COLONIC MICROBIOTA, NUTRITION AND HEALTH

Colonic Microbiota, Nutrition and Health

Edited by

Glenn R. Gibson and Marcel B. Roberfroid

KLUWER ACADEMIC PUBLISHERS

DORDRECHT / BOSTON / LONDON

Library of Congress Catalog-in-Publication data is available.

ISBN 0-4127-9880-8

Published by Kluwer Academic Publishers,
P.O. Box 17, 3300 AA Dordrecht, The Netherlands.

Sold and distributed in North, Central and South America
by Kluwer Academic Publishers,
101 Philip Drive, Norwell, MA 02061, U.S.A.

In all other countries, sold and distributed
by Kluwer Academic Publishers,
P.O. Box 322, 3300 AH Dordrecht, The Netherlands.

Printed on acid-free paper

Printed in the Netherlands.

Table of Contents

vi

INTRODUCTION

MARCEL B. ROBERFROID[1] AND GLENN R. GIBSON[2]
[1]*Université Catholique de Louvain, Department of Pharmaceutical Sciences, Avenue Mounier 73, B-1200 Brussels, BELGIUM*
[2]*Food Microbial Sciences Unit, Department of Food Science and Technology, The University of Reading, Reading, UK*

It is clear that diet fulfils a number of important human requirements. These include the provision of sufficient nutrients to meet the requirements of essential metabolic pathways, as well as the sensory (and social) values associated with eating. It is also evident that diet may control and modulate various body functions in a manner that can reduce the risk of certain diseases. This very broad view of nutrition has led to the development of foodstuffs with added "functionality".

Many different definitions of functional foods have arisen. Most of these complicate the simple issue that a functional food is merely a dietary ingredient(s) that can have positive properties above its normal nutritional value. Other terms used to describe such foods include vitafoods, nutraceuticals, pharmafoods, foods for specified health use, health foods, designer foods, etc. Despite some trepidation, the concept has recently attracted much interest through a vast number of articles in both the popular and scientific media.

There are five identifiable ways in which a foodstuff can have an improved function on human well-being:

1) Elimination of a component that causes a negative effect (*e.g.* an allergenic protein, toxin)
2) Increasing the concentration of a natural component to a more desirable level (*e.g.* fortification with a micronutrient above the recommended daily intake, but compatible with dietary guidelines for disease prevention)
3) Addition of a component for which beneficial effects have been demonstrated (*e.g.* non vitamin antioxidants)
4) Replacement of a component, usually a macronutrient, the intake of which is often excessive and causes deleterious effects (*e.g.* fats), by a component that is more benign (*e.g.* emulsified carbohydrates)
5) Improvement of the bioavailability of food components for which beneficial effects have been demonstrated.

The design and development of functional foods is a key nutritional issue which should rely on clear scientific guidelines. Some important aspects are given in Table 1.

G.R. Gibson and M.B. Roberfroid (eds.), Colonic Microbiota, Nutrition and Health, vii-viii.
© 1999 *Kluwer Academic Publishers. Printed in the Netherlands.*

Table 1. Scientific aspects pertinent to functional food development

---

1. Basic scientific knowledge relevant to functions:
   - sensitive to modulation by food components
   - pivotal to the maintenance of well-being and health, including disease prevention

2. Exploitation of this knowledge in the development of markers key to relevant functions

3. The generation of new hypothesis-driven human intervention studies which:
   - include the use of validated, relevant markers
   - allow the establishment of effective and safe intakes

4. Development, and exploitation, of advanced technologies for efficacious human studies:
   - minimally invasive
   - applicable on a large scale, including multiple centres

The first generation of functional foods largely involved the addition of certain components (*e.g.* vitamins, micronutrients) to the diet. However, one of the most promising current targets for functional food development is the gastrointestinal tract (GIT). The large intestine is, by far, the most densely populated areas of the human GIT. The organ is involved with control of transit time, bowel habit, absorptive and mucosal function - all of which may be associated with the resident gut microflora. Moreover, it appears that the resident microbiota may, through its normal metabolic state, exert other important physiological processes relevant to host health and disease. These are discussed in many of the following chapters.

The purpose of this book is to overview current knowledge of the activities and functions of the gut microflora. This is approached through the collation of recognised expertise in the areas of gut microbial function, molecular biology, clinical nutrition and industrial relevance. The following aspects of gastrointestinal microbiology are reviewed in terms of gut functionality:

- the GIT flora composition and activities
- the fermentation process
- luminal and biofilm bacterial processes
- gut flora modulation through diet (probiotics, prebiotics)
- the harnessing of molecular methodologies in gut microbiology
- applied relevance in terms of health ourcome
- consumer perspectives

It is implicit that the impact of gastrointestinal microorganisms on this contemporary area of nutritional sciences has much relevance.

CHAPTER 1

The Human Colonic Microbiota

GEORGE T. MACFARLANE AND ANDREW J. MCBAIN
Medical School, University of Dundee, Ninewells Hospital, Dundee, UK

1   Introduction

It has become increasingly evident over the last 20 years that the large intestine is a highly specialised digestive organ, which through the activities of its constituent microflora, rivals the liver in metabolic capacity, and in the diversity of its biochemical transformations. Approximately 90% of the 10^{14} cells associated with the human body are microorganisms, and the vast majority of these reside in the large bowel. The colonic microbiota consists of approximately 10^{13} cells which, with the possible exception of bacteria growing in the oral cavity, are physiologically very different to those associated with any other part of the host. Several hundred bacterial strains and species normally exist in the large intestine, with viable counts typically being in the region of 10^{12} per gram of intestinal contents. While host tissues and other substrates of endogenous origin (sloughed epithelial cells, mucins, pancreatic and other secretions) are continually being broken down and recycled by intestinal bacteria, the species composition and metabolic activities of the colonic microbiota are primarily determined by diet. Therefore, what we eat, particularly carbohydrate and protein, affects ecological, physiological and metabolic events in the large bowel. An outline of the main substrates available for intestinal bacteria is shown in Fig. 1. Through fermentation and the absorption and metabolism of short chain fatty acids (SCFA), the large intestinal microflora plays an important role in host digestive processes, enabling energy to be salvaged from unabsorbed dietary residues, as well as body tissues and secretions.

Intestinal microorganisms play a major role in health and disease, and affect human physiology in a multiplicity of ways (see Table 1), through obligate host requirements for bacterial fermentation products, maintenance of colonisation resistance to microbial pathogens, activation or destruction of genotoxins and mutagens, and modulation of immune system function. For example, the colonic microflora affects the cytokine network that regulates the effector arms of the immune response, as evidenced by the fact that reduction in IFN-γ and IFN-α production in aging mice is reversed with

G.R. Gibson and M.B. Roberfroid (eds.), Colonic Microbiota, Nutrition and Health, 1-25.
© 1999 *Kluwer Academic Publishers. Printed in the Netherlands.*

lactic acid bacteria probiotics (Muscettola *et al.,* 1994), while in human peripheral blood mononuclear cells, these organisms have been shown to induce formation of IL-1ß, TNF-α and IFN-γ, though not IFN-α or IL-2 (Pereyra and Lemmonier, 1993).

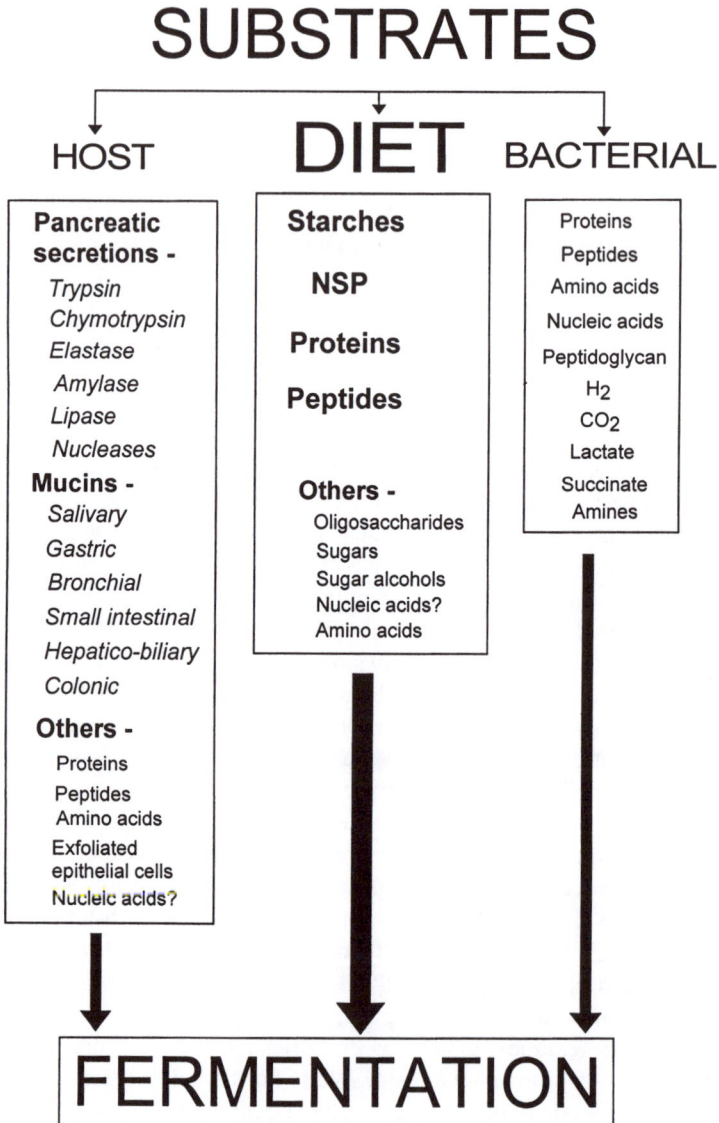

SUBSTRATES

HOST   DIET   BACTERIAL

Pancreatic secretions - *Trypsin Chymotrypsin Elastase Amylase Lipase Nucleases* **Mucins -** *Salivary Gastric Bronchial Small intestinal Hepatico-biliary Colonic* **Others -** Proteins Peptides Amino acids Exfoliated epithelial cells Nucleic acids?	**Starches** **NSP** **Proteins** **Peptides** **Others -** Oligosaccharides Sugars Sugar alcohols Nucleic acids? Amino acids	Proteins Peptides Amino acids Nucleic acids Peptidoglycan H_2 CO_2 Lactate Succinate Amines

FERMENTATION

Figure 1. Substrates available for fermentation by bacteria growing in the large intestine.
(NSP = non starch polysaccharide)

Table 1. Some health-related activities associated with bacteria growing in the large intestine

Processes	Examples	Effects on host
Carbohydrate fermentation	Digestion of starches, non-starch polysaccharides and oligosaccharides that escape digestion in the small bowel	SCFA formation supports epithelial cell growth, energy reclamation and other aspects of large bowel function. Increased bacterial growth and faecal output. Reduced absorption of toxic products of protein fermentation. Gas production
Proteolysis and amino acid fermentation	Recycling of C and N in proteins and peptides in dietary residues and pancreatic secretions. Production of ammonia, amines, HS⁻, thiols, phenols and indoles	Toxic metabolites associated with hepatic coma, other neurological symptoms, cytotoxicity and colon cancer. Gas production
Hydrogen disposal	Methane production, acetogenesis, HS⁻ formation	Reduction in colonic gas volume, HS⁻ cytotoxicity and ulcerative colitis?
Bile acid metabolism	Deconjugation and dehydroxylation of bile acids, desulphation of bile acid sulphates	Absorption of secondary bile acids. Possible promoting activity in colon cancer
Mutagen production	N-Nitrosation of secondary amines	Large bowel cancer
Metabolism of neutral steroids	Chemical modification of cholesterol, plant sterols and steroid hormones	Reabsorption and recycling of reduced corticosteroids, progesterone, oestrogens, role in breast cancer?
Transformations of xenobiotic substances	Desulphation of cyclamate to produce cyclohexylamine. Desulphation and deconjugation of drugs excreted in bile. β-Glucuronidase/methylazoxymethanol formation from cycasin. Conversion of azo bond in sulfasalazine to produce the active drug 5-aminosalicylic acid	Formation of acutely toxic substances, prolonged enteropathic circulation of foreign compounds
Metabolism of lignans and phytooestrogens	Conversion to enterodiol, enterolactone and equol by the microbiota	Oestrogenic and antioestrogenic effects: Related to fertility and breast cancer
Immune system development and modulation	Shown in probiotic studies, investigations with gnotobiotic animals	Enhanced resistance to infection
Colonisation resistance	Barrier effect of mature microbiota against invading species. Degraded during illness or antibiotic/drug treatments	Resistance to disease

2   The human large intestine

Due to gastric acid and the washout effect resulting from rapid passage of digestive substances through the stomach and small bowel, the principal areas of permanent colonisation of the human gastrointestinal tract are the terminal ileum and large intestine. This is primarily a result of the slowing down of movement of digestive material in the colon, which allows time for a complex and stable microbial ecosystem to develop (Cummings, 1978).

In adults, the large intestine is approximately 1.5 m long, and has a volume in excess of 500 ml. It daily receives about 1.5 kg of material from the small gut, the majority of which is water and is rapidly absorbed. The colon contains around 200 g of contents, of which approximately half consists of bacterial cells (Cummings *et al.*, 1990). Microorganisms are therefore a major component of faeces, comprising approximately 55% of solids in persons consuming Western-style diets (Stephen and Cummings, 1980). In the United Kingdom, the average daily output of material is 100-200 g (Cummings *et al.*, 1992).

After digestive materials enter the caecum, rapid breakdown of readily fermentable substrates occurs. Dietary residues form a pool of digesta in the proximal colon, and there is a significant degree of mixing from one day to the next (Wiggins and Cummings, 1976). Portions of this material are periodically transferred to the transverse and distal colon, where further water absorption increases the viscosity and reduces mixing. Due to utilisation of digestive substances by bacteria in the proximal colon, particularly carbohydrates, there is a progressive reduction in substrate availability towards the distal gut, affecting such factors as the type and amounts of fermentation products that are formed, as well as colonic pH (Cummings *et al.*, 1987). This can influence a number of metabolic processes in the bowel, including fermentation product formation (Blackwood *et al.*, 1956) and enzyme activities such as bile acid 7-α-dehydrogenase (Midvedt and Norman, 1968), proteases and peptidases (Macfarlane and Macfarlane, 1997; Macfarlane *et al.*, 1992a). Thus, some bacterial populations and activities may be restricted to, or predominate in, particular parts of the large gut (Macfarlane *et al.*, 1992b).

Transit of digestive substances through the large intestine consists of two chronological components, time spent in the mixing region (caecum and ascending colon), and time spent in passage and storage of partly solidified stool in the distal bowel (Wiggins and Cummings, 1976). The length of time digestive material spends in the large intestine is an important determinant of colonic bacterial metabolism. Long colonic transits affect the metabolism of carbohydrates, proteins and xenobiotic substances, and have been linked to the occurrence of large bowel cancer (Cummings *et al.*, 1992). Moreover, a significant correlation exists between transit time and bacterial mass in the large intestine: Stephen *et al.* (1986) used the drug senokot to speed up gut transit time from 64 to 25h in human volunteers. This increased mean stool weight from 148 to 285 g/d, with bacterial mass increasing from 18.9 to 20.3 g. In contrast, treatment with codeine/lopramide slowed colonic transit times from 47 to 88h and reduced bacterial cell mass from 18.9 to 16.1 g.

3   Acquisition and development of the colonic microflora

Humans are born with sterile colons, but bacteria begin to appear in excreta over the first few days of life. Breast and bottle-fed infants are inoculated by large numbers of Gram positive and Gram negative microorganisms during birth (Bullen *et al.*, 1976), and an outline of events relating to subsequent colonisation events in the large bowel during the first 2 years is shown in Fig. 2. Facultative anaerobes such as enterococci and enterobacteria are early colonisers, and they reduce pO_2 sufficiently to enable anaerobic organisms, particularly bacteroides and bifidobacteria, to establish. After weaning, the intestinal microbiota becomes more stable, as adult-type climax bacterial communities develop.

Considerable discrepancies occur in the literature concerning developing bacterial populations in the infant colon, which probably reflect inter-individual, cultural and environmental differences, and possibly, the methodologies employed in their study. For example, the work of Simhon *et al.* (1982) indicated that the faecal microflora of breast-fed and bottle-fed infants was similar, but many other investigations show that marked differences exist between these two groups, and that bifidobacteria, in particular, are the predominant organisms in formula-fed babies.

As shown in Fig. 3, a study of 35 breast-fed and 35 bottle-fed Japanese children (Benno *et al.*, 1984), indicated that bifidobacteria occurred high numbers in both groups, though some other types of intestinal bacteria including anaerobic Gram positive cocci, eubacteria, lactobacilli, enterobacteria, enterococci and bacteroides occurred in significantly lower numbers in the breast-fed infants.

In some European children, *Bifidobacterium bifidum* seems to predominate in the colons of breast-fed children (see Fig. 4), due to the presence of specific growth factors in human milk. Infant formulas, cows, sheeps or pigs milk milk do not contain these substances, and do not promote growth of this species, although they can support *B. infantis* and *B. longum* (Beerens *et al.*, 1980). Recent work indicates that *B. longum*, *B. adolescentis*, *B. pseudocatenulatum and B. parabifidum* can also occur in high numbers in some breast-fed children (Kleessen *et al.*, 1995). In Japan, *B. breve*, and *B. infantis* have been reported to be the numerically important bifidobacteria in infant faeces (Mitsuoka, 1989).

A variety of monosaccharides are present in human milk including lactose, glycoproteins and glycolipids, sialic acid, fucose and N-acetylglucosamine. Milk also contains over one hundred other more complex carbohydrates, including oligosaccharides (degree of polymerisation, DP, 3-11) based on lactose, and containing N-acetylglucosamine, fucose, galactose, sialic acid and glucose (Miller *et al.*, 1994). These oligomers are not digested to a significant extent in the small bowel of young infants, which has led to the suggestion that they have a multi-functional role in the gut, acting as ligands for pathogenic microorganisms and preventing their adherence to the small bowel mucosa, while serving as carbon sources for bifidobacteria in the colon (Brand Miller *et al.*, 1995).

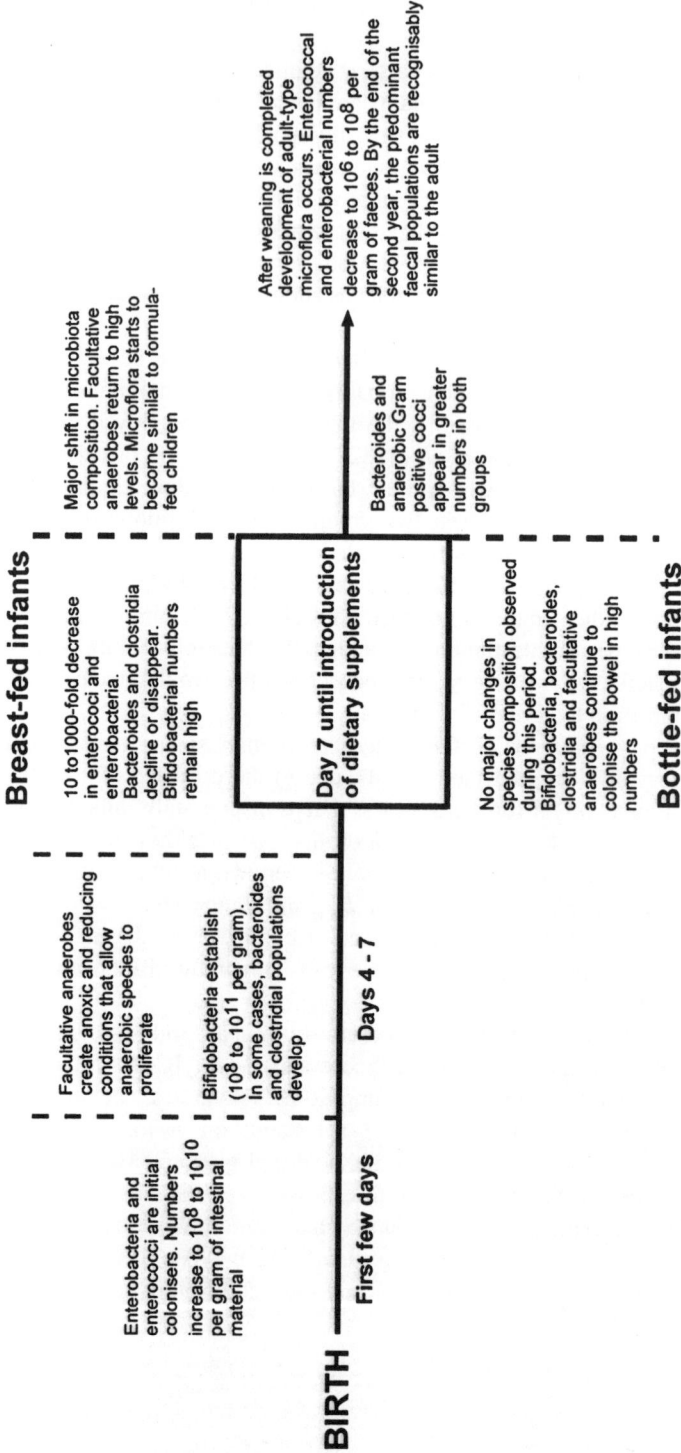

Figure 2. Outline of events during colonisation and bacterial succession in the infant colon. Summarised from Cooperstock and Zedd (1983).

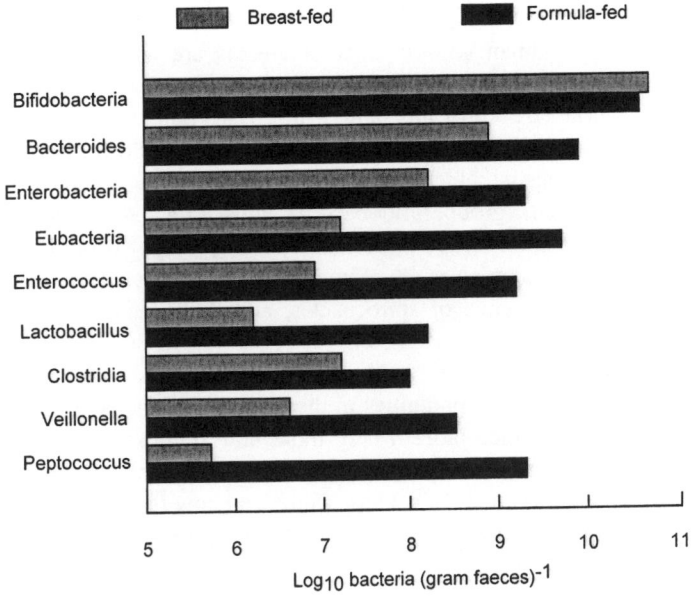

Figure 3. Comparison of bacterial populations in breast and bottle-fed children. Data are adapted from Benno *et al. (*1984).

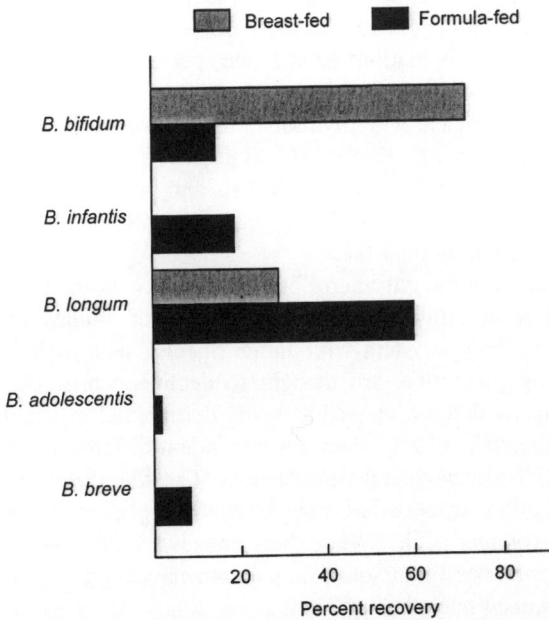

Figure 4. Bifidobacteria obtained from breast-fed (12 faecal samples, 81 isolates) and bottle-fed (39 faecal samples, 153 isolates) infants. Data are adapted from Beerens *et al. (*1980).

3.1 BACTERIAL POPULATIONS IN THE ADULT COLON

The vast majority of human colonic microorganisms are anaerobes, although they exhibit varying degrees of tolerance towards oxygen, ranging from the relatively oxygen tolerant bifidobacteria (Bezkorovainy and Miller-Catchpole, 1989) and bacteroides (Duerden, 1980), to very strictly anaerobic methanogenic archae. Anaerobic bacteria outnumber aerobic species by a factor of between 100 and 1000, and most studies show that the *Bacteroides fragilis* group, bifidobacteria, eubacteria and a variety of anaerobic Gram positive cocci are numerically important. While superficially similar to the faecal microfloras in other animal species (see Fig. 5) the human microbiota differs in many respects, including the absence of spirochaetes, and the relatively high numbers of clostridia and enterobacteria.

Except in very broad terms, little is known of the metabolic relationships that exist between individual bacterial communities in the colon, or of the ecology and multicellular organisation of the microbiota. Many molecular studies have shown that only a small fraction of bacterial species in natural communities are culturable (Hugenholtz and Pace, 1996; Dunbar *et al.*, 1997). Thus, while we know that the colonic ecosystem contains a very large number of different bacterial species, its actual composition is unknown.

The colonic microbiota is a stable entity within an individual (Bornside, 1978), but it is probable that at the level of individual species, considerable variations in cell populations occur. This was shown in measurements with human volunteers, where 10 individuals were studied over a 12-month period, and it was found that up to 1000-fold differences in counts of *Bacteroides fragilis* group organisms occurred during the course of the investigation (Meijer-Severs and Van Santen, 1986). The data of Finegold and his co-workers (Finegold *et al.*, 1983) show that large inter-individual variations in intestinal bacterial populations exist at the species level, and suggest that diet influences the composition of the microbiota.

A variety of host, microbiological and environmental factors affect colonisation of the large bowel (Table 2). Aries *et al.* (1969) and Hill *et al.* (1971) found higher faecal excretion of bifidobacteria and bacteroides by British and Americans compared to Ugandans, Japanese and Indians. The Ugandans characteristically had higher numbers of enterococci and enterobacteria in their faeces.

While metabolic variations in the gut microflora associated with the ageing process in the host have not been investigated, a limited number of studies indicate that structural changes occur in the ecosystem with aging. Species such as bifidobacteria, that are regarded as being protective, are thought to decline, whilst clostridia and enterobacterial populations, which are viewed as being detrimental to health, increase (Gorbach *et al.*, 1967; Mitsuoka, 1984). Other age-dependent differences in the faecal microflora were reported by Mitsuoka and Hayakawa (1972), who observed that lactobacilli and clostridia, in particular, occurred in significantly higher numbers in elderly people, in comparison to younger adults. While there are several possible explanations for these observations, age-related immunological phenomena and changes in the types and amounts of food consumed may be significant.

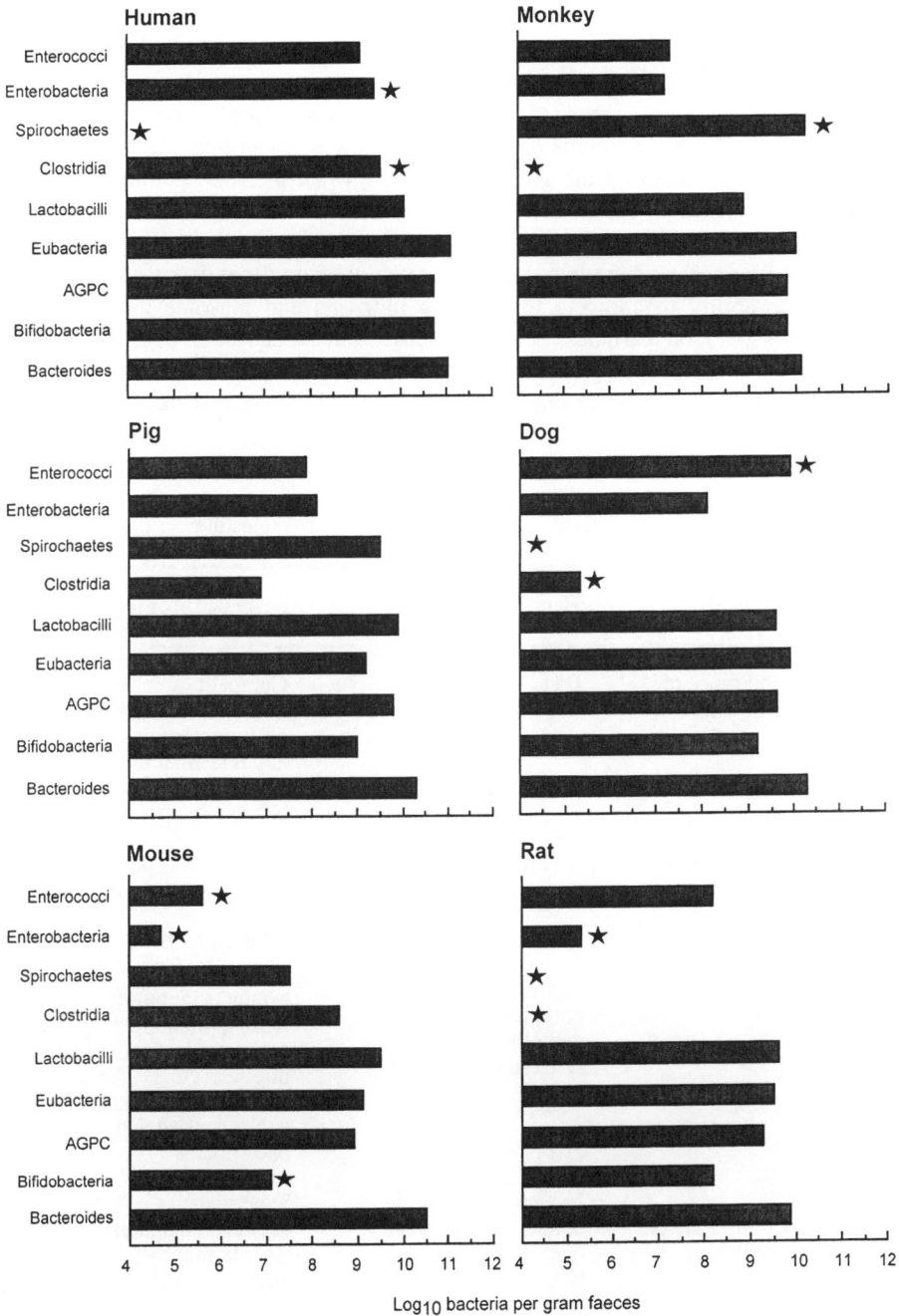

Figure 5. Climax intestinal bacterial communities in man and different animal species. Asterisks show significant characteristics associated with particular populations. Animal data are faecal wet weight values, and are adapted from Mitsuoka (1982). Human values are dry weight results (Moore and Holdeman, 1974).

Bacterial species diversity largely derives from the multiplicity of different carbon and energy sources available for fermentation in the colon, and the principal host factors regulating the microbiota are substrate availablity and colonic transit time. However, many different types of ecological interaction occur between intestinal microorganisms, including commensalism, where one species is stimulated by a second, which itself is unaffected by the growth and activities of the first; neutralism, where bacterial communities co-exist but have no significant metabolic effect on each other; antagonism, in which one population is repressed by inhibitory substances produced by another, and symbiosis, where two species have an obligate dependence on each other. However, the ability to compete for limiting nutrients and in some circumstances, adhesion sites on food particles, colonic mucus or the mucosa, is considered to be the most important microbiological factor determining the composition of the microbiota, with unsuccessful species being rapidly eliminated from the ecosystem.

Table 2. Factors affecting bacterial colonisation in the large intestine

Environmental	Host	Microbiological
Amounts and types of substrate available	Diet	Competition for limiting nutrients and adhesion sites on food particles, mucus and intestinal mucosa. Cooperative interactions between microorganisms
pH of intestinal contents	Colonic transit time, epithelial cell turnover rates	Generic and species composition of microbiota
Redox potential	Disease, drugs, antibiotic therapy, rates of mucus production and its chemical composition, pancreatic and other secretions, lysozyme at mucosa	Inhibition of allochthonous species by fermentation products including HS^-, SCFA, phenolic compounds, deconjugated bile salts etc. Bacterial secretion of antagonistic substances such as bacteriocins
Geographical residence/cultural factors associated with host	IgA production and defensin secretion at mucosal surface. Neuroendocrine system activity?	Synergistic effects of bacterial antagonism and local immunity in the mucus layer and on the colonic mucosa

3.2   PRODUCTION OF ANTAGONISTIC SUBSTANCES BY INTESTINAL BACTERIA

Some, but not all colonic microorganisms are able to form antagonistic substances such as bacteriocins. Bacteriocin formation is growth-associated in some species, and is dependent on carbon availability (Lejeune *et al.*, 1998). Many of these small peptides are highly species specific. For example, only one strain of 13 test bifidobacterial isolates tested for bacteriocin formation produced a protease-sensitive inhibitory substance, with broad spectrum activity against other bifidobacteria, streptococci, lactobacilli and clostridia, but not Gram negative species such as *Escherichia coli*, proteus, pseudomonas or klebsiella (Meghrous *et al.*, 1990).

Other studies (Gibson and Wang, 1994a) have shown that eight different bifidobacterial species produced similar inhibitory substances, that were antagonistic to various degrees against a range of intestinal pathogens, belonging to several bacterial genera (*Salmonella, Listeria, Campylobacter, Shigella, Vibrio*).

Unlike the bacteriocins of lactic bacteria, the antimicrobial secretory products of bifidobacteria have not been well characterised, although they generally appear to exhibit a much wider activity spectrum than is usually associated with conventional bacteriocins. Studies by O'Riordan *et al.*, (1995) indicate that *B. infantis* NCFB 2255 and *B. breve* NCFB 2258 may produce two different types of antimicrobial agent. One is thought to be proteinaceous in nature and largely affects Gram positive bacteria, whilst the other was not affected by proteolytic enzymes, and inhibited Gram negative organisms.

While many investigations have been made on bacteriocin production using pure cultures in the laboratory, the effectiveness of these substances in a highly proteolytic environment such as the intestinal tract is not clear. However, studies on an enterocin secreted by a ruminal strain of *Enterococcus faecium* showed that it could retain anti-microbial activity for long periods in rumen fluid (Laukova and Czikkova, 1998).

4   The large intestinal microbiota and homeostasis

The normal colonic microflora plays an important role in colonisation resistance (Hentges, 1983), a term defined as the mechanism whereby the intestinal microbiota protects itself against incursion by new and occasionally pathogenic microorganisms (Gorbach *et al.*, 1988). Colonisation resistance, otherwise known as the barrier effect, bacterial antagonism or bacterial interference, affects homeostasis in the large bowel, and may also be viewed as having a protective role against proliferation of potentially harmful elements present in the autochthonous microbiota.

Increasing species diversity in a microbial ecosystem such as the large intestine enforces metabolic homeostasis and structural stability, while degenerative changes in bacterial composition, through for example, antibiotic administration, are known to reduce the ability of the ecosystem to resist invading pathogens. Colonisation resistance is a multifactorial phenomenon in which anaerobes are more important than aerobes, and there is evidence that bacteroides in particular play a significant role in protection against *C. difficile* infection (Tvede and Rask-Madsen *et al.*, 1989). In animals, food deprivation leads to major changes in gut bacterial populations, and renders them more susceptible to enterobacterial infections (Tannock and Savage, 1974).

Due to deterioration of the normal ecological balance in the large bowel, resulting from either antibiotic treatment, alterations in diet, or physiological alterations in the gastrointestinal tract, profound changes can occur in the composition and activities of the colonic microbiota. Clinical symptoms often manifest in the form of diarrhoeal episodes of varying persistence and severity (Gibson *et al.*, 1997). While much information is available in the literature dealing with the mechanisms of infection and the inhibitory effects of various antibiotics on individual gut species, little is known of how gastrointestinal infections or antibiotic treatment affect either the stability or metabolic activities of the gut microbiota as a whole, or the physiological consequences for the host resulting from these phenomena.

5   Carbohydrate metabolism

The vast majority of colonic bacteria are saccharolytic, and carbohydrate fermentation is an important force driving the microecology and physiology of the large intestine. With respect to the efficiency of fermentation, bacterial cell yields have been calculated to be in the region of 25 to 36 g bacteria per 100 g carbohydrate fermented (Cummings, 1987). A wide variety of different carbohydrates are potentially available for fermentation in the large bowel, by many different types of anaerobic bacteria, yet these substances are catabolised in a relatively small number of metabolic pathways to produce a comparatively limited range of end products.

Carbohydrate metabolism is quantitatively more important than amino acid fermentation, particularly in the proximal colon, where substrate availability is greatest. Starches that for various reasons resist the action of pancreatic amylase, are hydrolysed by bacterial amylases (Macfarlane and Englyst, 1986) and these substances, together with plant cell wall polysaccharides (dietary fibre) are quantitatively of greatest significance. Dietary fibre comprises cellulose and non-cellulosic polysaccharides, such as pectins, hemicelluloses, gums and mucilages. Around 50% of the amorphous cellulose in vegetables is fermented in the colon, rising to 80% if all plant cell wall polysaccharides are considered (Cummings, 1983). The amounts reaching the large intestine vary markedly from diet to diet, but are not less than 20 g/d, reaching 60 to 80 g when carbohydrate foods form the major staple.

5.1   MECHANISMS OF CARBOHYDRATE UTILISATION
IN COLONIC ANAEROBES

Ecological experiments using chemostats show that bifidobacteria are adapted to compete for a plentiful supply of carbohydrate at high specific growth rates, whereas bacteroides adopt a different strategy, and are able to compete most effectively under C-limiting conditions, at low specific growth rates (Degnan and Macfarlane 1995a; Macfarlane *et al.*, 1995). Both competition for carbohydrate, and metabolic cooperation between bifidobacteria and bacteroides were observed in these continuous culture studies, and were primarily dependent on substrate availability, with obligate bifidobacterial cross-feeding of carbohydrate fragments produced by bacteroides occurring during C-limited growth.

Despite its ecological implications, few studies have addressed the problem of how carbohydrate is assimilated by intestinal microorganisms. This is of considerable importance, because many different types of (limiting) carbohydrates are available for bacteria growing in the large bowel, and strategies involved in substrate utilisation employed by individual species will strongly affect their growth and ability to compete.

Experiments with *Bacteroides thetaiotaomicron*, one of the most numerous species in the large intestine, demonstrated that hexose metabolism was energetically more efficient than pentose fermentation in this organism (Degnan and Macfarlane, 1995b). Carbohydrate utilisation was controlled by substrate-induced regulatory mechanisms, as shown by the fact that mannose inhibited uptake of glucose, galactose and arabinose, but had less effect on xylose transport. Arabinose and xylose were preferentially utilised at high dilution rates in C-limited chemostats grown on mixtures of arabinose, xylose, galactose and glucose. When mannose was also present, xylose was co-assimilated at all dilution rates. Under N-limited conditions, however, mannose repressed uptake of all sugars, demonstrating that its effect on xylose metabolism was strongly concentration dependent. Transport systems for glucose, galactose, xylose and mannose were shown to be inducible in these studies. Experiments using inhibitors of carbohydrate transport demonstrated the existence of a variety of different systems involved in sugar transport.

Studies on the abilities of bifidobacteria (*B. infantis, B. longum, B. adolescentis, B. bifidum, B. catenulatum, B. pseudolongum*) to grow on 15 different carbohydrate sources demonstrated that substrate utilisation was highly variable, and that considerable inter-species and inter-strain differences existed (Hopkins *et al.*, 1998). Galactooligosaccharides (GOS) and fructooligosaccharides (FOS), with a low DP supported best growth of the test microorganisms. In contrast, xylooligosaccharides and pyrodextrins were invariably poor substrates. In many species, maximum specific growth rates and bacterial cell yields were higher on oligosaccharides compared to their monosaccharide constituents, reflecting the activities of substrate transport systems, which appear to be more efficient with dimeric and oligomeric carbon sources. *Bifidobacterium pseudolongum, B. longum* and *B. catenulatum* were the most nutritionally versatile species.

5.2   REGULATION OF CARBOHYDRATE METABOLISM

Because saccharolytic bacteria exist in a multisubstrate-limited environment in the large bowel, particularly in the distal colon, they must be able to adapt rapidly to changing nutritional circumstances to ensure growth and survival. These organisms control the utilisation of polymerised C-sources through regulation of synthesis of hydrolytic enzymes and substrate uptake systems (catabolite repression), as well as by more responsive and rapid acting mechanisms such as catabolite inhibition and inducer exclusion, which operate at the level of substrate transport into the cells. Inducer exclusion mechanisms include inhibition of sugar uptake by intracellular sugar phosphates (Saier and Roseman, 1976) and competition for binding sites on transport proteins

(Dills *et al.*, 1980), while formation of a transmembrane potential by one sugar may selectively inhibit uptake of others (Reider *et al.*, 1979). Catabolite inhibition is similar to inducer exclusion. In this process, inhibition of sugar uptake results from translocation of another substrate across the cell membrane (McGinnis and Paigen, 1973).

Although the metabolic impact of catabolite regulatory mechanisms on fermentation reactions in the large bowel is dependent on the types and amounts of carbohydrate present in the ecosystem, as well as on the relative cell population densities of individual carbohydrate-utilising species, their operation enables intestinal bacteria to selectively use certain substrates to the exclusion of others, allowing some organisms to compete more effectively for a restricted range of preferred substrates in relatively specialised metabolic niches. Since different bacteria form distinct patterns of metabolic end products, catabolite regulatory mechanisms ultimately affect the types and amounts of these substances that can be formed from individual substrates (Macfarlane and Gibson, 1996).

These control processes are ecologically important in the large intestine, and are of great physiological significance to the host, due to the fact that different fermentation products are metabolised at various body sites. Thus, butyrate is the principal energy source for the colonic epithelium (Roediger and Nance, 1986), and plays an important role in regulating epithelial cell growth (Cummings, 1995), as well as immune cell growth and apotosis (Kurita-Ochiai *et al.*, 1997), propionate is cleared by the liver and affects cholesterol metabolism, while acetate is oxidised in brain, heart and peripheral tissues (Cummings, 1995).

While many studies have shown that fermentation of different polysaccharides gives rise to specific patterns of SCFA production, for example, butyrate and lactate are particularly associated with starch breakdown (Macfarlane and Englyst, 1986; Englyst *et al.*, 1987), little is known of how the microbiota deals with several substrates simultaneously, and fermentation experiments using a single carbohydrate source probably do not provide realistic information on their likely metabolism *in vivo*.

6   Proteolysis

The majority of nitrogen entering the large bowel occurs in the form of proteins and peptides (Chacko and Cummings, 1988) and bacteria must therefore depolymerise these macromolecules to take advantage of the carbon, energy and nitrogen that they contain.

The large intestine contains complex mixtures of bacterial and pancreatic endopeptidases, and is one of the most proteolytic natural environments known (Macfarlane *et al.*, 1986). The importance of microbial proteolysis is that while large quantities of pancreatic peptidases enter the colon daily, these serine enzymes have comparatively limited substrate specificities, and are relatively ineffective at digesting globular proteins. Bacterial proteases degrade these substances in the large bowel, as well as the pancreatic secretions, to short peptides and free amino acids, which then become available for fermentation.

Many nutritional and environmental factors regulate protease production in colonic microorganisms. For example, *Clostridium sporogenes* secretes at least six extracellular serine, thiol and metalloproteases (Allison and Macfarlane, 1992). Their synthesis is catabolite regulated, and is inhibited by glucose, ammonia and peptides. Protease secretion is primarily a response to nutrient limitation, which is consistent with the enzymes having a scavenging role. In *C. bifermentans*, a multiplicity of extracellular metalloproteases are produced. Their synthesis is principally controlled by cell growth rate, and is inversely related to dilution rate in C- and N- limited chemostats (Macfarlane and Macfarlane, 1992).

Although not as proteolytic as clostridia, bacteroides outnumber them by two or three orders of magnitude in intestinal contents, and are important proteolytic species by virtue of their great numbers. In *B. splanchnicus* and *B. fragilis*, production of cell-bound serine, thiol and metalloproteases is growth rate-associated, and stimulated by N limitation (Macfarlane *et al.*, 1992c; Macfarlane and Macfarlane, 1997). Unlike the clostridia, bacteroides proteases are inefficient at breaking down globular protein substrates, but are highly effective in digesting residual pancreatic endopeptidases (Macfarlane and Macfarlane, 1991), indicating that they play an important role in the reclaimation and recycling organic N-containing compounds in the large gut.

Wallace and McKain (1997) showed that the principal route of peptide breakdown in human intestinal bacteria occurs as result of dipeptidyl peptidase activity. This study indicated that bacteroides, in particular, express high levels of these enzymes, compared to amino acyl peptidases. The authors concluded that peptide hydrolysis in the large bowel is a two-stage process initiated by bacteroides dipeptidyl peptidases.

7   Peptide utilisation and dissimilatory amino acid metabolism

Peptides and amino acids are important carbon, nitrogen and energy sources for saccharolytic and asaccharolytic microorganisms. Some saccharolytic species are able to derive small amounts of energy from the carbon skeletons of peptides and amino acids, but this is insufficient for these nutrients to serve as energy sources by themselves (Russell, 1983). Other bacteria including enterococci and enterobacteria ferment carbohydrates and amino acids, whilst some organisms have an obligate dependency on amino acids for their energy requirements.

Generation of free amino acids from proteins and peptides in the large bowel probably has little nutritional significance to the host, due to bacterial competition for the substrate, and the fact that the colon does not have the absorptive machinery needed for their uptake (Wrong, 1988). Instead, amino acids are catabolised in a wide variety of deamination, hydroxylation, dehydroxylation, desaturation, decarboxylation, oxidative and reductive reactions. Many different types of electron donors and acceptors participate in these processes, including other amino acids, keto acids, unsaturated fatty acids and molecular H_2. Intestinal microorganisms use four direct pathways for deaminating single amino acids, together with the Stickland reaction, in which pairs of amino acids are fermented.

Deamination can occur by oxidation, with formation of an α-keto acid, by reduction in which a saturated fatty acid is produced, by hydrolysis with production of an α-hydroxy fatty acid, or alternatively, ammonia is removed to leave an unsaturated product, as in the deamination of aspartate to fumarate by aspartase. Amino acid fermenting clostridia largely use Stickland reactions, which consist of coupled oxidation-reductions be tween suitable pairs of amino acids, where one substrate is oxidatively deaminated and decarboxylated while the other is reductively deaminated. Saturated fatty acids (acetate, propionate, butyrate) and ammonia are the main products of protein breakdown and amino acid fermentation in the large bowel (Macfarlane and Allison, 1986), indicating that dissimilatory reactions are primarily effected through reductive deamination.

Short chain fatty acids are important products of amino acid fermentation in the colon, and their formation is dependent on the chemical compositions of the amino acid substrates. Recent studies (Smith and Macfarlane, 1997) showed that acetate and butyrate are formed from the acidic amino acid glutamate, while acetate and propionate are produced from aspartate. Intestinal bacteria rapidly ferment basic amino acids such as lysine and arginine to butyrate and acetate, and ornithine and citrulline, respectively. The major products of histidine deamination are also acetate and butyrate. These investigations further showed that fermentation of S-containing amino acids was slow and incomplete. Acetate, propionate and butyrate were formed from cysteine, whereas the main products of methionine metabolism were propionate and butyrate. The aliphatic amino acids alanine and glycine were respectively fermented to acetate, propionate and butyrate, and acetate and methylamine, whilst branched-chain amino acids were slowly metabolised by colonic bacteria, with the main acidic products being branched chain fatty acids (BCFA) one carbon atom shorter than the parent amino acid. Aliphatic-hydroxy amino acids were, however, rapidly deaminated, with serine being converted to acetate and butyrate, while threonine was primarily fermented to propionate. Proline was poorly utilised by intestinal microorganisms, although hydroxyproline was rapidly dissimilated to acetate and propionate.

Studies on branched chain amino acid fermentation, using material obtained from different regions of the large bowel of sudden death victims at autopsy, have shown that amino acid fermentation becomes quantitatively more important in the distal colon, where carbohydrate is depleted (Macfarlane *et al.*, 1992b). In these investigations, approximately 17% of SCFA in proximal bowel contents could be attributed to amino acid fermentation, rising to around 38% in the distal large intestine. Correspondingly, concentrations of phenolic and indolic compounds, which are products of dissimilatory aromatic amino acid metabolism, were found to be four-fold higher in the distal bowel. *In vitro* experiments using a three-stage continuous culture model of the colon confirmed that quantitatively and qualitatively, amino acid fermentation is regulated by carbohydrate availability, and inhibited in acid pH conditions characteristic of the caecum and ascending colon. Increasing system retention times from 27h to 67h in the model, resulted in a three-fold increase in production of phenols and phenol-substituted fatty acids, illustrating how long colonic transit times predispose towards production of putrefactive substances by the microflora (Macfarlane *et al.*, 1998).

Many products of amino acid fermentation are harmful to the host. For example, high concentrations of ammonia in the colonic lumen may select for neoplastic growth (Matsui *et al.*, 1995), while ammonia also contributes towards the onset of portal-systemic encephalopathy in patients with liver disease (Weber *et al.*, 1987). Butyrate and valerate have also been implicated in hepatic coma, where patients with cirrhosis have raised concentrations in blood and cerebro-spinal fluid (Vince, 1986). Formation of BCFA by intestinal bacteria is not usually associated with toxicity, however, confusion and stupor accompanies a rise in these metabolites in children with isovaleric acidaemia (Tanaka *et al.*, 1976). Phenolic and indolic compounds are believed to act as co-carcinogens (Dunning *et al.*, 1950), while amines serve as precursors of nitrosamine production (Shephard *et al.*, 1987). Amines, phenols and indoles have also been implicated in schizophrenia (Dalgliesh *et al.*, 1958), migraine (Anon, 1968) and the onset of hypertensive symptoms (Boulton *et al.*, 1970).

A wide range of amines are formed by human colonic anaerobes, including simple aliphatic compounds such as methylamine, dimethylamine, propylamine, butylamine and 2-methybutylamine, the polyamines putrescine and cadaverine, and their heterocyclic oxidation products pyrrolidine and piperidine, as well as histamine and the aromatic amines tyramine, tryptamine and phenylethylamine (Smith and Macfarlane, 1996). Large amounts of these metabolites occur in intestinal contents: Total concentrations of simple amines in colonic material obtained at autopsy were approximately 22 mmol/kg in the proximal bowel and 16 mmol/kg in the distal large intestine. Propylamine, piperidine and methylamine predominated in all regions. Faecal amine concentrations ranged from 1.8 to 41.5 mmol/kg. These differences may be due to variations in substrate (proteins, peptides, amino acids, carbohydrate) availability, gut transit times and intestinal pH, together with inter-individual differences in bacterial populations. Once formed, amines are further metabolised in the colonic microbiota, especially in the presence of a fermentable carbohydrate source. Amine formation by faecal bacteria was shown to be maximal at near neutral pH, while carbohydrate reduced net production by 80%.

A direct relationship exists between dietary protein intake and amine excretion, and this may in part explain epidemiological findings that relate high levels of dietary protein to increased incidence of large intestinal disease (Drasar and Irving, 1973). Feeding studies with human volunteers show that increased dietary protein intake results in greater faecal excretion of sulphydryl compounds and ammonia, and urinary excretion of p-cresol (Geypens *et al.*, 1997).

However, studies with rats fed thermolysed casein have suggested that protein fermentation products may not be important in the promotion of cancer (Corpet *et al.*, 1995). While the ultimate nutritional determinant affecting amino acid fermentation in the gut is availability of organic-N containing compounds (proteins, peptides, amino acids), production of these potentially toxic metabolites by intestinal microorganisms can be reduced by increasing the supply of fermentable carbohydrate to the colon. Thus, reduction of putrefactive processes in the large bowel is achievable through manipulation of diet.

8   Hydrogen

Fermentations are regulated by the need to maintain redox balance, principally through the reduction and oxidation of ferredoxins, flavins and pyridine nucleotides. This affects the flow of carbon through bacteria and energy yield obtained from the substrate, as well as the fermentation products that are generated. Formation of reduced metabolites, such as lactate, succinate, ethanol and H_2 is used to effect redox balance, whilst production of oxidised fermentation products such as acetate is associated with ATP synthesis. Estimates of fermentation stoichiometry suggest that 10 g carbohydrate fermented in the colon should yield about 100 mmol SCFA and 50-100 ml H_2 (Cummings and Macfarlane, 1997).

Physical removal of H_2 in the colon occurs through excretion in breath or flatus, and by its utilisation by H_2-consuming syntrophs such as dissimilatory sulphate reducing bacteria (SRB), methanogenic archae and acetogenic bacteria (Macfarlane and Gibson, 1996). Strocchi and Levitt (1992) showed that the extent of H_2 removal, and to some degree, its net production, was influenced by stirring of gut contents, which was directly related to the partial pressure of the gas. Stirring released H_2 from intestinal material to the adjacent gas head space at a more rapid rate than by diffusion alone. Due to its poor solubility, mass transfer resistance of H_2 between gas-liquid interfaces affects its rate of utilisation by syntrophic microorganisms. This emphasises the importance of close spatial proximity existing between microbial communities that produce and utilise H_2, as shown in studies on sewage sludge, where more than 90% of H_2 transfer reactions occurred between anaerobic populations existing in close proximity, or in microniches (Conrad et al., 1985).

9   Diet and the colonic microflora

It is still widely believed that while food intake may regulate the metabolic activities of intestinal microorganisms, changing diet has little effect on their relative population sizes. However, this concept is increasingly being challenged. For example, SRB can occur in high numbers in the colon (Gibson et al., 1988; Gibson et al., 1993). These organisms derive energy from the reduction of SO_4^{2-} and SO_3^{2-} to HS^-. Human intestinal SRB use a variety of electron-donors in SO_4^{2-} reduction, which can account for substantial consumption of H_2 in the large bowel. SRB occur predominantly in the distal colon (Macfarlane et al., 1992b), and their activities are primarily controlled by availability of oxidised S-compounds (terminal electron acceptors), especially SO_4^{2-} and SO_3^{2-}, which are used as food preservatives. Ileostomy studies show that up to 7 mmol SO_4^{2-} is absorbed per day in the bowel, before spillover into the colon occurs.

Sulphide is toxic to the colonic epithelium, and concentrations as low as 2 mM inhibit butyrate oxidation in human colonocytes (Roediger et al., 1993). This metabolic defect is similar to the lesion observed in ulcerative colitis, and is thought to result in intracellular energy deficiency. Using human epithelial cells, Christl et al. (1996) showed that HS^- at a concentration of 1 mM significantly increased the rate of colonocyte proliferation, and other cellular changes that occur in colitis. Dietary SO_4^{2-} may therefore selectively stimulate SRB growth in the large intestine with potentially harmful effects to the mucosa (Pitcher and Cummings, 1996).

9.1 BIFIDOBACTERIA AND PREBIOTICS

As described earlier, bifidobacteria are numerically important members of the normal human colonic microflora, comprising up to 25% of all cultivable bacteria, with *B. adolescentis* and *B. longum* predominating in adults (Mitsuoka, 1984; Scardovi, 1986). There is increasing evidence that these organisms play an important role in the eco-physiology of the large intestinal microbiota (Gibson and Wang, 1994a,b), as well as in direct interactions with host metabolism (Gibson *et al.*, 1997).

Given orally, bifidobacteria alter faecal bacterial enzyme activities (Marteau *et al.*, 1990), reduce antibiotic-related side effects (Colombel *et al.*, 1987), inhibit IQ-induced mammary and liver tumours in rats (Reddy and Rivenson, 1993), and in conjunction with fructooligosaccharides (FOS), reduce 1,2-dimethylhydrazine induced colonic carcinogenesis in mice (Koo and Rao, 1991). They have also been reported to stimulate the immune system against certain tumours (Sekine *et al.*, 1985) and bacterial invasion (Yamazaki *et al.*, 1985). Other putative health benefits include cholesterol lowering effects, increased lactose digestion, and relief from constipation. For these reasons, there is currently much interest in increasing their numbers in the large intestine, either through direct introduction as probiotics, or through the inclusion of short chain carbohydrate substrates in the diet.

Feeding studies with human volunteers have shown that bifidobacterial populations can be selectively stimulated by FOS and galactooligosaccharides (GOS) *in vivo* (Ito *et al.*, 1990; Gibson *et al.*, 1995), though the species involved were not identified in these investigations. The ability to target specific groups of organisms in the large intestine, for defined health-promoting purposes may be of significant health value. Two factors that are especially important in this respect are the rate at which an organism can grow on a particular carbon source, since this will influence its ability to compete with other bacteria in the colon, the other is the extent to which the substrate is converted into bacterial mass, because cell population sizes affect the degree of both prebiotic and probiotic activity.

To test the inhibitory effects that probiotics (*Bifidobacterium longum*) and prebiotics (FOS, GOS) might have on the formation of mutagenic substances in the large gut, studies were made using a three-stage continuous culture model of the colon (McBain and Macfarlane, 1997). These experiments investigated their effects on synthesis of hydrolytic and reductive enzymes associated with the formation of genotoxic bacterial metabolites. Results showed that adding *B. longum* to established communities of colonic microorganisms effected substantial increases in nitroreductase in the vessel reproducing the caecal ecosystem. However, both FOS and GOS suppressed synthesis of ß-glucosidase and ß-glucuronidase in the model, but strongly stimulated azoreductase and nitroreductase formation. These data indicate that indiscriminate use of either pro- or prebiotics whose side-effects have not been extensively investigated, might in some circumstances result in dysbiotic manifestations in the large intestine, through increased genotoxin production.

10   Conclusion

Through fundamental laboratory based experiments, as well as clinical studies with patients, and healthy volunteers, considerable work is required to further our understanding of the role of diet in regulating the composition and metabolic activities of the human colonic microbiota, as they relate to host health. The fact that the intestinal microbiota can be manipulated by inclusion of short chain carbohydrates in the diet, may have significant health implications. An advantage of using prebiotics is that the target bacteria are already commensal to the gut. However, for a variety of reasons, there is a fundamental lack of knowledge concerning the physiology of colonic anaerobes, and an increasing awareness of bacterial substrate preferences, in particular, is needed to facilitate future prebiotic developments.

References

Allison, C. and Macfarlane, G.T. (1992) Physiological and nutritional determinants of protease secretion by *Clostridium sporogenes*: Characterisation of six extracellular proteases. *Applied Microbiology and Biotechnology* **37**, 152-56.

Anon. (1968) Headache, tyramine, serotonin and migraine. *Nutrition Reviews* **26**, 40-44.

Aries, V.C., Crowther, J.S., Drasar, B.S., Hill, M.J. and Williams, R.E.O. (1969) Bacteria and the etiology of cancer of the large bowel. *Gut* **10**, 334-35.

Beerens, H., Romond, C. and Neut, C. (1980) Influence of breast-feeding on the bifid flora of the newborn intestine. *American Journal of Clinical Nutrition* **33**, 2434-39.

Benno, Y., Sawad, K. and Mitsuoka, T. (1984) The intestinal microflora of infants: composition of fecal flora in breast-fed and bottle-fed infants. *Microbiology and Immunology* **28**, 975-86.

Bezkorovainy, A. and Miller-Catchpole, R. (1989). *Biochemistry and Physiology of Bifidobacteria*, CRC Press, Boca Raton.

Blackwood, A.C., Neish, A.C. and Ledingham, G.A. (1956) Dissimilation of glucose at controlled pH values by pigmented and non-pigmented strains of *Escherichia coli. Journal of Bacteriology* **72**, 497-99.

Bornside, G.H. (1978) Stability of human fecal flora. *American Journal of Clinical Nutrition* **31**, 5141-44.

Boulton, A.A., Cookson, B. and Paulton, R. (1970) Hypertensive crisis in a patient on MAOI antidepressants following a meal of beef liver. *Canadian Medical Association Journal* **102**, 1395.

Brand Miller, J.C., McVeagh, P., McNeil, Y. and Gillard, B. (1995) Human milk oligosaccharides are not digested and absorbed in the small intestine of young infants. *Proceedings of the Nutrition Society of Australia.* **19**, A44.

Bullen, C.L., Tearle, P.V. and Willis, A.T. (1976) Bifidobacteria in the intestinal tract of infants: an in vivo study. *Journal of Medical Microbiology* **9**, 325-27.

Chacko, A. and Cummings, J.H. (1988) Nitrogen losses from the human small bowel: obligatory losses and the effect of physical form of food. *Gut* **29**, 809-15.

Christl, S.U., Eisner, H.D., Scheppach, W. and Kasper, H. (1996) Effect of sodium hydrogen sulfide on cell proliferation of colonic mucosa. *Gastroenterology* **106**, A664.

Colombel, J.F., Cortot, A. and Neut, C. (1987) Yoghurt with *Bifidobacterium longum* reduces erythromycin-induced gastrointestinal effects. *Lancet* **2**, 43.

Conrad, R., Phelps, T.J. and Zeikus, J.G. (1985) Gas metabolism evidence in support of the juxtaposition of hydrogen-producing and methanogenic bacteria in sewage sludge and lake sediments. *Applied and Environmental Microbiology* **50**, 595-601.

Cooperstock, M.S. and Zedd, A.J. (1983) Intestinal flora of infants, in *Human Intestinal Microflora in Health and Disease* (ed. D.J. Hentges), Academic Press, London, pp. 79-99.

Corpet, D.E., Yin, Y., Zhang, X-M., Remesy, C., Stamp, D., Medline, A., Thompson, L., Bruce, W.R. and Archer, M.C. (1995). Colonic protein fermentation and promotion of colon carcinogenesis by thermolyzed casein. *Nutrition and Cancer* **23**, 271-81.

Cummings, J.H. (1978) Diet and transit through the gut. *Journal of Plant foods* **3**, 83-95.

Cummings, J.H. (1987) Dietary fiber. *American Journal of Clinical Nutrition* **45**, 1040-43.

Cummings J.H. (1983) Dietary fibre and the intestinal microflora, *in Nutrition and the Intestinal Flora* (ed. B. Hallgren),Swedish Nutrition Foundation, Stockholm, pp. 77-86

Cummings, J.H. (1995) Short chain fatty acids, *in Human Colonic Bacteria: Role in Nutrition, Physiology and Pathology* (eds. G.R. Gibson and G.T. Macfarlane). CRC Press, Boca Raton. pp. 101-30.

Cummings, J.H., Banwell, J.G., Englyst, H.N., Coleman, N., Segal, I. and Bersohn, D. (1990) The amount and composition of large bowel contents. *Gastroenterology* **98**, A408.

Cummings, J.H., Bingham, S.A., Heaton, K.W. and Eastwood, M.A. (1992) Fecal weight, colon cancer risk, and dietary intake of nonstarch polysaccharides (dietary fiber). *Gastroenterology* **103**, 1783-89.

Cummings, J.H. and Macfarlane, G.T. (1997) Colonic Microflora: Nutrition and Health. *Nutrition* **13**, 476-78.

Cummings, J.H., Pomare, E.W., Branch, E.J., Naylor, C.P.E. and Macfarlane, G.T. (1987) Short chain fatty acids in human large intestine, portal, hepatic and venous blood. *Gut* **28** 1221-27.

Dalgliesh, C.E., Kelley, W. and Horning, E.C. (1958) Excretion of a sulphatoxyl derivative of skatole in pathological studies in man. *Biochemical Journal* **70**, 13P.

Degnan, B.A. and Macfarlane, G.T. (1995a) Arabinogalactan utilization in continuous cultures of *Bifidobacterium longum*: Effect of co-culture with *Bacteroides thetaiotaomicron*. *Anaerobe* **1**, 103-12.

Degnan, B.A. and Macfarlane, G.T. (1995b) Carbohydrate utilization patterns and substrate preferences in *Bacteroides thetaiotaomicron*. *Anaerobe* **1**, 25-33.

Dills, S.S., Apperson, A., Schmidt, M.R. and Saier, M.H. Jr. (1980) Carbohydrate transport in bacteria. *Microbiological Reviews* **44**, 385-418.

Drasar, B.S. and Irving, D. (1973) Environmental factors and cancer of the colon and breast. *British Journal of Cancer* **27**, 167-72.

Duerden, B.I. (1980) The isolation and identification of *Bacteroides* spp. from the normal human faecal flora. *Journal of Medical Microbiology* **13**, 69-78.

Dunbar, J., White, S. and Forney, L. (1997) Genetic diversity through the looking glass: Effect of enrichment bias. *Applied and Environmental Microbiology* **63**, 1326-31.

Dunning, W.F., Curtis, M.R. and Maun, M.E. (1950) The effect of added dietary tryptophane on the occurrence of 2-acetyl-aminofluorene-induced liver and bladder cancer in rats. *Cancer Research* **10**, 454-59.

Englyst, H.N., Hay, S. and Macfarlane, G.T. (1987) Polysaccharide breakdown by mixed populations of human faecal bacteria. *FEMS Microbiology Ecology* **95**, 163-71.

Finegold, S.M., Sutter, V.L. and Mathisen, G.E. (1983) Normal indigenous intestinal flora, in *Human Intestinal Microflora in Health and Disease* (ed. D.J. Hentges) Academic Press, London, pp. 3-31.

Geypens, B., Claus, D., Evenepoel, P., Hiele, M., Maes, B., Peeters, M., Rutgeerts, P. and Ghoos, Y. (1997) Influence of dietary protein supplements on the formation of bacterial metabolites in the colon. *Gut* **41**, 70-76.

Gibson, G.R., Beatty, E., Wang, X. and Cummings, J.H. (1995) Selective stimulation of bifidobacteria in the human colon by oligofructose and inulin. *Gastroenterology* **106**, 975-82.

Gibson, G.R., Macfarlane, G.T. and Cummings, J.H. (1988) Occurrence of sulphate-reducing bacteria in human faeces and the relationship of dissimilatory sulphate reduction to methanogenesis in the large gut. *Journal of Appllied Bacteriology* **65**, 103-11.

Gibson, G.R., Macfarlane, S. and Macfarane, G.T. (1993) Metabolic interactions involving sulphate-reducing and methanogenic bacteria in the human large intestine. *FEMS Microbiology Ecology* **12**,117-25.

Gibson, G.R., Saavedra, J.M., Macfarlane, S. and Macfarlane, G.T. (1997) Gastrointestinal microbial disease and probiotics, in *Probiotics: Therapeutic and Other Beneficial Effects* (ed. R. Fuller), Chapman & Hall, London, pp. 10-39.

Gibson, G.R. and Wang, X. (1994a) Regulatory effects of bifidobacteria on the growth of other colonic bacteria. *Journal of Applied Bacteriology* **77**, 412-20.

Gibson, G.R. and Wang, X. (1994b) Bifidogenic properties of different types of fructooligosaccharides. *Food Microbiology* **11**, 491-98.

Gorbach, S.L., Barza, M., Giuliano, M. and Jacobus, N.V. (1988) Colonization resistance of the human intestinal microflora: testing the hypothesis in normal volunteers. *European Journal of Clinical Microbiology and Infectious Diseases* **7**, 98-102.

Gorbach, S.L., Nahas, L., Lerner, P.I. and Weinstein, L. (1967) Studies of intestinal microflora. 1. Effects of diet, age, and periodic sampling on numbers of fecal microorganisms in man. *Gastroenterology* **53**, 845-55.

Hentges, D.J. (1983) Role of the intestinal microflora in host defense against infection, in *Human Intestinal Microflora in Health and Disease* (ed. D.J. Hentges), Academic Press, London, pp. 311-32.

Hill, M.J., Drasar, B.S., Aries, V., Crowther, S. and Williams, R.E.O. (1971) Bacteria and aetiology of cancer of the large bowel. *Lancet* 95-100.

Hopkins, M.J., Cummings, J.H. and Macfarlane, G.T. (1998) Interspecies differences in maximum specific growth rates and cell yields of bifidobacteria cultured on oligosaccharides and other simple carbohydrate sources. *Journal of Applied Microbiology* **85**, 381-86.

Hugenholtz, P. and Pace, N.R. (1996) Identifying microbial diversity in the natural environment: a molecular phylogenetic approach. *Trends in Biotechnology* **14**, 190-97.

Ito, M., Deguchi, Y., Miyamori, A., Matsumoto, K., Kikuch, H. *et al.* (1990) Effects of administration of galactooligosaccharides on the human faecal microflora, stool weight and abdominal sensation. *Microbial Ecolology in Health and Disease* **3**, 285-92.

Kleesen, B., Bunke, H., Tovar, K., Noack, J. and Sawatzki, G. (1995) Influence of two infant formulas and human milk on the development of the faecal flora in newborn infants. *Acta Paediatrica* **84**, 1347-56.

Koo, M. and Rao, A.V. (1991) Long-term effect of bifidobacteria and Neosugar on precursor lesions of colonic cancer in CF1 mice. *Nutrition Reviews* **51**, 137-46.

Kurita-Ochiai, T., Fukushima, K. and Ochiai, K. (1997) Butyric acid-induced apoptosis of murine thymocytes, splenic T-cells, and human Jurkat T-cells. *Infection and Immunity* **65**, 35-41.

Laukova, A. and Czikkova, S. (1998) Inhibition effect of enterocin CCM 4231 in the rumen fluid environment. *Letters in Applied Microbiology* **26**, 215-18.

Lejeune, R., Callewaert, R., Crabbe, K. and De Vuyst, L. (1998) Modelling the growth and bacteriocin production production by *Lactobacillus amylovorus* DCE 471 in batch cultivation. *Journal of Applied Microbiology* **84**, 159-68.

Macfarlane, G.T. and Allison, C. (1986) Utilization of protein by human gut bacteria. *FEMS Microbiology Ecology* **38**, 19-24.

Macfarlane, G.T., Cummings, J.H. and Allison, C. (1986) Protein degradation by human intestinal bacteria. *Journal of General Microbiology* **132**, 1647-56.

Macfarlane, G.T. and Englyst, H.N. (1986) Starch utilization by the human large intestinal microflora. *Journal of Applied Bacteriology* **60**, 195-201.

Macfarlane, G.T. and Gibson, G.R. (1996). Carbohydrate fermentation, energy transduction and gas metabolism in the human large intestine, in *Ecology and Physiology of Gastrointestinal Microbes Vol 1: Gastrointestinal Fermentations and Ecosystems*. (eds. R.I. Mackie and B.A. White), Chapman & Hall, New York, pp. 269-318.

Macfarlane, G.T., Gibson, G.R., Beatty, E.A. and Cummings, J.H. (1992a) Estimation of short chain fatty acid production from protein by human intestinal bacteria, based on branched chain fatty acid measurements. *FEMS Microbiology Ecology* **101**, 81-88.

Macfarlane, G.T., Gibson, G.R. and Cummings, J.H. (1992b) Comparison of fermentation reactions in different regions of the human colon. *Journal of Applied Bacteriology* **72**, 57-64.

Macfarlane, G.T. and Macfarlane, S. (1991) Utilization of pancreatic trypsin and chymotrypsin by proteolytic and non-proteolytic *Bacteroides fragilis*-type bacteria. *Current Microbiology* **23**, 143-48.

Macfarlane, G.T. and Macfarlane, S. (1992) Physiological and nutritional factors affecting the synthesis and secretion of extracellular metalloproteases by *Clostridium bifermentans* NCTC 2914. *Applied and Environmental Microbiology* **58**, 1195-200.

Macfarlane, S. and Macfarlane, G.T. (1997) Formation of a dipeptidyl arylamidase by *Bacteroides splanchnicus* NCTC 10582 with specificities towards glycylprolyl-x and valylalanine-x substrates. *Journal of Medical Microbiology* **46**, 1-9.

Macfarlane, G.T., Macfarlane, S. and Gibson, G.R. (1992c) Synthesis and release of proteases by *Bacteroides fragilis*. *Current Microbiology* **24**, 55-59.

Macfarlane, G.T., Macfarlane, S. and Gibson, G.R. (1995) Co-culture of *Bifidobacterium adolescentis* and *Bacteroides thetaiotaomicron* in arabinogalactan-limited chemostats: Effects of dilution rate and pH. *Anaerobe* **1**, 275-81.

Macfarlane, G.T., Macfarlane, S. and Gibson, G.R. (1998) Validation of a three-stage compound continuous culture system for investigating the effect of retention time on the ecology and metabolism of bacteria in the human colon. *Microbial Ecology* **35**, 180-87.

Marteau, P., Pochart, P. and Flourie, B. (1990) Effect of chronic ingestion of a fermented dairy product containing *Lactobacillus acidophilus* and *Bifidobacterium bifidum* on metabolic activities of the colonic flora in humans. *American Journal of Clinical Nutrition* **52**, 685-88.

Matsui, T., Matsukawa, Y., Sakai, T., Nakamura, K., Aoike, A. and Kawai, K. (1995) Effect of ammonia on cell-cycle progression of human gastric cancer cells. *European Journal of Gastroenterology and Hepatology* **7**, S79-81.

McBain, A.J. and Macfarlane, G.T. (1997) Investigations of bifidobacterial ecology and oligosaccharide metabolism in a three-stage compound continuous culture system. *Scandinavian Journal of Gastroenterology* **32**, 32-40.

McGinnis, J.F. and Paigen, K. (1973) Site of inhibition of carbohydrate metabolism. *Journal of Bacteriology* **114**, 885-87.

Meghrous, J., Euloge, P., Junelles, A.M. Ballongue, J. and Petitdemange, H. (1990) Screening of *Bifidobacterium* strains for bacteriocin production. *Biotechnology Letters* **12**, 575-80.

Meijer-Severs, G.J. and Van Santen, E. (1986) Variations in the anaerobic faecal flora of ten healthy human volunteers with special reference to the *Bacteroides fragilis* group and *Clostridium difficile*. *Zentralblatt fur Bakteriologie Mikrobiologie und Hygiene* **261**, 43-52.

Midvedt, T. and Norman, A. (1968) Parameters in 7-α dehydroxylation of bile acids by anaerobic lactobacilli. *Acta Pathologica Microbiologica Scandinavica* **72**, 313-29.

Miller, J.B., Bull, S., Miller, J. and McVeagh, P. (1994) The oligosaccharide composition of human milk: Temporal and individual variations in monosaccharide components. *Journal of Pediatric Gastroenterology and Nutrition* **19**, 371-76.

Mitsuoka, T. (1982) Recent trends in research on intestinal flora. *Bifidobacteria Microflora* **1**, 3-24.

Mitsuoka, T. (1984) Taxonomy and ecology of bifidobacteria. *Bifidobacteria Microflora* **3**, 11-28.

Mitsuoka, T. (1989) Taxonomy and ecology of the indigenous intestinal bacteria, in *Recent Advances in Microbial Ecology* (eds. T. Hattori. Y. Ishida, Y. Maruyama, R.Y. Morita and A. Uchida), Japan Scientific Societies Press, Tokyo, pp. 493-98.

Mitsuoka, T. and Hayakawa, K. (1972) The fecal flora of man 1. Communication: The composition of the fecal flora of ten healthy human volunteers with special reference to the *Bacteroides fragilis*-group and *Clostridium difficile. Zentralblatt fur Bakteriologie Mikrobiologie und Hygiene* **261**, 43-52.

Moore, W.E.C. and Holdeman, L.V. (1974) Human faecal flora: the normal flora of 20 Japanese-Hawaiians. *Applied Microbiology* **27**, 961-79.

Muscettola, M., Massai, L., Tanganelli, C. and Grasso, G. (1994) Effects of lactobacilli on interferon production in young and aged mice. *Annals of the New York Academy of Sciences* **717**, 226-32.

O'Riordan, K.C., Condon, S. and Fitzgerald, G.F. (1995) Bacterial interference by *Bifidobacterium* species and a comparative analysis of genomic profiles from strains of this genus. *Proceedings of the Lactic Acid Bacteria Conference, Cork, Ireland*, p. 207.

Pereyra, B.S. and Lemmonier, D. (1993) Induction of human cytokines by bacteria used in dairy foods. *Nutrition Research* **13**, 1127-40.

Pitcher, M.C.L. and Cummings, J.H. (1996) Hydrogen sulphide: a bacterial toxin in ulcerative colitis? *Gut* **39**, 1-4.

Reddy, B.S. and Rivenson, A. (1993) Inhibitory effect of *Bifidobacterium longum* on colon, mammary and liver carcinogenesis induced by 2-amino-3-methylimidazo [4,5 -f] quinoline, a food mutagen. *Cancer Research* **53**, 3914-18.

Reider, E., Wagner, F. and Schweiger, M. (1979) Control of phosphoenolpyruvate-dependent phosphotransferase-mediated sugar transport in *Escherichia coli* by energization of the cell membrane. *Proceedings of the National Academy of Science, U.S.A.* **76**, 5529-33.

Roediger, W.E.W., Duncan, A., Kapaniris, O. and Millard, S. (1993) Reducing sulfur compounds of the colon impair colonocyte nutrition: implications for ulcerative colitis. *Gastroenterology* **104**, 802-9.

Roediger, W.E.W. and Nance, S. (1986) Metabolic induction of experimental ulcerative colitis by inhibition of fatty acid oxidation. *British Journal of Experimental Pathology* **67**, 773-76.

Russell, J.B. (1983) Fermentation of peptides by *Bacteroides ruminicola* B14. *Applied and Environmental Microbiology* **45**, 1566-74.

Saier, M.H. Jr. and Roseman, S. (1976) Regulation of carbohydrate intake in gram positive bacteria. *Journal of Biological Chemistry* **251**, 893-94.

Scardovi, V. (1986) Genus *Bifidobacterium,* In *Bergey's Manual of Systematic Bacteriology Vol 2* (ed. N.S. Mair), Williams & Wilkins, New York, pp. 1418-34.

Sekine, K., Toida, T., Saito, M., Kuboyama, M., Hawashima, T. and Hashimoto, Y. (1985) A new morphologically characterized cell wall preparation (whole peptidoglycan) from *Bifidobacterium infantis* with a higher efficacy on the regression of an established tumor in mice. *Cancer Research* **45**, 1300-7.

Shephard, S.E., Schlatter, C. and Lutz, W.K. (1987) N-Nitrosocompounds: relevance to human cancer, in *IARC Scientific Publications No. 57.* (eds. H. Bartels, I.K. O'Neill, R.S. Herman), IARC Publications, Lyons, pp. 328-32.

Simhon, A., Douglas J.R., Drasar B.S., Soothill, J.F. (1982) Effect of feeding on infants' faecal flora. *Archives of Diseases in Childhood* **57**, 54-58.

Smith, E.A. and Macfarlane, G.T. (1996) Studies on amine production in the human colon: Enumeration of amine forming bacteria and physiological effects of carbohydrate and pH. *Anaerobe* **2**, 285-97.

Smith, E.A. and Macfarlane, G.T. (1997) Dissimilatory amino acid metabolism in human colonic bacteria. *Anaerobe* **3**, 327-37.

Stephen, A.M. and Cummings, J.H. (1980) The microbial contribution to human faecal mass. *Journal of Medical Microbiology* **13**, 45-56.

Stephen, A.M., Wiggins, H.S. and Cummings, J.H. (1986) Effect of changing transit time on colonic microbial metabolism in man. *Gut* **28**, 601-09.

Strocchi, A. and Levitt, M.D. (1992) Factors affecting hydrogen production and consumption by human fecal flora: The critical role of hydrogen tension and methanogenesis. *Journal of Clinical Investigation* **89**, 1304-11.

Tanaka, K., Budd, M.A., Efron, M.L. and Isselbacher, K.J. (1976) Isovaleric acidemia: a new genetic defect of leucine metabolism. *Proceedings of the National Academy of Science* **56**, 236-42.

Tannock, G.W. and Savage, D.C. (1974) Influence of dietary and environmental stress on microbial populations in the murine gastrointestinal tract. *Infection and Immunity* **9**, 591-98.

Tvede, M. and Rask-Madsen, J. (1989) Bacteriotherapy for chronic relapsing *Clostridium difficile* diarrhoea in six patients. *Lancet* **i**, 1156-60.

Vince, A.J. (1986) Metabolism of ammonia, urea, and amino acids, and their significance in liver disease, in *Microbial Metabolism in the Digestive Tract*. (ed. M.J. Hill), CRC Press, Boca Raton, pp. 83-105.

Wallace, R.J. and McKain, N. (1997) Peptidase activity of human colonic bacteria. *Anaerobe* **3**, 251-57.

Weber, F.L., Banwell, J.G., Fresard, K.M. and Cummings, J.H. (1987) Nitrogen in fecal bacteria, fiber and soluble fractions of patients with cirrhosis: effects of lactulose and lactulose plus neomycin. *Journal of Laboratory and Clinical Medicine*. **110**, 259-63.

Wiggins, H.S. and Cummings J.H. (1976) Evidence for the mixing of residue in the human gut. *Gut* **17**, 1007-11.

Wrong, O.M (1988) Bacterial metabolism of protein and endogenous nitrogen compounds, in *Role of the Gut Flora in Toxicity and Cancer* (ed. I.R. Rowland), Academic Press, New York. pp. 227-62.

Yamazaki, S., Machii, K., Tsuyuki, S., Momose, H., Kawashima, T. and Ueda, K. (1985) Immunological responses to monoassociated *Bifidobacterium longum* and their relation to prevention of bacterial invasion. *Immunology* **56**, 43-50.

CHAPTER 2

Growth Substrates for the Gut Microflora

HENRIK ANDERSSON AND ANNA MARIA LANGKILDE
Department of Clinical Nutrition, Institute of Internal Medicine, Göteborg University, SWEDEN

1   Introduction

Material from the ileum constitutes the major substrate for colonic fermentation. It is comprised of a mixture of digested and non-digested nutrients, endogenous fluids, cells and mucus. The human large bowel normally contains about 200 g wet weight of material (Banwell *et al.,* 1981; Cummings *et al.,* 1980) with bacteria contributing over half of this (Stephens and Cummings, 1980).

Whilst a number of factors can influence the gut microflora composition and activities, bacteria are dependent on the type and amount of substrate that is excreted from the small bowel and it is therefore of considerable importance to quantify and characterize energy-donating nutrients that reach the large bowel.

2   Methods to estimate small bowel excretion

2.1   BREATH HYDROGEN TEST

An early, and frequently used, technique is to measure breath H_2 excretion after the consumption of test meals (Anderson *et al.,* 1981; Wolever *et al.,* 1986; Levitt *et al.,* 1987; Flourié *et al.,* 1988). Although this test was the first to show that carbohydrates are fermented in the gut, the method is only semi-quantitative and suffers from several problems (Cummings and Englyst, 1991). These include:
1) the proportion of H_2 excreted in the breath is not constant
2) different types of carbohydrates do produce different amounts of H_2 per gram of carbohydrate
3) substantial amounts of H_2 is disposed of by other routes, *e.g.* bacterial activities.

G.R. Gibson and M.B. Roberfroid (eds.), Colonic Microbiota, Nutrition and Health, 27-35.
© 1999 *Kluwer Academic Publishers. Printed in the Netherlands.*

Consequently, the breath test is now considered to be an unreliable way of obtaining good quantitative data (Cummings and Englyst, 1991).

2.2   THE INTUBATION TECHNIQUE

A quantitative assessment of ileal flow can be made by using the intubation technique and estimating quantity with an unabsorbable marker (Phillips and Giller, 1973; Levitt and Bond, 1977; Stephen *et al.*, 1983; Flourié *et al.*, 1988). Volunteers are intubated with a triple lumen tube led by a mercury bag, which can be inflated with air to accelerate progression of the tube (Beaugerie *et al.*, 1990). The tip of the catheter is confirmed by X-ray with the subject in a semi-recumbent position. The method is, however, time consuming, invasive and expensive. Consequently, only a few studies have been performed with this technique. The intestinal tube also affects gastrointestinal function (Holgate and Read, 1983), but the main problem of the method is through assessing the ileal flow rate of solids with precision. This is because only a portion of the intestinal flow can be aspirated and quantitative estimation is dependent on the amount of a marker from the intestinal flow.

2.3   THE ILEOSTOMY MODEL

Although studies in ileostomists have been performed earlier (Werch and Ivy, 1940), reliable results of specific analyses of the ileostomy contents could not be obtained until the ileostomy model was introduced in the late 1970's (Andersson and Sandberg, 1979). Subjects were given a controlled diet and efforts were also made to minimise bacterial degradation by frequent collection of contents and immediate freezing.

The main advantage of the ileostomy model is that because intestinal transit time is short, the effluent corresponding to one day's intake is excreted before the next morning (Sandberg *et al.*, 1981; Englyst and Cummings, 1986). The intra-patient and diet daily variations are small, making short-term balance studies feasible (Tornquist *et al.*, 1986). The coefficient of variation for dry matter excretion in ileostomy subjects is about 5% (Ellegård and Bosaeus, 1991). A complete quantitative collection of ileostomy contents is relatively easy to perform for ileostomy subjects, as handling of the excreta is a routine matter for these subjects.

It may be queried if the amount of substrate passing to the ileostomy bag of proctocolecomized subjects is the same as what would normally pass from the small to the large bowel. Comparisons between the amount of energy found in the immediate postoperative period (6.4% of intake) and several months later (7% in ileosotomy subjects given an enteral feed show little sign of "adaptation" (Andersson *et al.*, 1984a, b). Furthermore, the bacterial flora in the distal ileum of proctocolectormized subjects differs from that in the normal distal ileum. Numbers of bacteria in the terminal ileum of ileostomy subjects have been estimated to be about 10^7-10^8 per gram compared to 10^5-10^6 per gram in the normal ileum (Finegold *et al.*, 1970), but there will only be a small microbial degradation if the ileostomy bags are handled properly. Less then 5 mmol/L of short chain fatty acids (SCFA) are found in samples direct from the ileum (Cummings and Englyst, 1991) and between 5-30 mmol/L in ileostomy bags.

Degradation of bile acids and neutral steroids is minimal or absent (Bosaeus *et al.,* 1986; Bosaeus and Andersson, 1987). Moreover, there is limited degradation of non-starch polysaccharide (NSP) components from pectin, bran or starchy foods (Schweizer *et al.,* 1990; Englyst and Cummings, 1987a; Englyst and Kingman, 1990). Transit-time through the stomach and small intestine of ileostomy subjects is similar to that observed in healthy subjects (Holgate and Read, 1983).

Two studies have been performed to show that there is no difference between the immediate response on ileal excretion to dietary change and the excretion pattern after some weeks on the same diet (Zhang *et al.,* 1992; Zhang *et al.,* 1994). The immediate response to diet thus appears to remain in long term studies.

A comparison between the amount calculated from intubation studies and the ileostomy model, showed that the former technique gave higher values for resistant starch (Langkilde *et al.,* 1994). The results could partly be explained by different flow rates in the small intestine, with a more rapid transit in the intubation study.

3   Total amount of substrate

The total amount of substrate needed to maintain the bacterial flora in the large intestine has been estimated to be 50-70 g/d (Smith and Bryant, 1979; Bolin *et al.,* 1981; McNeil, 1984; Cummings and Macfarlane, 1991).

Ileostomy studies have shown similar figures, with 50-60 g/d of dry matter excreted from the small intestine when subjects were consuming a typical Western diet (Table 1). It can be noted that on a low fibre and low resistant starch diet, the amount of dry matter is about 50 g/d. As soon as dietary fibre or resistant starch is added to the diet the amount of dry matter increases. On high-fibre or starch diets excretion of dry matter from the small intestine reaches 80-90 g/d (Table 1, Silvester *et al.,* 1995).

As can be seen from Table l, there is a fairly good correspondence between the dry weight and energy content. The energy excreted on low-fibre, low-resistant starch diets is around 800-1000 kJ/d, and on high-fibre or high-resistant starch diets is about 1400-1700 kJ/d.

3.1   CARBOHYDRATES

3.1.1 *Starch*

Various studies have shown that part of the ingested starch, known as resistant starch (RS), escapes digestion and absorption in the small intestine of humans. This has been confirmed in breath H_2 analyses (Anderson *et al.,* 1981; Levitt *et al.,* 1987; Christl *et al.,* 1992), intubation studies (Stephen *et al.,* 1983) and with ileostomy models (Sandberg *et al.,* 1981, 1983; Englyst and Cummings 1985, 1986, 1987a,b). The total amount of resistant starch in the human diet differs according to the food items consumed. In ileostomy studies (Table 1), with diets low in fibre and resistant starch, the amount of excreted starch is around 3-4% of the daily ingested starch, but as soon as high-RS products, such as beans, are included in the diet this increases. Calculations from in vitro

Table 1. Amounts of dry weight, energy and nutrients excreted from the small bowel in groups of ileostomy subjects (n= 7-11) on different diets. Mean figures per 24h

	Dry weight (kJ/24h)	Energy (g/24h)	Nitrogen (g/24h)	Fat (g/24h)	Total starch (g/24)	Dietary fibre (g/24h)	Reference
Low fat, low-fibre, (15 g/d) diet	46	790	1.8	1	3	13	Ellegård and Bosaeus 1991
Low fat, high-fibre, (35 g/d) diet	81	1430	2.4	3	4	33	Ellegård and Bosaeus 1991
Low-fibre, low resistant starch diets with addition of:							
Wheat flour, 187 g/d	47	870	1.8	0.8	5.3	11.6	Lia et al., 1996
Potato flakes, 102 g/d	52	1000	2.1	3	3	21.6	Langkilde et al., 1990
							Schweizer et al., 1990
Ordinary corn starch, 100 g/d	55	10899	2.4	2.5	5	11	Langkilde et al., 1998
Retrograded high amylose corn starch, 100 g/d	93	1547	2.3	2.8	39	11	Langkilde et al., 1998
Bean flakes, 174 g/d	89	1590	2.7	3.5	13.2	44.3	Langkilde et al., 1990
							Schweizer et al., 1990
Oat bran, 135 g/d + wheat flour, 90 g/d + wheat gluten, 20 g	91	1690	3.6	5.5	4.8	31.2	Lia et al., 1996

measurements of RS in food items have shown figures from 3 to 6 g/d of daily RS intake in 10 different countries and in a separate study on the Italian diet 7-9 g/d (Dysseler and Hoffem, 1994, Brighenti,1997). In countries where the diets are based on starch-rich foods amounts are probably considerably higher (Cummings and Englyst, 1991).

3.1.2 *Non-starch polysaccharides (NSP)*

A major group of polysaccharides fermented by the colonic microflora is the non-starch polysaccharides (NSP). They comprise the main components of plant cell walls and include cellulose, hemicellulose and pectins but also substances like gums. Non-starch polysaccharide products added to the diet are recovered by 80-100% in the ileostomy content (Sandberg *et al.*, 1981).

The average NSP intake of in European diets is estimated to be 12-25 g/d (Cummings, 1995) and is probably similar in other Western diets. In other parts of the world, this amount can be considerably higher. Interestingly however, only less than half of the amount of energy in the ileostomy content is NSP, which means that main components are derived from other substrates.

3.1.3 *Sugars and oligosaccharides*

Lactose (in lactose-intolerant subjects), stachyose and raffinose escape digestion in the small bowel. In a recent study (Sandberg *et al.*, 1993), 65-78% raffinose and 67-70% stachyose were recovered in the ileostomy content. Sugars are recovered only in very small amounts (Schweizer *et al.*, 1990).

The amount of inulin in the large gut varies dependent on that available in foods. Wheat flour contains 1-4% (w/v) fructans in the solid matter, artichoke 20-65% (w/v), asparagus 30% (w/v) and onion up to 50% (w/v). Ileostomy studies have shown that about 90% of inulin and oligofructose in the diet can be recovered in ileostomy effluent (Bach Knudsen and Hessov, 1995; Ellgård *et at.*, 1997). Addition of these purified substances does not influence the absorption of other nutrients from the small bowel (Ellegård *et al.*, 1997).

3.2   PROTEIN AND FAT

Daily amounts of N_2 excreted from the small bowel in ileostomy studies vary from 1-2 g/d on low-fibre, low-RS diets to almost 3 g/d on high-fibre or high-RS diets (Table 1). The major part of the N_2 excreted is protein, (48-51%), and peptides, (20-30%), with only small amounts of urea, ammonia and nitrate (Gibson *et al.*, 1976; Florin *et al.*, 1990). There is a correlation between the amount of N_2 excreted and that in the diet (Silvester *et al.*, 1995). According to estimates given by Silvester and Cummings (199S), N_2 losses amount to between 1g/d and 3g/d from a nitrogen intake of 0-20 g/d. This is in accordance with the figures given in Table 1. It has been suggested that dietary fibre increases N_2 excretion. However, it was shown that intake of the viscous fibre pectin only increased excretion by less than 1 g/d (Sandberg *et al.*, 1983).

3.3 MUCUS

The amount of available mucus (acidic glycoprotein from the goblet cells) in the intestine has been estimated at 2-5 g/d (Stephen *et al.*, 1983; Englyst and Cummings 1 985; Wolever *et al.*, 1986; Levitt *et al.*, 1987). This has been analysed in just one ileostomy subject (Schweizer *et al.*, 1990). Quantitative determination of key mucus constituent sugars revealed the presence of 0.45 g N-acetyl-glucosamine, 0.2 g N-acetyl-galactosamine, 0.2 g fructose and 0.4 g sialic acids.

4 Summary

A variety of different substrates are excreted from the small bowel as possible substrates for bacterial fermentation in the large intestine. Ileostomist studies have shown that between 50-90 g/d of dry weight, which yields 800-1700 kJ/d, is excreted from the small bowel. Although the amount of dietary fibre, (NSP), in the diet, causes considerable changes in excretion patterns, NSP may not be the dominant part of the energy source provided to the large bowel. Resistant starch amounts to 3-4 % of the total starch intake. Oligosaccharides, sugars and fat also contribute, but at lower amounts. The excretion of protein can be estimated to 10-15 g/d. Calculation of substrate and energy shows that a considerable amount of energy equivalent to 15-40 g of carbohydrate is metabolised in the large bowel.

References

Anderson, I.H., Levine, A.S. and Levitt, M.D. (1981) Incomplete absorption of the carbohydrate in all-purpose wheat flour. *New England Journal of Medicine* **304**, 891-92.

Andersson, H., Hultén, L., Magnusson, O. *et al.* (1984a) Energy and mineral utilization from a peptide-based elemental diet and a polymeric enteral diet given to ileostomists in the early postoperative course. *Journal of Parenteral and Enteral Nutrition* **8**, 497-500.

Andersson, H., Bosaeus, I., Ellegår, L., *et al.* (1984b) Comparison of an elemental and two polymeric diets in colectomized patients with or without intestinal resection. *Clinical Nutrition* **3**, 183-89.

Andersson, H. and Sandberg, A-S. (1979) En försöksmodell för kostfiberstudier. *Livsmedelsteknik* **21**, 194-95.

Bach Knudsen, K.E. and Hessov, I. (1995) Recovery of inulin from Jerusalem artichoke (Helianthus tuberosus L.) in the small intestine of man. *British Journal of Nutrition* **74**, 101-13.

Banwell, J.G., Bransch, W.J. and Cummings, J.H. (1981) The microbial mass in the human large intestine. *Gastroenterology* **80**, 1104.

Beaugerie, L., Flourié, B., Marteau, P., *et al.* (1990) Digestion and absorption in the human intestine of three sugar alcohols. *Gastroenterology* **99**, 717-23.

Bolin, T., Sjödahl, R., Sundqvist, T., *et al.* (1981) Passage of molecules through the wall of the gastrointestinal tract. Increased passive permeability in rat ileum after exposure to lysolecithin. *Scandinavian Journal of Gastroenterology* **16**, 897-901.

Bosaeus, I. and Andersson, H. (1987) Short-term effect of two cholesterol-lowering diets on sterol excretion in ileostomy patients. *American Journal of Clinical Nutrition* **45**, 54-59.

Bosaeus, I., Carlsson, N.-G., Sandberg, A.-S. *et al.* (1986) Effect of wheat bran and pectin on bile acid and cholesterol excretion in ileostomy patients. Human Nutrition: *Clinical Nutrition* **40C**, 429-40.

Brighenti, F., Casiraghi, M.C. and Baggio, C. (1997) Resistant starch in the Italian diet. Abstracts. 16th International Congress of Nutrition, Montreal (1997), PW12.5, p 71.

Christl, S., Murgatroyd, P.R., Gibson, G.R. *et al.* (1992) Production, metabolism, and excretion of hydrogen in the large intestine. *Gastroenterology* **102**, 1424-26.

Cummings, J.H. (1995) Dietary fibre intakes in Europe: overview and summary of European research activities, conducted by members of the Management Committee of COST 92. *European Journal of Clinical Nutrition* **49**, Suppl 3, S5-9.

Cummings, J.H. and Englyst, H.N. (1991) Measurement of starch fermentation in the human large intestine. *Canadian Journal of Physiology and Pharmacology* **69**, 121-29.

Cummings, J.H. and Macfarlane, G.T. (1991) The control and consequences of bacterial fermentation in the human colon. *Journal of Applied Bacteriology* **70**, 443-59.

Cummings, J.H., Banwell, J.G., Segal, I., *et al.* (1980) The amount and composition of large bowel contents in man. *Gastroenterology* **98**, A408.

Dysseler, P. and Hoffem, D. (1994) Estimation of resistant starch intake in Europe. In *Proceedings of the Concluding Plenary Meeting of* EURESTA. (eds. N-G. Asp, J.M.M. van Amelsvoort, J.G.A.J. Hautvast). European Flair-Concerted Action No. 11 (COST 911), pp. 84-86.

Ellegård, L. and Bosaeus, I. (1991) Sterol and nutrient excretion in ileostomists on prudent diets. *European Journal of Clinical Nutrition* **45**, 451-57.

Ellegård, L., Andersson, H. and Bosaeus, I. (1997) Inulin and oligofructose do not influence the absorption of cholesterol, or the excretion of cholesterol, Ca, Mg, Zn, Fe, or bile acids but increases energy excretion in ileostomy subjects. *European Journal of Clinical Nutrition* **51**, 1-5.

Englyst, H.N. and Cummings, J.H. (1985) Digestion on the polysaccharides of some cereal foods in the human small intestine. *American Journal of Clinical Nutrition* **42**, 778-87.

Englyst, H.N. and Cummings, J.H. (1986) Digestion of the carbohydrates of banana (*Musa paradisiaca sapientum*) in the human small intestine. *American Journal of Clinical Nutrition* **44**, 42-50.

Englyst, H.N. and Cummings, J.H. (1987a) Digestion of polysaccharides of potato in the small intestine of man. *American Journal of Clinical Nutrition* **45**, 423-31.

Englyst, H.N. and Cummings, J.H. (1987b) Resistant starch, a 'new' food component: A classification of starch for nutritional purposes. In *Cereals in a European Context* (ed. I.D. Morton) Ellis Horwood, Chichester, pp. 221-33.

Englyst, H.N. and Kingman, S.M. (1990) Dietary fiber and resistant starch: A nutritional classification of plant polysaccharides. In *Dietary Fiber, Chemistry, Physiology and Health Effects* (eds. D. Kritchevsky, J.A. Anderson) Plenum Publ Co, New York, pp. 49-65.

Finegold, S.M., Sutter, V.L., Boyle, J.D. *et al.* (1970) The normal flora of ileostomy and transverse colostomy effluents. Journal of Infectious Diseases **122**, 376-81.

Florin, T.H.J., Neale, G. and Cummings, J.H. (1990) The effect of dietary nitrate on nitrate and nitrite excretion in ileal effluents and urine in man. *British Journal of Nutrition* **64**, 387-97.

Flourié, B., Leblond, A., Florent, Ch., *et al.* (1988) Starch malabsorption and breath gas excretion in healthy humans consuming low-and high-starch diets. *Gastroenterology* **95**, 356-63.

Gibson, J.A., Sladen, G.E. and Dawson, A.M. (1976) Protein absorption and ammonia production: the effects of dietary protein and removal of the colon. *British Journal of Nutrition* **35**, 61-65.

Holgate, A.M. and Read, N.W. (1983) Relationship between small bowel transit time and absorption of a solid meal; influence of metoclopramide, magnesium sulfate and lactulose. *Digestive Diseases and Sciences* **28**, 812-19.

Langkilde, A.M., Andersson, H., Schweizer, T.F. *et al.* (1990) Nutrients excreted in ileostomy effluents after consumption of mixed diets with beans or potatoes. I. Minerals, protein, fat and energy. *European Journal of Clinical Nutrition* **44**, 559-66.

Langkilde, A.M., Andersson, H., Faisant, N. and Champ M. (1994). A comparison between the intubation technique and the ileostomy model for in vivo measurement of RS. In *Proceedings of the Concluding Plenary Meeting of* EURESTA. (eds. N-G. Asp, J.M.M.van Amelsvoort, J.G.A.J. Hautvast) European Flair-Concerted Action No. 11 (COST 911), pp. 28-30.

Langkilde, A.M., Ekwall, H., Björck, I., *et al.* (1998) Retrograded high amylose corn starch reduces cholic acid excretion from the small bowel in ileostomy subjects. *European Journal of Clinical Nutrition* In press.

Levitt, M.D. and Bond, J.H. (1977) Use of the constant perfusion technique in the nonsteady state (editorial). *Gastroenterology* **73**, 1450-54.

Levitt, M.D., Hirsch, P., Fetzer, C.A., *et al.* (1987) H2 excretion after ingestion of complex carbohydrates. *Gastroenterology* **92**, 383-9.

Lia, Å., Sundberg, B., Åman, P., *et al.* (1996) Substrates available for colonic fermentation from oat, barley and wheat bread diets. A study in ileostomy subjects. *British Journal of Nutrition* **76**, 797-808.

McNeil, N.I. (1984) The contribution of the large intestine to energy supplies in man. *American Journal of Clinical Nutrition* **39**, 338-42.

Phillips, S.F. and Giller, J. (1973) The contribution of the colon to electrolyte and water conservation in man. *Journal of Laboratory and Clinical Medicine* **81**, 733-46.

Sandberg, A.-S., Andersson, H., Hallgren, B., *et al.* (1981) Experimental model for in vivo determination of dietary fibre and its effect on the absorption of nutrients in the small intestine. *British Journal of Nutrition* **45**, 283-94.

Sandberg, A-S., Ahderinne, R., Andersson, H., *et al.* (1983)The effect of citrus pectin in the absorption of nutrients in the small intestine. *Human Nutrition: Clinical Nutrition* **37C**, 171-83.

Schweizer, T.H., Andersson, H., Langkilde, A.M., *et al.* (1990) Nutrients excreted in ileostomy effluents after consumption of mixed diets with beans or potatoes. II. Starch, dietary fibre and sugars. *European Journal of Clinical Nutrition* **44**, 567-75.

Silvester, K.R. and Cummings, J.H. (1995) Does digestibility of meat protein help explain large bowel cancer risk? *Nutrition and Cancer* **24**, 279-88.

Silvester, K.R., Englyst, H.N. and Cummings, J.H. (1995) Real recovery of starch from whole diets containing resistant starch measured in vitro and fermentation of ileal effluent. *American Journal of Clinical Nutrition* **62**, 403-11.

Smith, C.J. and Bryant, M.P. (1979) Introduction to metabolic activities of intestinal bacteria. *American Journal of Clinical Nutrition* **32**, 149-57.

Stephen, A.M. and Cummings, J.H. (1980) The microbial contribution to human faecal mass. *Journal of Medical Microbiology* **13**, 45-56.

Stephen, A.M., Haddad, A.C. and Phillips, S.F. (1983) Passage of carbohydrate into the colon. Direct measurements of humans. *Gastroenterology* **85**, 589-95.

Tornquist, H., Rissanen, A. and Andersson, H. (1986) Balance studies in patients with intestinal resection, how long is enough? *British Journal of Nutrition* **56**, 11-16.

Werch, S.C. and Ivy, A.C. (1940) On the fate of ingested pectin. *American Journal of Clinical Nutrition* **8**, 101-5.

Wolever, T.M., Cohen, Z., Thompson, L.U., *et al.* (1986) Ileal loss of available carbohydrate in man: comparison of a breath hydrogen method with direct measurement using a human ileostomy model. *American Journal of Gastroenterology* **81**, 115-22.

Zhang, J.-X., Hallmans, G., Andersson, H., *et al.* (1992) Effects of oat bran on plasma cholesterol and bile acid excretion in nine subjects with ileostomies. *American Journal of Clinical Nutrition* **56**, 99-105.

Zhang J.-X., Lundin, E., Hallmans, G., *et al.* (1994) Effect of rye bran on excretion of bile acids, cholesterol, nitrogen, and fat in human subjects with ileostomists. *American Journal of Clinical Nutrition* **59**, 389-94.

CHAPTER 3

Biochemistry of Fermentation

ANNICK BERNALIER[1], JOËL DORE[2] AND MICHELLE DURAND[3]
[1]*Laboratoire de Microbiologie, INRA, Centre de Recherches de Clermont-Ferrand-Theix, 63122 Saint-Genès-Champanelle, FRANCE*
[2]*Unité d'Ecologie et Physiologie du Système Digestif, INRA, Domaine de Vilvert, 78352 Jouy-en-Josas, FRANCE*
[3]*Laboratoire de Nutrition et Sécurité Alimentaire, INRA, Domaine de Vilvert, 78352 Jouy-en-Josas, FRANCE*

1   Introduction

The human large intestine is a complex anaerobic ecosystem, composed of numerous different species, which degrade and ferment substrates that have either escaped the digestion in the upper digestive tract or are produced by the host. It is recognised that a significant daily quantity of undigested dietary carbohydrate enters the colon (Edwards and Rowland, 1992; see Chapter 2). In contrast, the amount of carbohydrate fermented from endogenous sources like mucus remains undefined (Cummings and Macfarlane, 1991; Flourié *et al.*, 1991). The microbial degradation of this organic matter in the colon constitutes a fundamental process which requires the contribution of different groups of microorganisms linked in a trophic chain (Wolin and Miller, 1983). These food-chain reactions break macromolecules such as complex polysaccharides down to short-chain fatty acids (mainly acetate, propionate and butyrate) and gases (H_2, CO_2 and in some case CH_4). Polysaccharide degrading bacteria hydrolyse polymers into smaller fragments that can be used by saccharolytic bacteria. This cross-feeding allows maintainence of bacterial diversity in the ecosystem. The fermentation products of hydrolytic and saccharolytic bacteria include intermediates, such as lactate or succinate, that are metabolised by other species and do not accumulate to any significant extent in the colon. Hydrogen, which derives enterely from these fermentative processes, can be re-utilized in situ by hydrogenotrophic microorganisms.

G.R. Gibson and M.B. Roberfroid (eds.), Colonic Microbiota, Nutrition and Health, 37-53.
© 1999 *Kluwer Academic Publishers. Printed in the Netherlands.*

2   Degradation of carbohydrates to monomers

Dietary carbohydrates are variably susceptible to bacterial attack. Their rates and intensities of fermentation differ widely according to structure, physicochemical properties, types of monomeric linkages, etc. The bacterial population in the large bowel is well adapted to grow on these different types of polysaccharide. The main enzymes and bacterial species involved in the hydrolysis of the carbohydrates available in the colon have been reviewed and are listed in Table 1. It is clear that one polysaccharide substrate can be fermented by different microorganisms.

Intestinal bacteria may control the metabolism of polymerised carbohydrates in a variety of manners (Macfarlane *et al.*, 1990). They can regulate the synthesis of enzymes involved in initial hydrolysis of the substrates and control subsequent transport of carbohydrates into the cells. Salyers (1985) has reported that most of polysaccharide-degrading enzymes are inducible.

Main polysaccharide degraders belong to the genera *Bacteroides*, *Bifidobacterium*, *Ruminococcus* and some *Eubacterium* and *Clostridium*. *Bacteroides* spp. are the most versatile polysaccharide utilisers in the colon. Most of the polysaccharidases produced appear to be cell-associated, with the location of these enzymes being in the cytoplasm, periplasm and cell envelope (Macfarlane *et al.*, 1990). Membrane proteins seem to be involved in bringing the polysaccharides into contact with the enzymes (McCarthy *et al.*, 1985; Fonty and Gouet, 1989).

3   Fermentation of monomers

Monomeric fermentation allows conservation of part of the energy as the energy-rich phosphate-bond of ATP. The energy of ATP is further used for the biosynthesis of cell material. A proportion of the carbon skeletons and N compounds released during carbohydrate and protein degradation enter the biosynthetic processes. The remaining products are excreted and form major end-products of fermentation.

The formation of ATP from ADP and inorganic phosphate (Pi) is coupled to oxido-reduction reactions. Two main types of phosphorylation are recognised, substrate level phosphorylation (SLP) and electron transport phosphorylation (ETP).

The SLP pathway forms intermediate products, rich in energy, by reactions involving dehydrogenases or lyases and energetic transfer using kinases. As a consequence, there is a release of reduced cofactors such as $NADH + H^+$ or $NADPH + H^+$. These reduced cofactors must be regenerated, either by means of hydrogenases with the release of H_2 or by reduction of electron acceptors with a final formation of H_2-rich end products such as ethanol, lactate or succinate (Figs. 1 and 2).

In the ETP pathway, free energy is first conserved as a transmembrane electrochemical gradient which then drives ATP synthesis via membrane-associated ATP synthases. Energy can be conserved in the form of transmembrane proton or sodium ion gradients (Gottschalk, 1988). This pathway of ATP formation is involved in the reduction of fumarate to succinate and in methanogenesis, sulphate and nitrate reduction as well as in the autotrophic pathway of reductive acetogenesis.

Table 1. Enzymatic activities of human colonic bacteria

Dietary carbohydrates	Enzymes	Bacterial species
Cellulose	Cellulase, glycosidase	*Bacteroides, Ruminococcus*[1]
Xylan	Xylanase, arabinofuranosidase, xylosidase	*Bacteroides, Bifidobacterium*
Arabinogalactan	Arabinogalactanase, arabinofuranosidase, galactosidase	*Bacteroides, Bifidobacterium*
Pectin	Pectinase, exopectate lyase[2], glucuronidase, arabinofuranosidase	*Bacteroides, Bifidobacterium, Eubacterium, Clostridium, Lachnospira*
Starch	Amylase, glycosidase	*Bacteroides, Bifidobacterium, Eubacterium, Clostridium, Fusobacterium*
Guar gum	Galactomannanase, mannosidase, galactosidase	*Bacteroides, Ruminococcus*
Gum arabic	Arabinogalactanase, pullulanase, galactosidase, glucuronidase	*Bifidobacterium*
Laminarin	Laminarinase (ß-glucanase) ß-glucosidase	*Bacteroides, Peptostreptococcus*
Galacto-oligosaccharides	ß-galactosidase	*Bifidobacterium, Lactobacillus, Bacteroides*
Fructooligosaccharides	Fructooligosaccharidases	*Bifidobacterium, Eubacterium, Clostridium*
Endogenous carbohydrates		
Mucin	Galactosidase, glycosidase, fucosidase, glucosulphatase, sialata o-acetyl-esterase[3], hyaluronidase, sialidase	*Bacteroides, Ruminococcus, Bifidobacterium*
Chondroitin sulphate	Pullulanase, glucuronidase, galactosidase	*Bacteroides*

Based on: Salyers, 1985; Cummings *et al.*, 1989; Macfarlane and Cummings, 1991; Edwards and Rowland, 1992; Macfarlane and Macfarlane, 1993; Szylit and Andrieux, 1993

---

[1] Wedeking *et al.*, 1988

[2] Jensen and Canale-Parola, 1986

[3] Corfield *et al.*, 1992

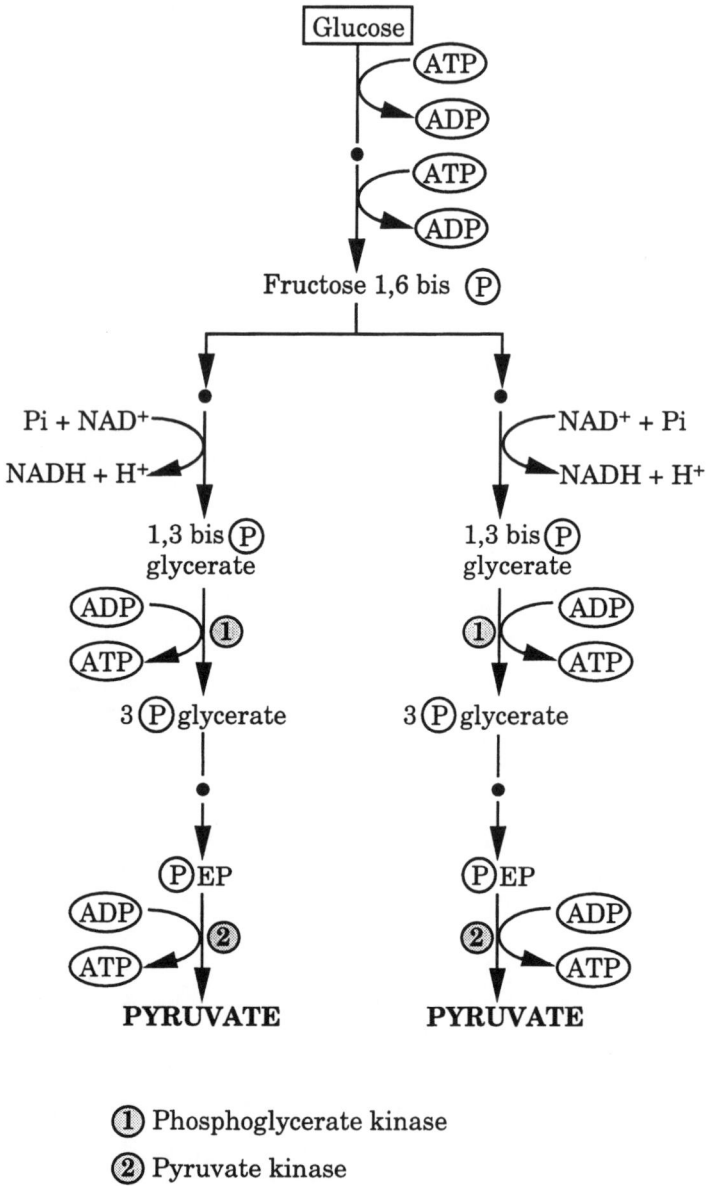

Figure 1. Simplified scheme of the EMP pathway showing substrate-linked phosphorylation.

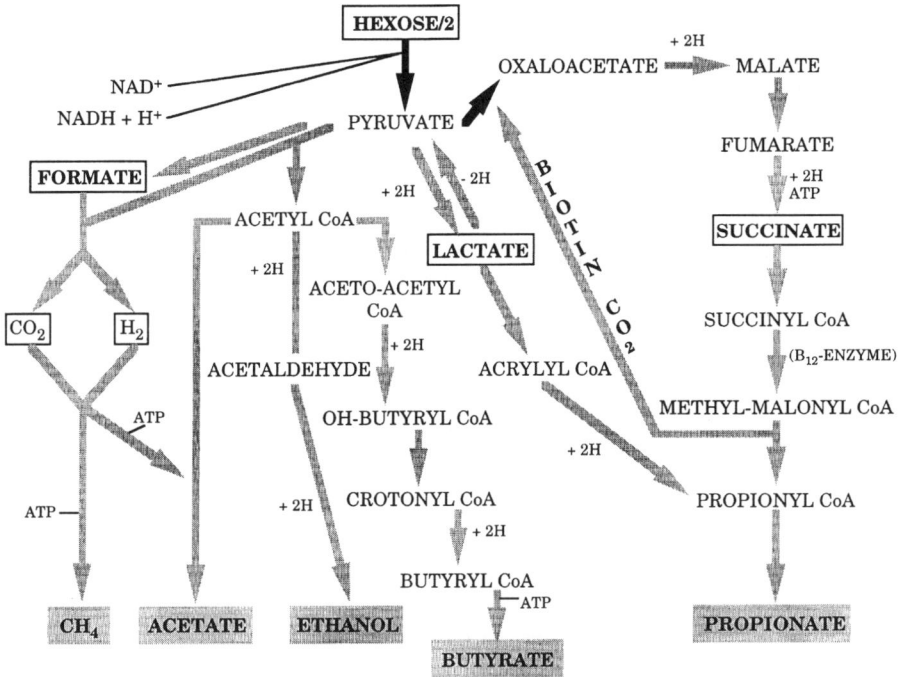

• 2H represents reduced equivalents or H_2

Figure 2. Main pathways of pyruvate conversion to end-products (adapted from Southgate, 1990).

3.1   GLYCOLYSIS AND PENTOSE METABOLISM

The Embden-Meyerhof-Parnas (EMP) pathway (Fig. 1) maximises the yield of ATP through SLP. It is a most common pathway in intestinal bacteria. Two ATP are formed during the oxidation of one glucose to two pyruvates, together with the release of two NADH + H$^+$. Alternative catabolic routes employing the Entner-Doudoroff, pentose and pentose phosphoketolase pathways can also be used to metabolise hexoses. These pathways have in common the enzyme glucose-6-phosphate dehydrogenase. The Entner-Doudoroff pathway yields 2 NADPH + H$^+$ and only 1 ATP per 2 pyruvate. This pathway, not advantageous for anaerobic bacteria, might be important when gluconate, mannonate or hexuronates serve as substrates (Gottschalk, 1988).

Pentose fermentation can proceed by the transketolase and transaldolase reactions of the pentose cycle or by a phosphorolytic cleavage (phosphoketolase). The pentose cycle is also important in the interconversion of sugars and fulfils the requirements for precursors of nucleotide synthesis.

3.2   PYRUVATE METABOLISM

Pyruvate is the central intermediate in fermentation. $NADH + H^+$ formed during glycolysis must be regenerated in pyruvate metabolism so that fermentation can continue (Fig. 2).

3.2.1   *Conversion of pyruvate into formate, acetate, butyrate or ethanol via acetyl-CoA formation*

Pyruvate can be oxidised to acetyl-CoA with the formation of either formate or CO_2 plus a reduced electron carrier. Formate can be produced by the pyruvate formate lyase or formate dehydrogenase pathways by many pure cultures of bacterial species including *Bacteroides*, *Clostridium*, *Eubacterium*, *Ruminococcus*, *Lachnospira*, *Butyrivibrio*, *Succinivibrio* and *Bifidobacterium*. In mixed culture, little formate is observed under steady-state conditions, which suggests a reutilisation of formate, or reducing equivalents derived from its oxidation to CO_2.

With pyruvate oxidoreductase, the reduced electron carrier can be either the iron-sulphur protein ferredoxin or a flavin nucleotide. When ferredoxin is the electron carrier, in this reaction, the electrons can readily reduce protons to H_2. Thus, H_2 can be formed as observed for *Ruminococcus* and *Clostridium*. Reduced flavins participate in the reduction of fumarate to succinate or the reduction of crotonyl-CoA to butyryl-CoA in *Bacteroides* and *Butyrivibrio* and in the reduction of acrylyl-CoA to propionate in *Megasphœra elsdenii*. The overall reaction of acetate formation is:

$$\text{Hexose} \longrightarrow 2 \text{ acetate} + 4\ H_2 + 2\ CO_2 + 4ATP$$

Acetate is produced by many intestinal bacteria including *Ruminococcus* and *Propionibacterium*, and is the major acid product of nearly all species of *Bacteroides*. In ruminococci, acetate formation requires collaboration between H_2-producing and H_2-utilising organisms such as methanogens, acetogens or sulphate reducers (Wolin and Miller, 1994). The partial pressure of H_2 must be low enough (at or below 1 kPa) for the oxidation of $NADH + H^+$ to proceed with production of H_2. Otherwise, instead of producing acetate, the organisms produce lactate or ethanol in order to utilise $NADH + H^+$.

Acetyl-CoA serves as a precursor for butyrate and ethanol. For butyrate, two moles of acetyl-CoA are condensed. Acetoacetyl-CoA is then converted to butyryl-CoA with two reducing steps by a series of reactions similar to a reversal of ß-oxidation. One mole of ATP is formed, by the conversion of acetoacetyl-CoA in butyrate.
The overall reaction is:

$$\text{Hexose} \longrightarrow \text{butyrate} + 2\ H_2 + 2\ CO_2 + 3\ ATP$$

Fusobacterium, *Eubacterium* and *Clostridium* spp. are major butyrate formers in human intestine (Wolin and Miller, 1983). Acetyl-CoA is an electron acceptor in ethanol formation. It is first reduced to acetaldehyde which is then reduced to ethanol with the overall oxidation of $2NADH + H^+$ per mole of ethanol formed. All the H_2 equivalents produced from hexose fermentation are utilised, but no more ATP is formed from acetyl-CoA and the overall reaction is:

$$\text{Hexose} \longrightarrow 2 \text{ ethanol} + 2\ ATP + 2\ CO_2$$

When grown in pure culture, ethanol can be formed by *Lactobacillus*, *Bifidobacterium*, *Clostridium*, *Eubacterium* and by a few *Bacteroides* species. Yet, in mixed culture it is only observed in trace quantities. Ethanol can be fermented by other microorganisms such as the clostridia.

3.2.2 *Conversion of pyruvate into propionate, valerate and caproate*

Propionate is formed by a variety of anaerobic bacteria, either from glucose or lactate fermentation. It is synthesised by two different pathways (Fig. 2) *i.e.* through succinate, or direct reductive process involving formation of lactate and acrylyl-CoA.

The succinate-propionate pathway. In this pathway, succinate can be an intermediate but also an end-product of fermentation in some species. *Bacteroides* from the human large intestine are able to decarboxylate succinate to propionate. The methyl-malonyl-CoA mutase is a vitamin B_{12} dependent enzyme. Vitamin B_{12} stimulates the formation of propionate from succinate in *Bacteroides*, whereas succinate accumulates in its absence. In mixed culture, vitamin B_{12} is probably produced by other microorganisms.

During the formation of propionate, methylmalonyl-CoA can act as CO_2 donor and biotin as a CO_2-carrier in transcarboxylation reaction of pyruvate or phosphoenolpyruvate to form oxaloacetate. When CO_2 is limiting, *Bacteroides fragilis* forms more propionate than succinate in order to increase CO_2 recycling (Caspari and Macy, 1983).

The lactate-acrylate pathway. Pyruvate can be reduced to lactate by many gut anaerobes. In the direct reductive pathway, lactate is converted into a CoA ester (lactyl-CoA) which is dehydrated into acrylyl CoA and subsequently reduced into propionyl-CoA. The reduction of acrylyl-CoA involves an electron transport chain employing ferredoxin and an electron-transferring flavoprotein. Russell and Wallace (1988) quoted that ATP synthesis had not been demonstrated. However, Gottschalk (1988) estimated that 1 mol ATP / 3 mol lactate was formed. The overall reaction for propionate formation is:

$$\text{3 Hexose} \longrightarrow \text{4 propionate} + \text{2 acetate} + \text{2 } CO_2 + \text{x ATP}$$

The number of moles of ATP produced depends on the pathway of propionate formation and may vary from 8 (acrylate pathway) to a maximum of 12 (succinate pathway). In humans, the predominant route of propionate formation would be the succinate pathway as suggested by Miller and Wolin (1979). However, this aspect requires further attention.

Formation of valerate and caproate. Valerate and/or caproate can be produced in reduced amounts from pyruvate by some species of *Clostridium* or *Megasphœra*. Valerate is formed from the condensation of propionate and acetate. Caproate is formed by the condensation of 3 acetate units or from butyrate and a C_2-unit (Prins, 1977).

4   Lactate formation and utilisation

Many lactate-producing bacteria are present in the gut (*Bacteroides*, *Eubacterium*, *Clostridium*, *Streptococcus*, *Peptostreptococcus*, *Lactobacillus*, *Bifidobacterium*). Lactic acid bacteria may exist in relatively low numbers but their numbers can increase considerably under suitable conditions (Czerkawski, 1986).

Pathways of lactate production involves two separate enzymes producing the two lactate isomers D (-) and L (+) (Prins, 1977; Gottschalk, 1988). *Lactobacillus* spp. can produce D (-), DL or L (+) lactic acid, whereas *Bifidobacterium* and *Streptococcus* produce only the L (+) form.

Lactobacilli use either of two pathways to form lactate:
- The homofermentative pathway yields 2 mol of lactate per mol of glucose. The overall reaction is:

$$\text{glucose} \longrightarrow \text{2 lactate + 2 ATP}$$

- The heterofermentative pathway generally yields 1 mol each of lactate, ethanol and CO_2 per mol of glucose. As in the oxidative pentose phosphate cycle, ribulose-5-phosphate is formed. Xylulose-5-phosphate is cleaved into glyceraldehyde-3-phosphate and acetylphosphate by a phosphoketolase. Acetyl phosphate is converted into acetyl-CoA and reduced to ethanol, while glyceraldehyde-3-P gives pyruvate and then lactate. The overall reaction is:

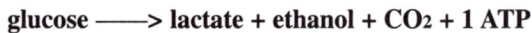

$$\text{glucose} \longrightarrow \text{lactate + ethanol} + CO_2 + \text{1 ATP}$$

Bifidobacteria utilise a distinct pathway of glucose fermentation. The classical bifidus pathway as described by Gottschalk (1988) yields 2 lactate and 3 acetate from 2 hexoses. Here, fructose-6-phosphate is split by phosphoketolases into acetyl phosphate and erythrose-4-phosphate. Erythrose-phosphate is converted to 2 acetylphosphate and 2 glyceraldehyde phosphate. The formation of acetate from acetylphosphate is coupled to the formation of ATP from ADP. Thus, 2.5 mol of ATP are produced per mol of hexose.

Lactate can be utilised by *Propionibacterium acnes*, *Veillonella* spp., *Megasphæra elsdenii*, *Clostridium* spp. and dissimilatory sulphate-reducers. It is converted into acetate, propionate (see above), butyrate and longer chain fatty acids. In the human colon, information on the specific role of different lactate utilisers and their pH-tolerance are lacking. However, effects of D-lactate absorption on the physiopathology of the host may be significant.

5   Hydrogen utilisation by hydrogenotrophic flora

The calculation of H_2 balance between production (2A+P+4B+3V+L) and utilisation (2P+2B+4V+L+4M+H_2) in methanogenic human flora gives recoveries (utilisation/ production) ranging between 80 and 100% (Durand *et al.*, 1994). Conversely, H_2 recoveries are much lower when the microflora harbours a low concentration of methanogens (Duncan and Henderson, 1990; Durand *et al.*, 1994). This means that in the absence of methanogenesis, alternatives other than the formation of H_2-rich acids exist for disposal of the gas (Gibson *et al.,* 1990). The other likely pathways are reduction of CO_2 to acetate, reduction of sulphate to sulphide, and reduction of nitrate to ammonia or nitrogen.

5.1 METHANOGENESIS

Human colonic methangens have been characterised by isolation and thorough identification from a limited number of stool samples (Nottingham and Hungate, 1968; Miller and Wolin, 1986). Only two species, belonging to two different genera have been described: *Methanobrevibacter smithii* and *Methanosphæra stadtmaniæ* (Miller and Wolin, 1986). These faecal methanogens belong to the domain Archae and are characterised by a fairly restricted substrate range.

Methanobrevibacter smithii is the predominant species of methanogen found in human colonic samples. It derives energy from the production of CH_4 by reduction of CO_2 with H_2. Hydrogen and CO_2, or formate, are the sole substrates. This colonic methanogen has a complete set of specific coenzymes including: Coenzyme M, methanofuran, tetrahydromethanopterin, 7-heptanoylthreonine phosphate and coenzymes F430 and F420 (Rouvière and Wolfe, 1988; Wolfe, 1993). These coenzymes, and associated enzymes, allow stepwise reduction of CO_2 to the formyl-, methenyl, methylene and methyl-levels of oxidation. After a methyl transfer yielding methyl-SCoM, methane is released by the methyl-reductase, a central enzymatic complex common to all methanogens (Wolfe, 1993). The overall stoichiometry of reaction is:

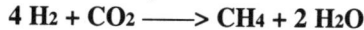

$$4 H_2 + CO_2 \longrightarrow CH_4 + 2 H_2O$$

Methanosphæra stadtmaniæ is the other reported species. It is widespread, yet consistently observed at lower population levels than *M. smithii*. *Methanosphæra stadtmaniæ* exclusively derives energy from the production of methane by reduction of methanol with H_2 (Miller and Wolin, 1985). *Methanosphæra stadtmaniæ* does not possess methanofuran nor active enzymes involved in the reduction of formaldehyde to methyl-SCoM (Van de Wijngaard *et al.,* 1991). The overall stoichiometry of metabolism is:

$$CH_3OH + H_2 \longrightarrow CH_4 + H_2O$$

It is most probable that this metabolic specificity is associated to the utilisation of methanol liberated by pectinolytic microorganisms (Miller and Wolin, 1985).

Enumeration of methanogens have shown that poplation levels may vary markedly between individuals (Miller and Wolin, 1983, 1986; Pochart *et al.,* 1992).

The dominant methanogenic Archae characterised from human faeces represent an ecological niche restricted to H_2 utilisation. However, it is most likely, that the actuel biodiversity of methanogens in the human colon is far from complete.

5.2   DISSIMILATORY SULPHATE-REDUCTION

Sulphate-reducing bacteria (SRB) can be characterised by the ability to utilise sulphate as their major terminal electron acceptor and to couple the generation of ATP (via ETP pathway) to this reduction. Most substrates are intermediate breakdown products of fermentation. Hydrogen is an energy source for many strains. Metabolism of a particular substrate involves either an incomplete oxidation to a C_2-unit, or a complete oxidation to CO_2. Amongst the substrates they can utilise as electron donors, in addition to H_2 and formate, are lactate, pyruvate and ethanol all of which are degraded to acetate.

The reduction of sulphate to hydrogen sulphide, proceeds through a number of intermediate stages. Sulphate is activated by means of ATP. The enzyme ATP sulfurylase catalyses sulphate attachment to a phosphate of ATP leading to the formation of adenosine phosphosulfate (APS). The sulphate ion of APS is directly reduced to sulphite (SO_3^{2-}) and subsequent reductions proceed readily. Hydrogen is evolved by a cytoplasmic hydrogenase and is utilised for sulphate reduction by periplasmic hydrogenases (Gottschalk, 1988). The equation for H_2 as electron donor is:

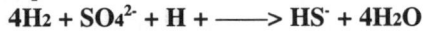

$$4H_2 + SO_4^{2-} + H + \longrightarrow HS^- + 4H_2O$$

A range of nutritionally and physiologically distinct SRB, belonging to different genera, have been identified in human faeces. The numerically predominant genus of SRB is *Desulfovibrio* (Macfarlane and Gibson, 1994). Sources of sulphate are represented by sulphated dietary components and endogenous secretions like mucins. As sulphate availability will vary according to diet or mucus secretion, it is likely that colonic SRB face a wide range of sulphate concentrations. In the human colon, this could in part be responsible for the conflicting and inconsistent results obtained on the interaction between sulphate reduction and methanogenesis. Whereas, Macfarlane and Gibson (1994) demonstrated that colonic SRB can outcompete methanogens for H_2 in the colon, Strocchi *et al.* (1994) have reported the opposite. No complete exclusion of one flora to another could be shown by Pochart *et al.* (1992). In the absence of sulphate or when its concentrations were very low, and methanogens are present, SRB can grow fermentatively on ethanol or lactate. The H_2 released is used by methanogens which, by maintaining a low pH_2, facilitate the process. In the gut, this particular type of interspecies H_2 transfer may explain the coexistence of both methanogens and SRB.

5.3   REDUCTIVE ACETOGENESIS

Reductive acetogenesis involves the reduction of two mol CO_2 to one mol of acetate (Fig. 3). In the autotrophic process, exogenous CO_2 and molecular H_2 are coupled in an energy yielding reaction ($2CO_2 + 4H_2 \longrightarrow CH_3COOH + 2H_2O$) whilst the heterotrophic pathway leads to the production of three acetates from one hexose. This homoacetate fermentation can be viewed as a partial oxidation of hexose to acetate and CO_2 accompanied by the reduction of CO_2 to acetate (Wood and Ljundahl, 1990; Ragsdale, 1991). The methyl group of acetate is formed by the reduction of CO_2 via formate and tetrahydrofolate bound C1 intermediates to 5'methyltetrahydrofolate (Fig. 3). The first enzyme of this pathway is a formate dehydrogenase (tungsten-selenoprotein) which catalyses the reduction of CO_2 to formate. The electron donor is only known for certain acetogenic

Hexose

4H

2 pyruvate **OXIDATIVE PATHWAY** **2 ACETATE**

4H

CO_2 CO_2

2H

Formate **REDUCTIVE PATHWAY** 2H

4H

Methyltetrahydrofolate **CO**
derivates

ACETATE

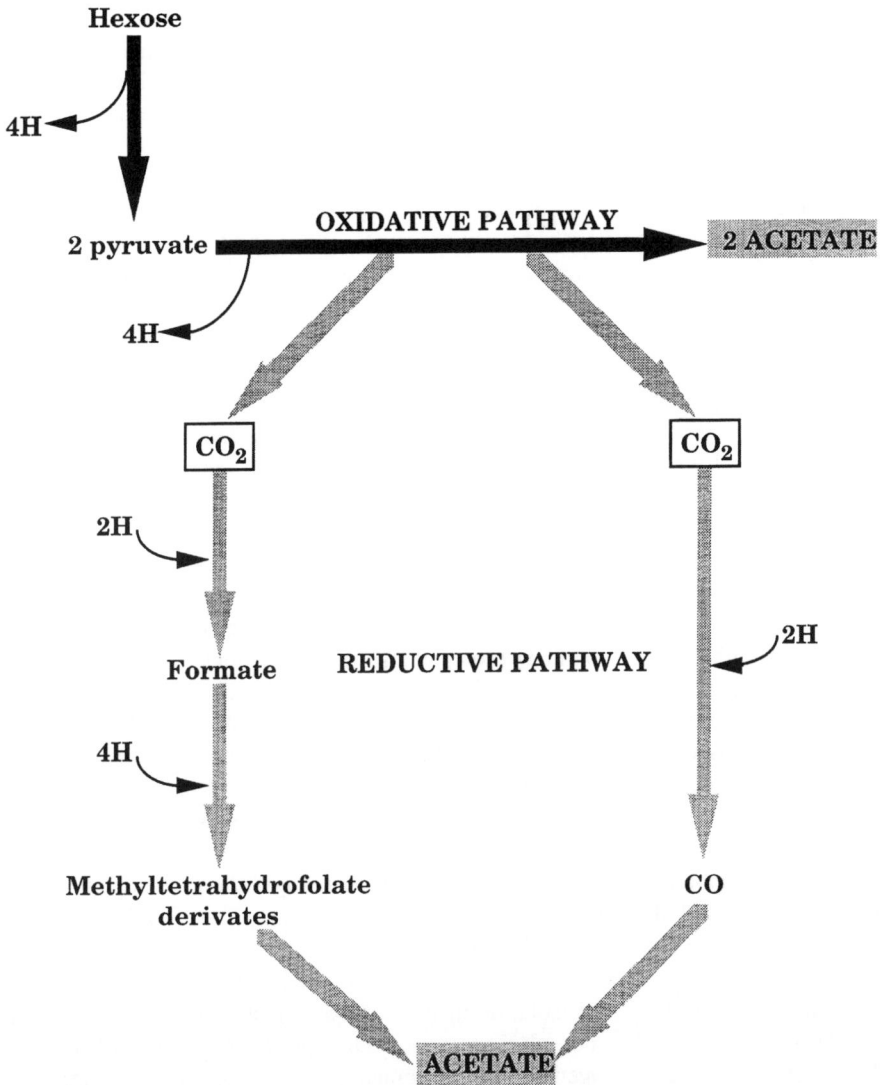

- 2H or 4H represent reduced equivalents or H_2

Figure 3. Simplified scheme of the reductive acetogenesis pathway.

bacteria and may be NADPH + H⁺ (*e.g. Clostridium thermoaceticum*). Formyltetra-
hydrofolate is then formed from formate in an ATP consuming reaction and the following
reactions lead to 5'methyltetrahydrofolate (Gottschalk, 1988). The reduction of CO_2 to

a bound carbonyl (CO) is catalysed by carbon monoxide dehydrogenase and leads to formation of the carboxyl group of acetate. This reaction requires an input of energy when H_2 is used as the electron donor (Diekert, 1990). The methyl group is then condensed with the carbonyl and CoA, to form acetylCoA which is converted to acetate with formation of one ATP.

Studies on the stoichiometry of colonic fermentation have suggested that acetogenesis could be a substantial source of acetate in the colon of subjects harbouring low concentrations of methanogens ($< 10^8$/g dry wet of faeces) (Wolin and Miller, 1983; Durand and Bernalier, 1994). An inverse relationship between acetogenesis and methanogenesis has been demonstrated in the human colon (Lajoie et al., 1988; Bernalier et al., 1996a). This is in agreement with the bacterial enumeration of acetogens and methanogens (Doré et al., 1995). Wolin and Miller (1983, 1994) have proposed that in non-methanogenic subjects, the equation of homoacetate fermentation ($C_6H_{12}O_6 \longrightarrow 3\ CH_3COOH$) should be used in place of the equation for methanogenic subjects ($C_6H_{12}O_6 \longrightarrow 2\ CH_3COOH + CH_4 + CO_2$). Wolin and Miller (1994) demonstrated that 35% of the acetate produced in non-methanogenic fæcal samples should be formed by reduction of CO_2.

Little is known about the bacterial species involved in acetogenesis in the human colon. Wolin and Miller (1993, 1994) have reported the isolation of four different acetogenic species from faeces of a non-methane excreting subject. These isolates were obtained using vanillate, methanol or glucose + H_2 as substrate. All the strains co-metabolised H_2 in the presence of organic compound, but only one strain could maintain its growth with H_2/CO_2 as a sole energy source (Wolin and Miller, 1993). More recently, Bernalier et al. (1996b) have isolated acetogenic strains from faeces of five non-methanogenic subjects, which were able to use H_2/CO_2 as the energy source to produce acetate, following the stoichiometric equation of reductive acetogenesis (Bernalier et al., 1996b; Leclerc et al., 1997). These strains can also grow on a variety of organic compounds (Bernalier et al., 1996b). Amongst these acetogenic isolates, different species of the genus *Clostridium* have been identified, whilst other Gram positive cocci were phylogenetically closely related to the genus *Streptococcus*. A new species of the genus *Ruminococcus*, named *R. hydrogenotrophicus*, was described (Bernalier et al., 1996c). The isolation of two new acetogenic isolates, closely related to *Clostridium coccoides*, has also recently been reported (Kamlage et al., 1997). These data show that acetogenesis in the human colon was a property exhibited by several different bacterial taxa. The area is at an early stage of research and it will clearly require more effort to clarify the taxonomic and physiological diversity among human colonic acetogens. Moreover, competitive interactions for H_2 between methanogens and acetogens should be more clearly defined.

6 Influence of types and amounts of substrates on biomass and end products

The nature of fermentation products will depend on the type and amount of available and bacteria involved in the fermentative chains. For example, endogenous mucin polysaccharides, like pectin, produce a relatively high proportion of acetate (Gibson *et al.*, 1988). The acetyl group of its component acetyl-glucosamine is apparently removed and excreted as acetate (Kotarski and Salyers, 1981).

Because of its role in colonic mucosal integrity, much attention has been given to butyrate. In vitro, starch seems the preferred substrate for butyrate formation. However, the in vivo significance remains unclear. The type of resistant starch (rate of fermentation), bacteria involved in its degradation, combination with other substrates and rate of passage may all be major factors responsible for discrepancies in vitro and in vivo.

Increasing the amount of fermentable substrate alters products of fermentation from in vitro pure cultures (Macfarlane and Macfarlane, 1993), and mixed faecal flora (Ducros *et al.*, 1993). A decrease in acetate and rise in lactate or in other H_2 sink acids allows for the oxidation of reduced cofactors.

An extreme situation is the development of lactic acidosis due to a large supply of highly fermentable oligosaccharides. From in vivo (Florent *et al.*, 1985) and in vitro culture (Ducros *et al.*, 1993) studies, high doses of lactulose induce, during adaptation, changes in metabolic pathways that result in decreased H_2 and an increase in lactate and acetate, probably due to the activities of lactobacilli and bifidobacteria. These reactions involved are shown in Fig. 4.

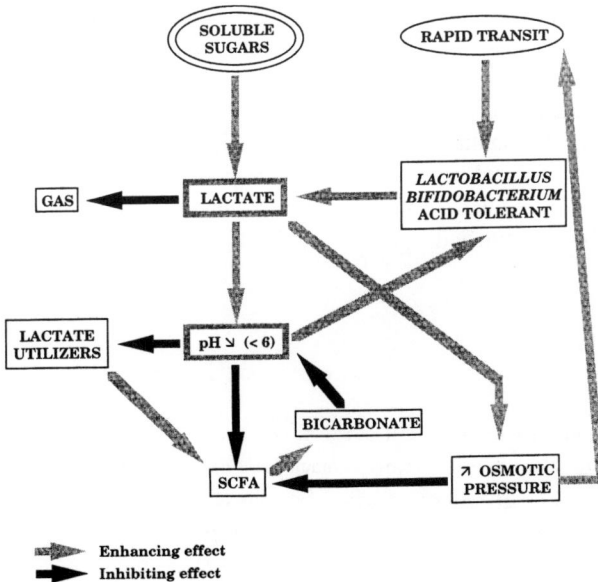

Figure 4. Chain reactions in lactic acidosis.

7   Future perspectives

Understanding the biochemistry of fermentation in the human colon is based principally on experiments involving both pure cultures or co-cultures and total faecal microflora incubations. These approaches are complementary, but not necessarily relevant, to the processes that occur in vivo. In humans, parameters such as region of the colon, supply of endogenous substrates (proteins, urea and mucin) and bicarbonate secretion, can greatly interact with bacterial metabolic processes. The following aspects are of significance:

1) With the increasing use of prebiotics and probiotics, a better understanding of the formation and fermentation of the two lactate isomers D (-) and L (+) as well as the mechanism of their colonic absorption is required.

2) Butyrate is thought to play a key role in the prevention of colonic cancer, therefore the conditions required to stimulate the in vivo formation should be investigated.

3) The role of mucin in fermentation is poorly understood. Its supply is not defined and its fermentation seems to follow specific pathways. Due to its metabolism and high sulphate content, it may be involved in sulphide production which reportedly has deleterious effects on the metabolism of mucosal cells.

4) Hydrogen utilisation by hydrogenotrophic microorganisms as well as interactions between these microorganisms can play an important role both in harmful sulphide production and in the general welfare of digestion and fermentation by avoiding flatulence and abdominal pains caused by excessive gas. Both taxonomic and physiological diversity as well as regulation of H_2 uptake within each group (methanogens, acetogens and sulphate reducers) need to be determined.

5) Microbial protein synthesis, that is dependent on the energy formed during carbohydrate fermentation, allows for the removal of putative deleterious products such as sulphides, ammonia, phenolic compounds, etc. The conditions necessary for maximising this should be investigated.

6) Optimisation of fermentation in the colon, which requires a continuous supply of SCFA up to the distal region, will only be possible when a better knowledge of the mechanisms of regulation of hydrolytic enzymes related to the nature of polysaccharides has been identified.

References

Bernalier, A., Lelait, M., Rochet, V. *et al.* (1996a) Acetogenesis from H_2 and CO_2 by methane- and non methane-producing human colonic bacterial communities. *FEMS Microbiology Ecology* **19**, 193-202.

Bernalier, A., Rochet, V., Leclerc, M. *et al.* (1996b) Diversity of H_2/CO_2-utilizing acetogenic bacteria from feces of non methane-producing humans. *Current Microbiology* **33**, 94-99.

Bernalier, A., Willems, A., Leclerc, M. *et al.* (1996c) *Ruminococcus hydrogenotrophicus* sp. nov., a new H_2/CO_2-utilizing acetogenic bacterium. *Archives of Microbiology* **166**, 176-83.

Caspari, D. and Macy, J.M. (1983) The role of carbon dioxide in glucose metabolism of *Bacteroides* fragilis. *Archives of Microbiology* **135**, 16-24.

Corfield, A.P., Wagner, S.A., Clamp *et al.* (1992) Mucin degradation in the human colon - production of sialidase, sialate O-acetylesterase, N-acetylneuraminate lyase, arylesterase and glycosulfatase activities by strains of fecal bacteria. *Infection and Immunity* **60** (10), 3971-78.

Cummings, J.H. and Macfarlane, G.T. (1991) The control and consequences of bacterial fermentation in the human colon. *Journal of Applied Bacteriology* **70**, 443-49.

Cummings, J.H., Macfarlane, G.T. and Drasar, B.S. (1989) The gut microflora and its significance, in *Gastrointestinal and Oesophageal Pathology*, (ed. R. Whitehead), Churchill Livingstone Edinburg, London, Melbourne, and New-York, pp. 201-29.

Czerkawski, J.W. (1986) *An Introduction to Rumen Studies,* Oxford, New-York Pergamon Press.

Diekert, G. (1990) CO_2 reduction to acetate in anaerobic bacteria. *FEMS Microbiology Reviews* **87**, 391-96.

Doré, J., Pochart, P., Bernalier, A. *et al.* (1995) Enumeration of H_2-utilizing methanogenic archae, acetogenic and sulfate-reducing bacteria from human feces. *FEMS Microbiology Ecology* **17**, 279-84.

Ducros, V., Durand, M., Beaumatin, P. *et al.* (1993) Adaptation of two doses of lactulose by human colonic flora in continuous culture. *Proceedings of the Nutrition Society* **52** (2), 156A.

Duncan, A.J. and Henderson, C. (1990) A study of the fermentation of dietary fibre by human colonic bacteria grown in vitro in semi-continuous culture. *Microbial Ecology in Health and Disease* **3**, 87-98.

Durand, M. and Bernalier, A. (1994) Reductive acetogenesis in human and animal gut, in *Physiological and Clinical Aspects of Short Chain Fatty Acids Metabolism* (eds. J.H. Cummings, J.L. Rombeau and T.L. Sakata), Cambridge University Press, pp. 107-17.

Durand, M., Beaumatin, P., Hannequart, G. *et al.* (1994) Effects of methanogenesis inhibition on human flora fermentation pattern in continuous culture, in *Nouvelles Tendances en Microbiologie Anaérobie* (ed. Société Française de Microbiologie), Paris, pp. 410-13.

Edwards, C.A. and Rowland, I.R. (1992) Bacterial fermentation in the colon and its measurement, in *Dietary Fibre - A Component of Food Nutritional Function in Health and Disease* (eds. T.F. Schweizer and C.A. Edwards), Springer-Verlag, pp. 119-50.

Florent, Ch., Flourié, B., Leblond, A. *et al.* (1985) Influence of chronic lactulose ingestion on the colonic metabolism of lactulose in man (an in vivo study). *Journal of Clinical Investigation* **75**, 608-13.

Flourié, B., Pellier, P., Florent, C. *et al.* (1991) Site and substrates for methane production in human colon. *American Journal of Physiology* **260**, G752-57.

Fonty, G. and Gouet, Ph. (1989) Fibre degrading microorganisms in the monogastric digestive tract. *Animal Feed Science and Technology* **23**, 91-107.

Gibson, G.R., Cummings, J.H. and Macfarlane, G.T. (1988) Use of a three-stage continuous culture system to study the effect of mucin on dissimilatory sulfate reduction and methanogenesis by mixed populations of human gut bacteria. *Applied and Environmental Microbiology* **54**, 2750-55.

Gibson, G.R., Cummings, J.H., Macfarlane, G.T. *et al.* (1990) Alternative pathways for hydrogen disposal during fermentation in the human colon. *Gut* **31**, 679-83.

Gottschalk, G. (1988) Bacterial Metabolism (Second Edition), Springer-Verlag, New-York.

Jensen, N.S. and Canale-Parola, E. (1986) *Bacteroides pectinophilus* sp. nov. and *Bacteroides galacturonicus* sp. nov.: two pectinolytic bacteria from the human intestinal tract. *Applied and Environmental Microbiology* **52** (4), 880-87.

Kamlage, B., Gruhl, B. and Blaut, M. (1997) Isolation and characterization of two new homoacetogenic hydrogen-utilizing bacteria from the human intestinal tract that are closely related to *Clostridium coccoides*. *Applied and Environmental Microbiology* **63**, **5**, 1732-38.

Kotarski, S.F. and Salyers, A.A. (1981) Effect of long generation times on growth of *Bacteroides thetaiotaomicron* in carbohydrate-limited continuous culture. *Journal of Bacteriology* **146** (3), 553-60.

Lajoie, S.F., Bank, S., Miller, T.L. *et al.* (1988) Acetate production from hydrogen and [^{13}C] carbon dioxide by the microflora of human feces. *Applied and Environmental Microbiology* **54**, 2723-7.

Leclerc, M., Bernalier, A., Donadille, G. *et al.* (1997) H$_2$/CO$_2$ metabolism in acetogenic bacteria isolated from the human colon. *Anaerobe* in press.

Macfarlane, G.T. and Cummings, J.H. (1991) The colonic flora, fermentation and large bowel digestive function, in *The Large Intestine: Physiology, Pathophysiology and Disease* (eds. S.F. Phillips, J.H. Pemberton and R.G. Shorter), Raven Press, New-York, pp. 51-92.

Macfarlane, G.T., Hay, S., Macfarlane, S. *et al.* (1990) Effect of different carbohydrates on growth, polysaccharidase and glycosidase production by *Bacteroides* ovatus, in batch and continuous culture. *Journal of Applied Bacteriology* **68** (2), 179-87.

Macfarlane, G.T. and Gibson, G.R. (1994) Metabolic activities of the normal colonic flora, in *Human health: the contribution of microorganisms* (ed. S.A.W. Gibson), Springer-Verlag, London, pp. 17-52.

Macfarlane, G.T. and Macfarlane, S. (1993) Factors affecting fermentations reactions in the large bowel. *Proceedings of the Nutrition Society* **52** (2), 367-73.

McCarthy, R.E., Kotarski, S.F. and Salyers, A.A. (1985) Location and characteristics of enzymes involved in the breakdown of polygalacturonic acid and by *Bacteroides thetaiotaomicron*. *Journal of Bacteriology* **161**, 493-99.

Miller, T.L. and Wolin, M.J. (1979) Fermentation by saccharolytic intestinal bacteria. *American Journal of Clinical Nutrition* **32**, 164-72.

Miller, T.L. and Wolin, M.J. (1983) Stability of *Methanobrevibacter smithii* populations in the microbial flora excreted from the human large bowel. *Applied and Environmental Microbiology* **51**, 429-31.

Miller, T.L. and Wolin, M.J. (1985) *Methanosphaera stadtmaniae* gen. nov., sp. nov.: a species that forms methane by reducing methanol with hydrogen. *Archives of Microbiology* **141**, 116-22.

Miller, T.L. and Wolin, M.J. (1986) Methanogens in human and animal intestinal tracts. *Systematic and Applied Microbiology* **7**, 223-29.

Nottingham, P.M. and Hungate, R.E. (1968) Isolation of methanogenic bacteria from feces of man. *Journal of Bacteriology* **86**, 2178-79.

Pochart, P., Doré, J., Lémann, F. *et al.* (1992) Interrelations between populations of methanogenic archæ and sulphate-reducing bacteria in the human colon. *FEMS Microbiology Letters* **98** (1-3), 225-28.

Prins, R.A. (1977) Biochemical activities of gut microorganisms, in *Microbial Ecology of the Gut* (eds. R.T.J. Clarke and T. Bauchop), Academic Press, pp. 73-183.

Ragsdale, S.W. (1991) Enzymology of the acetyl-CoA pathway of CO$_2$ fixation. *Critical Reviews in Biochemistry and Molecular Biology* **26**, 261-300

Rouvière, P.E. and Wolfe, R.S. (1988) Novel biochemistry of methanogenesis. *Journal of Biological Chemistry* **263** (17), 7913-16.

Russell, J.B. and Wallace, R.J. (1988) Energy yielding and consuming reactions, in *The Rumen Microbial Ecosystem* (ed. P.N. Hobson), Elsevier Applied Science, London and New-York, pp. 185-215.

Salyers, A.A. (1985) Breakdown of polysaccharides by human bacteria. *Journal of Environmental Toxicology and Oncology* **5** (6), 211-31.

Southgate, D.A.T. (1990) The role of the gut microflora in the digestion of starches and sugars: with special reference to their role in the metabolism of the host, including energy and vitamin metabolism, in *Dietary Starches and Sugars in Man: a Comparison.* (ed. J. Dodding), Springer-Verlag, London, Berlin, Heidelberg, New-York, Paris, Tokyo, pp. 67-87.

Strocchi, A., Furne, J., Ellis, C. *et al.* (1994). Methanogens outcompete sulphate-reducing bacteria for H$_2$ in the human colon. *Gut* **35**, 1098-101.

Szylit, O. and Andrieux, C. (1993) Physiological and pathophysiological effects of carbohydrate fermentation. *World Review of Nutrition and Dietetics* **74**, 88-102.

Van de Wijngaard, W.M.H., Creemers, J., Vogels, G.D. *et al.* (1991) Methanogenic pathways in Methanosphaera stadtmanifl. *FEMS Microbiology Letters* **80**, 207-12.

Wedeking, K.J., Mansfield, H.R. and Montgomery, L. (1988) Enumeration and isolation of cellulolytic and hemicellulolytic bacteria from human feces. *Applied and Environmental Microbiology* **54** (6), 1530-35.

Wolfe, R.S. (1993) Biochemistry of methanogenesis. *Biochemical Society Symposia* **58**, 41-49.

Wolin, M.J. and Miller, T.L. (1983) Carbohydrate fermentation, in *Human Intestinal Microflora in Health and Disease* (ed. D.J. Hentges), Academic Press, New-York, pp. 147-65.

Wolin, M.J. and Miller, T.L. (1993) Bacterial strains from human feces that reduce CO_2 to acetic acid. *Applied and Environmental Microbiology* **59** (11), 3551-56.

Wolin, M.J. and Miller, T.L. (1994) CO_2 acetogenesis in the human colonic ecosystem, in *Acetogenesis* (ed. H.L. Drake), Chapman et Hall, New-York, London, pp. 365-86.

Wood, H.G. and Ljundahl, G. (1990) Autotrophic character of the acetogenic bacteria, in *Variations in Autotrophic Life* (eds. J.M. Shively and L.L. Barton), Acadmic Press, London, pp. 201-50.

CHAPTER 4

Short Chain Fatty Acids

CHRISTIAN DEMIGNE, CHRISTIAN REMESY AND
CHRISTINE MORAND
Laboratoire des Maladies Metaboliques et des Micronutrients,
INRA de Centre de Recherches de Clermont-Ferrand-Theix,
63122 Saint-Genès-Champanelle, FRANCE

1   Introduction

Many carbohydrates present in plant foods exhibit complex structures. Thus, whilst the small intestine has the capacity to breakdown a limited number of polysaccharide compounds, different dietary carbohydrates still reach the colon. These include cell wall polysaccharides and related compounds (10-30 g/d), resistant starch (5-20 g/d) and varying amounts of oligosaccharides (inulin, α-galactosides, synthetic oligosaccharides). As a general rule, soluble fibres are more readily degraded than insoluble forms. In addition, there is a permanent supply of endogenous compounds like digestive secretions, mucus and sloughed epithelial cells, which all add to the pool of available carbon. Despite the diversity of substrates fermented by the colonic microflora, end-products of carbon metabolism are mostly represented by a limited number of carboxylic acids, especially short-chain fatty acids (SCFA) such as acetate, propionate and butyrate. Lactate, succinate and branched-chain fatty acids are more minor compounds under most conditions in monogastric species. Short chain fatty acid production in the large intestine constitutes a process which allows the recovery of carbon units and chemical energy from fermented compounds.

2   Intestinal absorption of short-chain fatty acids

It is thought that over 95% of SCFA formed from bacterial fermentation in the human large intestine can be absorbed across the colonic epithelium. The fact that SCFA are present in high concentrations in large intestinal contents (up to 80-150 mM) has been

G.R. Gibson and M.B. Roberfroid (eds.), Colonic Microbiota, Nutrition and Health, 55-69.
© 1999 *Kluwer Academic Publishers. Printed in the Netherlands.*

used as an argument to support the view that SCFA are essentially absorbed by a passive process, along a high concentration gradient. However, it is now recognised that several transport systems are probably involved in the digestive absorption of SCFA.

Weak carboxylic acids, such as SCFA, may exist as two forms, *i.e.* protonated neutral (AH) and ionised negatively charged (A⁻). The relative proportion of these two forms depends, for a given pH, on the pKa of the acid which is in the range 4.7-5.0 for SCFA. The percentage as a protonated form varies considerably between the proximal colon (pH around 6.0, with about 10% protonated SCFA) and the distal colon (pH close to neutrality, and around 1% protonated SCFA). Studies with animal models suggest that SCFA absorption in the proximal colon depends on three processes:

(i)   passive absorption of the lipid-soluble protonated form, probably coupled to an
         Na^+ - H^+ exchange
(ii) $SCFA^-$ <—> HCO^{3-} exchange
(iii)   diffusion of the ionised form through a paracellular pathway (Fig. 1).

Although luminal pH is generally more acidic in the proximal than distal region, only about 50% of the SCFA would be transported as a protonated form in the proximal area, whereas most SCFA absorption distally depends on diffusion of this form (Engelhardt, 1995). The non-ionic diffusion is probably promoted by Na^+ <—> H^+ and K^+ <—> H^+ exchange mechanisms which take place on the apical membrane, especially in the distal colon (a notable percentage of H^+ may arise from the bulk medium in the proximal colon). Carbonic anhydrase, which is more active in the proximal colon, also plays a substantial role in supplying H^+ (Reckemmer, 1994).

The rate of SCFA absorbed as a lipid-soluble protonated form should be higher as the carbon chain increases. However, no clear relationship with chain length has been found in the proximal colon. Similarly, in the distal colon, permeability seems higher for acetate than butyrate (Engelhardt, 1995; Velasquez *et al.*, 1996a).

Absorption of SCFA also has a stimulatory effect on water and electrolyte uptake in the large intestine. In the presence of SCFA, the proximal colon has double the capacity to absorb Na^+ (and secrete H^+), compared with the distal colon (Velasquez *et al.*, 1996b). Furthermore, the large intestine is a significant site of Mg^{++} and Ca^{++} absorption, and this is promoted by SCFA (Lutz and Scharrer, 1991; Younes *et al.*, 1996). An interactive mechanism (absorption under a $[Ca(Ac)]^+$ form) has also been proposed (Trinidad, 1993).

3   Intestinal metabolism of SCFA

Colonocytes exhibit many metabolic features that are characteristic of short life span tissues, such as a high rate of glucose and glutamine oxidation. More specifically, colonocytes also metabolise a substantial part of absorbed SCFA, especially butyrate. Glucose and glutamine are not totally oxidised to CO_2, and their metabolism results in cycling to products such as lactate, alanine or glutamate (Livesey and Elia, 1995). In the same manner, butyrate is not completely oxidised, with a proportion being recycled as ketone bodies (Henning and Hird, 1972; Clausen and Mortensen, 1994). Interactions between different energy fuels, such as a sparing effect of butyrate on

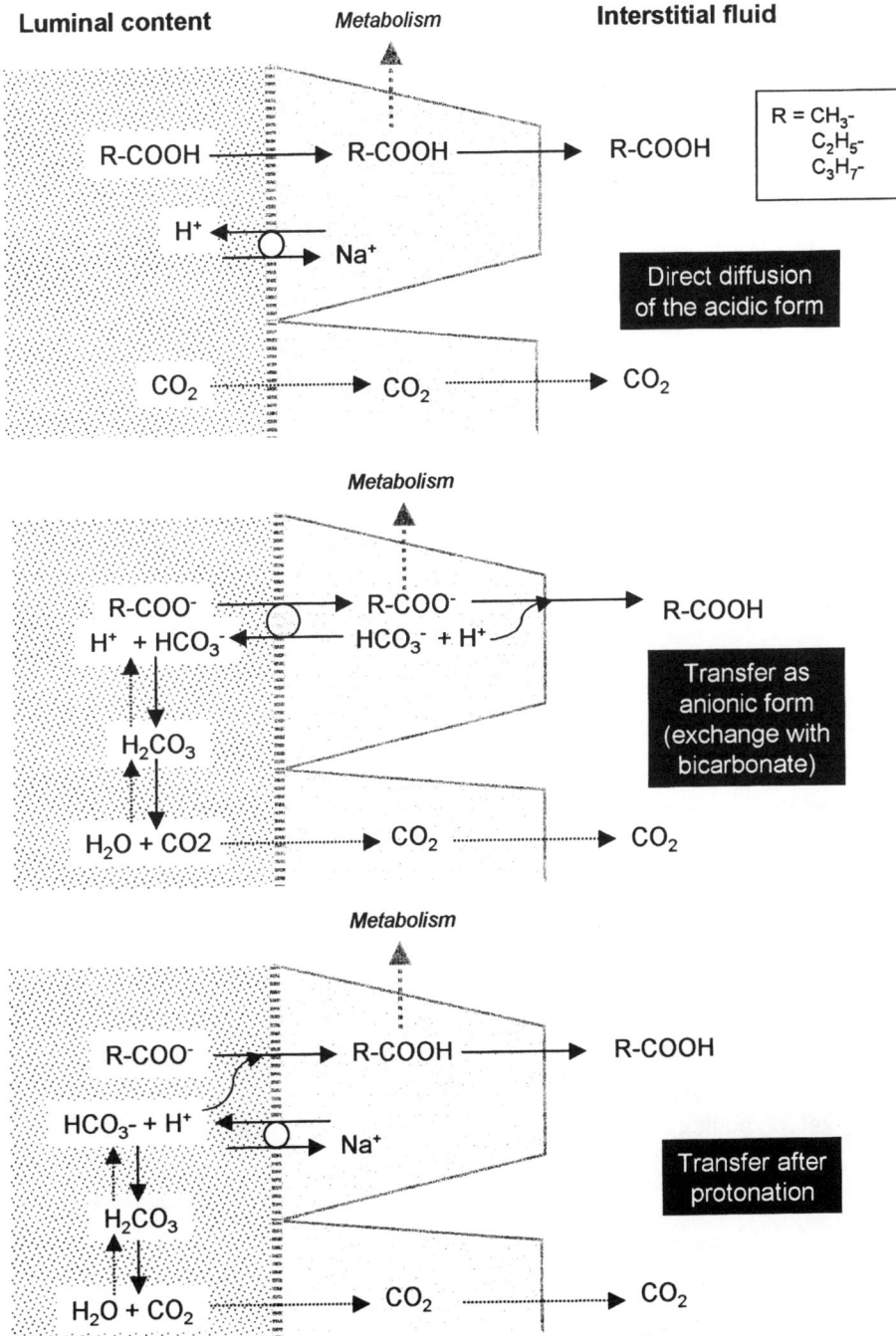

Figure 1. Pathways of chort-chain fatty acid transport in colonocytes.

glucose oxidation, probably take place in colonocytes (Clausen and Mortensen, 1994). Butyrate availability in the colon is generally lower than that of acetate or propionate. In colonocytes, the Vmax for oxidation is similar for the various acids, whereas the apparent Km is markedly lower for butyrate than for other substrates (Clausen and Mortensen, 1994). Moreover, butyrate has been reported as inhibitory for acetate and propionate utilisation (Gordon and Crabtree, 1992). Since the Km for butyrate metabolism has been found to be around 0.2 mM, whilst the luminal concentrations are frequently as high as 5-10 mM, it is clear that butyrate is generally metabolised at a high rate under most conditions. Nevertheless, competitive inhibition of butyrate oxidation by acetate and propionate may be effective when these SCFA are present in excess. Moreover, NH_4^+ may also inhibit butyrate utilisation (Darcy-Vrillon et al., 1996).

Altered capacities in the mucosal metabolism of butyrate have been postulated in the pathogenesis of ulcerative colitis (Christl et al., 1996) (Fig. 2). The metabolic alterations observed during this pathology do not seem to be the result of enzymatic deficiencies in the metabolic pathway of butyrate metabolism (Allan et al., 1996). In fact, HS^- anions produced by bacterial sulphate reduction have been suspected to disturb energy metabolism in colonocytes (Roediger 1982; Roediger et al., 1993).

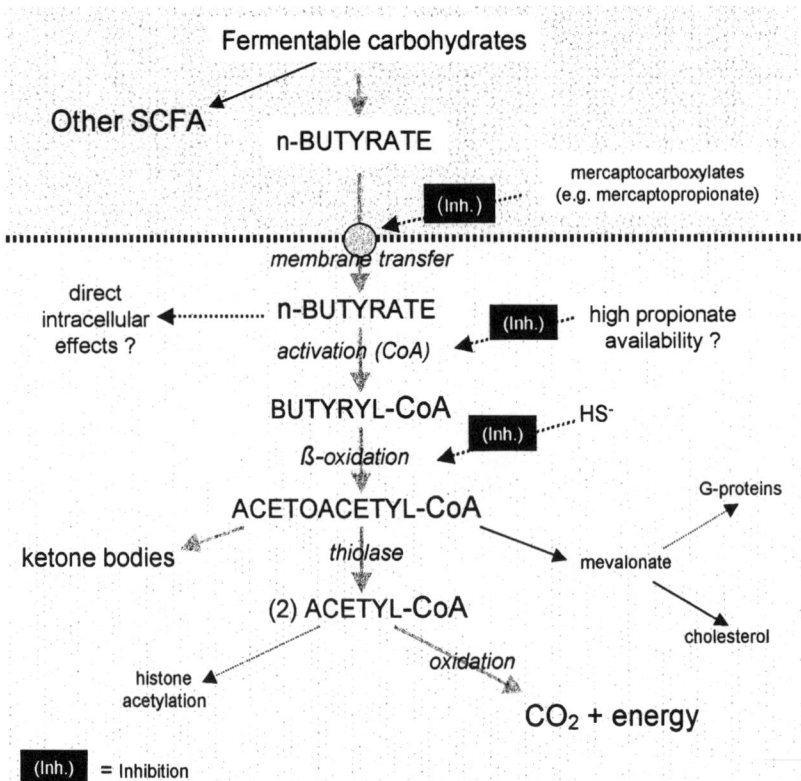

Figure 2. Factors that may effect butyrate metabolism in colonic epithelial cells.

Butyrate metabolism could be affected at the level of transfer across the apical membrane by mercapto-derivatives of SCFA (mercaptopropionate for example) formed by sulphate-reducing bacteria (Stein *et al.*, 1995, 1996). These alterations have been shown to induce mucosal hyperproliferation, and the consequences of impaired butyrate metabolism may be overcome by increased levels of SCFA. Furthermore, SCFA may have a stimulatory effect on mucus secretion (Sakata and Setoyama, 1995) and a stabilising effect on the integrity of the colon extracellular matrix. In the latter case, butyrate is liable to induce trans-glutaminase activity in factor Xlll (D'Argenio *et al.*, 1994). This factor stabilises fibrin and has the capacity to bind fibronectin, collagen, actin and myosin. In parallel, butyrate can inhibit urokinase activity, probably by acting at the transcriptional level (Scheppach *et al.*, 1995). This inhibition impairs plasminogen conversion to plasmin, a protease directed against fibrin, as well as various constituents of the basal membrane (*e.g.* laminin, fibronectin).

4　Trophic effects of SCFA on normal enterocytes

In vitro, a positive correlation between SCFA (mainly butyrate) and crypt development in the colonic mucosa has been reported (Lupton and Kurtz, 1993). Crypt cell proliferation in the caecum is stimulated by butyrate, as well as by propionate albeit to a smaller extent (Gamet *et al.*, 1992). The proliferative effect, accompanied by a stimulation of DNA synthesis, essentially affects the deepest zones of the crypts. This may have physiological relevance, since extension of the proliferative zone to upper part of the crypts may generally preceed colorectal cancer appearance. In a model in which upper proliferation was induced in the colonic crypt by deoxycholate, it was shown that butyrate reversed this effect (Velasquez *et al.*, 1996a).

5　Effects of SCFA on colonic cancer cells

Butyrate has the potential to arrest growth of neoplastic colonocytes and inhibits preneoplastic proliferation induced by certain tumour promoters in vitro (Fig. 3). In most cases, arrest takes place in the early G1 phase of the cell cycle (Kim *et al.*, 1994). With various carcinoma cell lines, butyrate can inhibit cell proliferation at a relatively low concentration (*e.g.* 1-5 mM), without affecting viability. However, at higher concentrations, butyrate is cytotoxic for these cells whilst normal cells remain relatively unaffected. Butyrate is not the only SCFA capable of exerting antiproliferative effects, since propionate (but not acetate) has also been shown to be effective (Gamet *et al.*, 1992, Scheppach *et al.*, 1995). Propionate is 2-3 fold less potent than butyrate at similar concentrations, but its antiproliferative importance may be significant as the luminal concentration is generally higher than that of butyrate. Moreover, the antiproliferative effects of butyrate may be potentiated by SCFA (Marsman and McBurney, 1996). Hague *et al.* (1993) have shown an induction of apoptosis in adenoma cell lines by butyrate, and this effect has also been observed with colonic carcinoma cell lines (Heerdt and Augenlicht, 1991).

Effects on the normal epithelium

-   Energy fuel

-   Trophic effects (↑ cell proliferation, ↑ DNA content, ↑ crypt height)

-   Counteracting effect on some crypt proliferation factors (e.g. DCA*)

Effects on preneoplastic or neoplastic colonocytes

-   Induction of systems involved in the potency or cellular availability of active compounds (e.g. ↑ mdr-1 gene, ↑ glutathione-S-transfererase Π)

-   Induction of differentiation markers (e.g. ↑ brush border hydrolases, ↑ EGF* receptor expression, ↑ SCLC* cluster-1 antigen)

-   Regulatory effects on systems involved in cellular adhesion and/or migration (e.g. ↑ LBP*, ↓ HLA-1*, ↓ urokinase secretion, ↑ G$_{M3}$ ganglioside synthesis)

-   Modulation of oncogene expression : inhibition of ras and src protooncogen and of p21 and pp60, inhibition of c-myc protooncogen, induction of c-fos and c-june.

-   Induction of apoptosis

Abbreviations :

↑ : increase	↓ : decrease
DCA : deoxycholic acid	EGF : epithelial growth factor
SCLC : small cell lung cancer...	mdr : multidrug resistance (factor)
LBP : laminin-binding protein	HLA-1 : human lymphocyte antigen class 1

Figure 3. Butyrate and colonocytes.

6   Mechanisms of action for butyrate effects

The complex nature of butyrate's biological effects raises the question of its mechanisms of impact. These could occur by binding to transregulatory proteins, interfering with the interaction of these proteins with DNA, or by affecting other regulatory enzymes like protein kinases. Butyrate modifies the degree of acetylation of histones, probably by inhibition of histone deacetylase (Boffa *et al.*, 1978). It also affects phosphorylation of histones (depressed for H1 and H2, enhanced for H3: Boffa *et al.*, 1981; Kruh *et al.*, 1994).

This could influence chromatin structure because of altered binding ot histones to DNA which, as a result, may therefore be more exposed to transcriptional factors or nucleases (Morrow *et al.*, 1994). Butyrate can promote hypermethylation of DNA (whether pre-existing or newly synthesised), particularly on cytosine residues (Kruh *et al.*, 1994). This alteration of DNA is generally correlated with decreased gene expression. As such, hypomethylation of colonic epithelial DNA has been reported in subjects that are prone to colonic adenoma (Bartram *et al.*, 1994).

It has also been proposed that butyrate may act at the membrane-cytoplasmic interface by interacting with G-proteins, which are key components of the signal transduction pathway and are also products of certain oncogenes such as *ras*. It has been speculated that butyrate may also interfere with mevalonate binding to G-protein, which is required for membrane translocation and subsequent *ras* protein activation (Velasquez *et al.*, 1996a). Recent data have also indicated that protein kinase CKII down-regulation could be involved in signal transduction initiated by butyrate, CKII being able to phosphorylate transcriptional factors such as *jun* and *myc* (Russo *et al.*, 1997).

7   Liver metabolism of SCFA

Short chain fatty acids that are absorbed through the portal vein are channelled directly to the liver. Quantification of the SCFA portal flux has been determined in various animal species, and represents up to 2 µmol/min/g liver in the rat (Remesy *et al.*, 1995). In humans, comparable data are scarce and rely on post mortem or surgery samples. In Western societies, the portal SCFA flux is around 300-400 mmol/24h, which corresponds to about 5% of daily caloric supply (Rambaud and Flourie, 1995). Propionate and butyrate are almost completely taken up by the liver, whereas quantitative acetate balance across this organ depends on differing nutritional and physiological conditions, as shown in Fig. 4.

7.1   ACETATE METABOLISM

Acetate uptake by the liver is approximately proportional to its portal concentration, and is significant only when the portal levels exceed 0.2-0.3 mM (Remesy *et al.*, 1995). There is a possibility of some recycling for acetate in the liver since this SCFA may be activated or produced in situ (from ethanol or the endogenous acetyl CoA pool). Acetate activation is possible in the cytoplasm and in mitochondria, but various experimental arguments suggest that acetate activation is mainly mitochondrial, with acetyl CoA hydrolysis being largely cytosolic (Crabtree *et al.*, 1990). Propionate and butyrate are potent inhibitors of acetate utilisation in the liver (Gordon and Crabtree, 1992; Remesy *et al.*, 1995). This inhibition is physiologically significant during high-propionic acid fermentations, which may be observed with readily fermented substrates such as certain oligosaccharides (Levrat *et al.*, 1994). A proportion of acetate from the colonic fermentation escapes hepatic uptake, and is thus available for peripheral tissues such as muscles or adipose tissue. This may be biologically relevant during situations of impaired acetate utilisation by the liver, *e.g.* ethanol ingestion or ketotic situations. In diabetic subjects, an increase of plasma acetate also reflects impaired glucose utilisation by peripheral tissues, which in turn alters acetate oxidation (Akanji *et al.*, 1989).

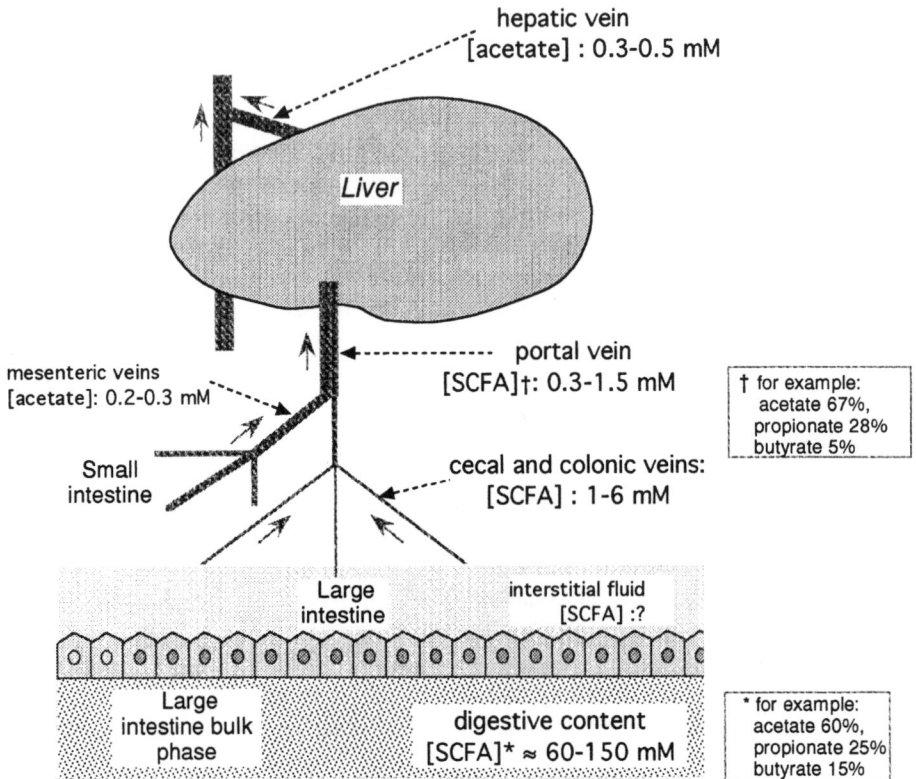

Figure 4. Approximate levels of short-chain fatty acids in the splanchnic region.

7.2   PROPIONATE AND BUTYRATE METABOLISM

The hepatic activation of these substrates occurs in mitochondria. Thus, only trace amounts of propionate or butyrate are present in extrasplanchnic blood. The rate limiting step of propionate metabolism in mitochondria is catalysed by methylmalonyl CoA mutase, which is affected by the vitamin B12 status (Fig. 5).

In herbivores, it has been established that propionate is a very effective neoglucogenic substrate. However, in omnivores, propionate availability is generally maximal under conditions that correspond to a low activity of the gluconeogenesis pathway. It is thus conceivable that propionate is essentially diverted towards glycogen, or triacylglycerol, synthesis. A variety of biological effects have been ascribed to propionate and its liver metabolites (especially the CoA esters derivatives), affecting glycolysis and gluconeogenesis, ureogenesis, fatty acid oxidation, cholesterol synthesis and oxidation (Remesy *et al.*, 1995). The propionate metabolism pathway is also used for catabolism of some amino acids (*e.g.* methionine, valine, isoleucine), but this supply is significant only with relatively high protein diets.

Butyrate is essentially channelled towards acetyl CoA in mitochondria. Butyrate is potentially ketogenic but this catabolism is observed only in the presence of a glucose shortage. Certain specific effects of butyrate have been reported in the liver, such as induction of apo B synthesis (Kaptein *et al.*, 1992). However, the physiological relevance of these observations remain uncertain, as absorbed butyrate is predominantly metabolised by the colonic epithelium. Thus, quantities reaching the liver are limited.

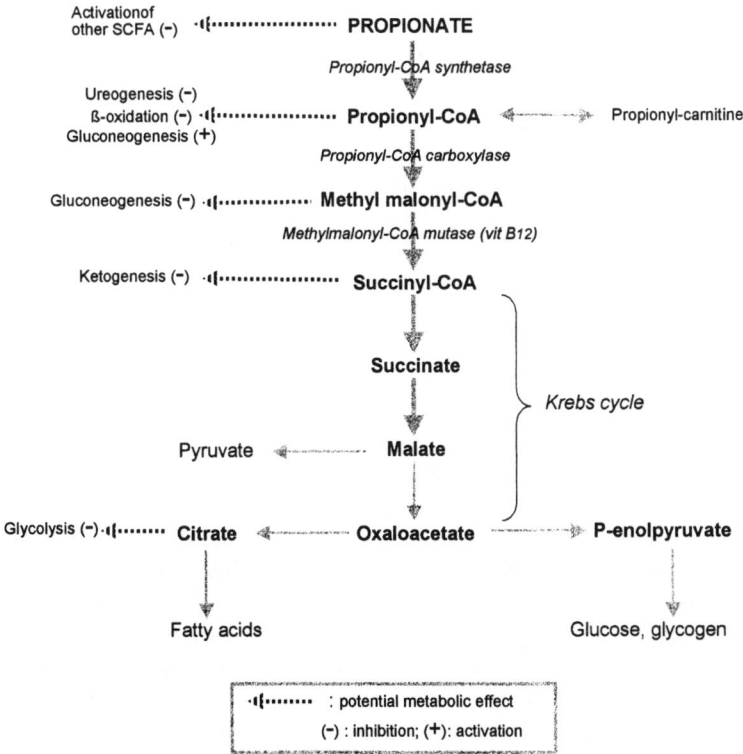

Figure 5. Pathway for propionate metabolism in the liver.

8   Metabolic effects of SCFA

8.1   LIPID METABOLISM

Fibres, especially the water-soluble types, may constitute effective cholesterol lowering agents. This has been hypothesised through an alteration of cholesterol absorption and enterohepatic cycling of bile acids, in the distal small intestine as well as in the large gut. In the latter case, microbial fermentation of soluble fibres or oligosaccharides yields an acidic pH which can contribute towards insolubilisation of bile acids and also inhibit biotransformation by the gut microflora. It has also been speculated that SCFA may constitute a mediator of the lipid-lowering effect of fibres. This could be particularly relevant for compounds devoid of sterol/bile acid binding capacities such as oligosaccharides, which have been shown effective in lowering plasma triglycerides, phospholipids and cholesterol in rats (Levrat *et al.*, 1994; Fiodaliso *et al.*, 1995).

In vitro experiments have suggested that acetate can stimulate fatty acid synthesis by rat liver cells, whilst inhihiting cholesterogenesis (Beynen *et al.*, 1982). However, the effects of propionate on lipid metabolism are more compelling. In vitro experiments have shown that propionate is an effective inhibitor of both fatty acid and cholesterol synthesis (Anderson, 1995; Demigne *et al.*, 1995). On the contrary, in vivo investigations with the rat have been less consistent (Levrat *et al.*, 1994; Berggren *et al.*, 1996). Nevertheless, investigations in humans have shown a short-term cholesterol-lowering effect of propionate (Wolever *et al.*, 1991). Venter *et al.* (1990) administered 7.5 g/d propionate to healthy subjects for 7 weeks. In this study, propionate did not affect overall serum cholesterol levels but did increase HDL cholesterol by 11%. Using a comparable approach (with 9.9 g propionate daily given in bread), Todesco *et al.* (1991) reported a slight lowering of total serum cholesterol and of LDL and HDL cholesterol. A significant inhibitory effect of propionate on [^{13}C] acetate recovery in triglycerides, but not in cholesterol, has also been observed (Wolever *et al.*, 1995).

A striking induction of HMG CoA reductase has been reported in rats fed diets promoting high propionic acid fermentations in the caecum (Levrat *et al.*, 1994). Nevertheless, it seems that propionate is a potent inhibitor of acetate utilisation by liver cells in both vitro and in vivo (Gordon and Crabtree, 1992; Remesy, 1995). It is conceivable that propionate may only be a significant inhibitor of lipid synthesis when acetate is a major precursor of cytosolic acetyl CoA (Demigne *et al.*, 1995). In this context, with isolated hepatocytes obtained from rats fed a control or pectin diet, it has been reported that acetate incorporation into cholesterol was accelerated considerably (about 9 fold; Stark and Madar, 1993). Propionate is also capable of inhibiting the metabolism of other acetyl CoA precursors, especially lactate or long chain fatty acids (Shaw *et al.*, 1985; Morand *et al.*, 1994; Remesy *et al.*, 1995). It has also been reported that propionate can accelerate the conversion of cholesterol to bile acids in hepatocytes (Imaizumi *et al.*, 1992). In work performed on rat and human liver cells (Lin *et al.*, 1995), propionate seemed to be ineffective in inhibiting cholesterogenesis in human hepatocytes.

8.2   CARBOHYDRATE METABOLISM

Studies in humans suggest that diets complemented with propionate may depress postabsorptive glycaemia, improve glucose tolerance and dampen postprandial hyperinsulinemia (Venter and Vorster, 1989). However, more recent data failed to demonstrate any significant influence of propionate on hepatic production of glucose and its utilisation in the bodies of normal or diabetic rats (Boillot *et al.*; 1995, Cameron-Smith *et al.*, 1994). Likewise, administration of SCFA to human subjects (corresponding to the fermentation of about 30 g fibre) did not significantly affect liver production of glucose (Laurent *et al.*, 1995).

In vitro, propionate and butyrate are effective modulators of lactate utilisation by isolated liver cells (Anderson 1995). Propionate tends to promote the release of lactate by liver cells, through an increased concentration threshold for the utilisation of lactate (Morand *et al.*, 1994). Propionate is also an inhibitor of pyruvate oxidation (Brass *et al.*, 1986), *e.g.* effects on pyruvate carboxylase by various CoA derivatives have been postulated (Remesy *et al.*, 1995).

8.2.1   *Interactions of propionate with carnitine*
Propionate can combine with carnitine to give propionyl-carnitine, which tends to increase when there is a high availability of propionate in the portal vein. It has been proposed that sequestration of a proportion of the carnitine pool as propionyl-carnitine, could elicit a signal that transfer carnitine from muscle to the liver, so sustaining fatty acid oxidation (Stephen, 1994). L-propionylcarnitine is thought to be protective in terms of functional recovery and restored high-energy phosphate of isolated perfused heart (Leipala *et al.*, 1991). Data also suggest that L-propionyl carnitine protects cell membrane and circulating lipoproteins from peroxidation (Bertelli *et al.*, 1994).

9   Summary

Despite a relatively limited flux from the large intestine, compared to other major nutrients, SCFA can have noticeable biological effects in tissues exposed to significant concentrations. This includes the colonic epithelium and, to a lesser extent, the liver. In the colon, the role of butyrate as a major energy fuel and control factor of cell proliferation is now established. Nevertheless, There are many aspects requiring further investigation:
(i)   situations by which butyrate bioavailability is limiting for colonocyte metabolism
(ii)   a lack of untoward effects of an elevated concentration of butyrate in the colonic bulk medium, as may be obtained by manipulation of the microflora and/or the supply of readily fermented substrates
(iii)   the importance of the other compounds liable to affect butyrate metabolism.

With low-fibre diets, the SCFA portal flux is probably too low to markedly affect metabolism in the liver. However, with diets rich in plant foods or other fermentable compounds, SCFA flux may be sufficient to affect certain hepatic metabolic pathways. This may be significant for glucose metabolism, since the presence of fermentable substrates in the diet, generally at the expense of more digestible carbohydrates, can lead to an SCFA/glucose substitution. This is probably less relevant for lipids, although it seems that dietary fibres can suppress cholesterol absorption and affect the kinetic of lipid digestion. As such, SCFA metabolism could take place during altered lipid metabolism (affecting remnant uptake, in situ lipogenesis and cholesterogenesis). It is relevant that some studies which assess the metabolic role of SCFA (especially by infusing propionate or by its addition to the diet) have been inconclusive, in terms of SCFA/glucose and alteration of lipid digestion.

References

Akanji, A.O., Peterson, D.B., Humphreys, S. (1989) Change in plasma acetate level in diabetic subjects on mixed high fiber diets. *American Journal of Gastroenterology* **84**, 1365-70.

Allan, E.S., Winter, S., Light, A.M. (1996) Mucosal enzyme activity for butyrate oxidation: no defect in patients with ulcerative colitis. *Gut* **38**, 886-93.

Anderson, J.W. (1995) Short-chain fatty acids and lipid metabolism: human studies, in *Physiological and Clinical Aspects of Short-Chain Fatty Acids* (eds. Cummings, J.H., Rombeau, J.L. and Sakata, T.), Cambridge University Press, Cambridge, pp. 509-23.

Bartram, H.P., Scheppach, X., Englert, S.*et al.* (1994) Effects of deoxycholic acid and butyrate on mucosal prostaglandin E2 release and cell proliferation in the human sigmoid colon. *Journal of Parenteral and Enteral Nutrition* **19**, 182-86.

Berggren, A.M., Nyman, M.G.L., Lundquist, I. *et al.* (1996) Influence of orally and rectally administered propionate on cholesterol and glucose metabolism in obese rats. *British Journal of Nutrition* **76**, 287-96.

Bertelli, A., Conte, A., Ronca, G. (1994) L-propionyl camitine protects erythrocytes and low density lipoproteins against peroxidation. *Drugs in Experimental Clinical Research* XX, 191-97.

Beynen, A.C., Buechler, K., Van Der Molen, A.J. (1982) The effects of lactate and acetate on fatty acid and cholesterol biosynthesis by isolated rat hepatocytes. *International Journal of Biochemistry* **14**, 165-69.

Boffa, L., Vidali, G., Mann, R., et al. (1978) Suppression of histone deacetylation in vivo and in vitro by sodium butyrate. *Journal of Biological Chemistry* **253**, 3364-66.

Boffa, L., Vidali, G., Mann, R., et al. (1981) Manifold effects of sodium butyrate on nuclear function. Selective and reversible inhibition of phosphorylation of histones H1 and H2A and impaired methylation of lysine and histidine residues in nuclear protein fractions. *Journal of Biological Chemistry* **256**, 9612-21.

Boillot, J., Alamovitch, C., Berger, A.M., et al. (1995) Effect of dietary propionate on hepatic glucose production, whole-body glucose utilization, carbohydrates and lipid metabolism in normal rats. *British Journal of Nutrition* **73**, 241-51.

Brass, E.P., Fennessey, P.V., Miller, L.V. (1986) Inhibition of oxidative metabolism by propionic acid and its reversal by carnitine in isolated rat hepatocytes. *Biochemical Journal* **236**, 131-36.

Cameron-Smith, D., Collier, G.R., O'Dea, K. (1994) Effect of propionate on in vivo carbohydrate metabolism in streptozotocin-induced diabetic rats. *Metabolism* **43**, 728-34.

Christl, S.U., Eisner, H-D., Dusel, G. *et al.* (1986) Antagonistic effects of sulfide and butyrate on proliferation of colonic mucosa. *Digestive Disease and Sciences* **41**, 2477-84.

Clausen, M.R. and Mortensen, P.B. (1994) Kinetic studies on the metabolism of short-chain fatty acids and glucose in isolated enterocytes. *Gastroenterology* **106**, 423-32.

Crabtree, B., Gordon, M.J. and Christie, S.L. (1990) Measurement of the rates of acetyl-CoA hydrolysis and synthesis from acetate in rat hepatocytes and the role of tehse fluxes in substrate cycling. *Biochemical Journal* **270**, 219-25.

D'Argenio, G., Cosenza, V. and Sorentini, I. (1994) Butyrate, mesalamine, and factor XIII in experimental colitis in the rat: effects of transglutaminase activity. *Gastroenterology* **106**, 399-404.

Darcy-Vrillon, B., Cherbuy, C. and Morel, M.T. (1996) Short chain fatty acids and glucose metabolism is isolated pig colonocytes: modulation by NH_4^+. *Molecular and Cellular Biochemistry* **156**, 154-61.

Demigne, C., Morand, C., Levrat, M.-A., *et al.* (1995) Effect of propionate on fatty acid and choelsterol synthesis and acetate metabolism in isolated rat hepatocytes. *British Journal of Nutrition* **74**, 209-19.

Engelhardt v, W. (1995) Absorption of short chain fatty acids from the large intestine, in *Physiological and Clinical Aspects of Short-Chain Fatty Acids* (eds. Cummings, J.H., Rombeau, J.L. and Sakata, T.), Cambridge University Press, Cambridge, pp. 149-70.

Fiordaliso, M., Kok, N., Desafer, J.P. *et al.* (1995) Dietary oligofructose lowers triglycerides, phospholipids and cholesterol in serum and very low density lipoprotein of rats. *Lipids* **30**, 163-67.

Gamet, L., Daviaud, D., Denis-Pouxciel, C. *et al.* (1992) Effects of short chain fatty acids on growth and differentiation of the human colon cell line HT29. *International Journal of Cancer* **52**, 286-89.

Gordon, M.J. and Crabtree, B. (1992) The effect of propionate and butyrate on acetate metabolism in rat hepatocytes. *International Journal of Biochemistry* **24**, 1029-31.

Hague, A., Manning, A.M., Hanlon, K.A., *et al.* (1993) Sodium butyrate induces apoptosis in human colonic tumor cell lines in a p53-independent pathway: implications for the possible role of dietary fibre in the prevention of large bowel cancer. *International Journal of Cancer* **55**, 498-505.

Heerdt, B.G. and Augenlicht, L.H. (1991) Effects of fatty acid on expression of genes encoding subunits of cytochrome c oxidase in HT29 human colonic adenocarcinoma cells. *Journal of Biological Chemistry* **266**, 19120-26.

Henning, S.J. and Hird, F.R.J. (1972) Ketogenesis from butyrate and acetate by the caecum and colon of rabbits. *Biochemical Journal* **130**, 785-90.

Imaizumi, K., Hirata, K., Yasni, S. *et al.* (1992) Propionate enhances synthesis and secretion of bile acids in primary cultures of rat hepatocytes via succinyl CoA. *Bioscience, Biotechnology and Biochemistry* **56**, 557-64.

Kaptein, A., Roodenburg, L. and Princen, H.M. (1992) Butyrate stimulates the secretion of apolipoprotein (apo) A-I and apo B100 by the human hepatoma cell line Hep G2. *Biochemical Journal* **278**, 557-64.

Kim, Y.S., Gum, J.R., Ho, S.B. *et al.* (1994) Colonocyte differentiation and proliferation. In *Short Chain Fatty Acids* (eds. Binder, H.J., Cummings, J., Soergel, K.H.), Kluwer Academic Press, Lancaster, pp. 119-34.

Kruh, J., Tichonicky, L. and Defer, N. (1994) Effect of butyrate on gene expression. In *Short Chain Fatty Acids* (eds. Binder, H.J., Cummings, J., Soergel, K.H.), Kluwer Academic Press, Lancaster, pp. 135-47.

Laurent, C., Simoneau, C., Marks, L. *et al.* (1995) Effect of acetate and propionate on fasting hepatic glucose production in humans. *European Journal of Clinical Nutrition* **49**, 484-91.

Leipala, J.A., Bhatnagar, R., Pineda, E. *et al.* (1991) Protection of reperfused heart by L-propionyl carnitine. *Journal of Applied Physiology* **71**, 1518-24.

Levrat, M.-A., Favier, M.-L., Moundras, C. *et al.* (1994) Role of dietary propionic acid and bile acid excretion in the hypocholesterolemic effects of oligosaccharides in rats. *Journal of Nutrition* **124**, 531-38.

Livesey, G. and Elias, M. (1994) Short chain fatty acids as energy source in the colon: metabolism and clinical implications. In *Physiological and Clinical Aspects of Short Chain Fatty Acids* (eds. Cummings, J.H., Rombeau, J.L., Sakata, T.), Cambrige University Press, Cambridge, pp. 427-81.

Lin, Y., Vonk, R.J., Maarten, J.H.S. *et al.* (1995) Differences in propionate-induced inhibition of cholesterol and triacylglycerol synthesis between human and rat hepatocytes in primary culture. *British Journal of Nutrition* **74**, 197-207.

Lupton, J.R. and Kurtz, P.P. (1993) Relationship of colonic luminal short-chain fatty acids and pH to in vivo cell proliferation in the rat. *Journal of Nutrition* **123**, 1522-30.

Lutz, T. and Scharrer, E. (1991) Effect of short-chain fatty acids on calcium absorption by the rat. *Experimental Physiology* **76**, 615-18.

Marsman, K.E. and McBurney, M.I. (1996) Dietary fiber and short-chain fatty acids affect cell proliferation and protein synthesis in isolated rat colonocytes. *Journal of Nutrition* **126**, 1429-37.

Morand, C., Besson, C., Demignè, C., *et al.* (1994) Importance of the modulation of glycolysis in the control of lactate metabolism by fatty acids in isolated hepatocytes from fed rats. *Archives of Biochemistry and Biophysics* **309**, 254-60.

Morrow, C.S., Nakagawa, M. and Goldsmith, M.E. (1994) Reversible transcriptional activation of mdn by sodium butyrate treatment of human colon cancer cells. *Journal of Biological Chemistry* **269**, 10739-46.

Rambaud, J.C. and Flouriè, B. (1995) Actions physiopathologiques des acides gras chaÓne courte. *Cahiers de Nutrition et Dietetique* **30**,150-58.

Rechkemmer, G. (1994) In vitro studies of short chain fatty acid transport with intact tissue. In *Short Chain Fatty Acids* (eds. Binder, H.J., Cummings, J., Soergel, K.H.), Kluwer Academic Press, Lancaster, pp. 83-92.

Remesy, C., Demigne, C. and Morand, C. (1995) Metabolism of short-chain fatty acids, in *Physiological and Clinical Aspects of Short-Chain Fatty Acids* (eds. Cummings, J.H., Rombeau, J.L., Sakata, T.), Cambridge University Press, Cambrige, pp. 171-90.

Roediger, W.E.W. (1982) Utilization of nutrients by isolated epithelial cells of the rat colon. *Gastroenterology* **83**, 424-29.

Roediger, W.E.W., Duncan, A. and Kipaniris, O. (1993) Reducing sulfur compounds in the colon impairs colonocyte nutrition: implication for ulcerative colitis. *Gastroenterology* **104**, 802-09.

Russo, G.L., Della Pietra, V., Mercurio, C., *et al.* (1997) Down-regulation of protein kinase CKII activity by sodium butyrate. *Biochemistry and Biophysical Research Communications* **233**, 673-77.

Sakata, T. and Setoyama, H. (1995) Local stimulatory effect of short-chain fatty acids on the mucus release from the hind-gut mucosa of rats. *Comparative Biochemistry and Physiology* **111 A**, 429-32.

Scheppach, W., Bartram, H.P. and Richter, A. (1995) Role of short chain fatty acids in the prevention of colorectal cancer. *European Journal of Cancer* **31A**, 1077-80.

Shaw, L. and Engel, P.C. (1985) The suicide inactivation of ox liver short chain acyl-CoA dehydrogenase by propionyl-CoA. *Biochemical Journal* **230**, 723-31.

Staecker, J.L. and Pitot, H.C. (1988) The effect of sodium butyrate on tyrosine aminotransferase induction in primary cultures of normal adult rat hepatocytes. *Archives of Biochemistry and Biophysics* **261**, 291-98.

Stark, A.H. and Madar, Z. (1993) In vitro production of short-chain fatty acids by bacterial fermentation of dietary fiber compared to the effects of those fibers on hepatic sterol synthesis. *Journal of Nutrition* **123**, 2166-73.

Stein, J., Schroder, O. and Bonk, M. (1996) Induction of glutathion-S-transferase-pi by short chain fatty acids in the intestinal cell line Caco-2. *European Journal of Clinical Investigation* **26**, 84-92.

Stein, J., Schroder, O. and Milovic, V. (1995) Mercaptopropionate inhibits butyrate uptake in isolated membrane vesicles of the rat distal colon. *Gastroenterology* **108**, 673-79.

Stephen, A. (1994) Propionate- sources and effects on lipid metabolism. In *Short Chain Fatty Acids*, (eds. Binder, H.J., Cummings, J. and Soergel, K.H.)., Kluwer Academic Press, Lancaster, pp. 250-71.

Todesco, T., Rao, A., Bosello, O., *et al.* (1991) Propionate lowers blood glucose and alters lipid metabolism in healthy subjects. *American Journal of Clinical Nutrition* **54**, 860-65.

Trinidad, P.T., Wolever, T.M.S. and Thompson, L.U. (1993) Interactive effects of calcium and short chain fatty acids on absorpbon in the distal colon of man. *Nutrition Research* **13**, 417-25.

Velasquez, O.C., Lederer, H.M. and Rombeau, J.L. (1996a) Butyrate and the colonocyte. Implications for neoplasia. *Digestive Diseases and Sciences* **41**, 727-39.

Velasquez, O.C., Seto, R.W. and Rombeau, J.L. (1996b) The scientific rationale and clinical application of short-chain fatty acids and medium-chain triacylglycerols. *Proceeding of the Nutrition Society* **55**, 49-78.

Venter, C.S. and Vorster, H.H. (1989) Possible metabolic consequences of fermentation in the colon for humans. *Medical Hypotheses* **29**, 161-66.

Venter, C.S., Vorster, H.H. and Cummings, J.H. (1990) Effects of dietary propionate on carbohydrate and lipid metabolism in healthy volunteers. *American Journal of Clinical Nutrition* **85**, 549-53.

Wolever, T.M.S., Spadafora, P and Eshuis, H. (1991) Interaction between colonic acetate and propionate in humans. *American Journal of Clinical Nutrition* **53**, 681-87.

Wolever, T.M.S., Spadafora, P.J., Cunnane, S.C., *et al.* (1995) Propionate inhibits incorporation of colonic [1,2-13C] acetate into plasma lipids in humans. *American Journal of Clinical Nutrition* **61**, 1241-47.

Younes, H., Demigne, C. and Remesy, C. (1996) Acidic fermentations in the caecum increases absorption of calcium and magnesium in the large intestine of the rat. *British Journal of Nutrition* **75**, 301-14.

CHAPTER 5

Bacterial Colonisation of Surfaces in the Large Intestine

SANDRA MACFARLANE, JOHN H. CUMMINGS AND
GEORGE T. MACFARLANE
Medical School, University of Dundee, Ninewells Hospital, Dundee, UK

1   Introduction

While the human large intestinal microbiota is widely perceived as being a homogeneous entity, individual species and groups of microorganisms exist in a multiplicity of different microhabitats and metabolic niches associated with the mucosa, the mucus layer or in the colonic lumen, where bacteria exist as 'free-living' organisms, as microcolonies, or grow in intimate associations on the surfaces of particulate materials (Englyst *et al.*, 1987; Macfarlane *et al.*, 1997a).

Where surfaces are available, bacteria form biofilms, which may be composed of a single species, but are usually microbial consortia that develop in response to the chemical composition of the substratum and other environmental constraints. Initial colonisation occurs through the attachment of single cells or small groups of organisms, followed by an increase in bacterial cell mass, leading to biofilm formation. Microscopic examination of human large intestinal contents shows that most of the bacteria are not freely dispersed, but occur in large clumps, and in aggregates attached to plant cell structures, resistant starch granules and other solids. Sessile microorganisms in biofilms often behave very differently from their planktonic, or free-living forms, and in particular, the nature and efficiency of their metabolism is changed, while the organisms exhibit greater resistance to antibiotics, and other environmental factors that have deleterious effects on planktonic cells (Anwar *et al.*, 1990; Van Loosdrecht *et al.*, 1990; Mozes and Rouxhet, 1992). Particle-associated and mucosal bacteria in the large bowel are members of complex multi-species consortia, which may be highly evolved assemblages, similar to those seen in oral biofilm communities, where partner recognition appears to be very specific during the formative stages of co-aggregation (Kolenbrander 1989). Biofilm communities often exhibit highly coordinated multicellular behaviour, within and between species, and many biofilm properties are dependent on local cell population densities, for example, quorum sensing transcriptional activation in Gram negative species (Salmond *et al.*, 1995).

G.R. Gibson and M.B. Roberfroid (eds.), Colonic Microbiota, Nutrition and Health, 71-87.
© 1999 *Kluwer Academic Publishers. Printed in the Netherlands.*

Microbial biofilm communities have been extensively studied in a variety of natural environments, such as sediments, soils, the oral cavity and the gastrointestinal tracts of animals, but they have hitherto been neglected niches in the human large intestine. Bacteria colonising surfaces in the gut lumen are likely to be more directly involved in the breakdown of complex insoluble polymeric substances than planktonic organisms, giving them a competitive advantage in the ecosystem. Close spatial relationships between bacterial cells on surfaces are important in metabolic communication between microorganisms in the microbiota, and are ecologically significant in that they minimise potential growth limiting effects on secondary cross-feeding populations, that are associated with mass transfer resistance (Conrad et al., 1985).

2   Mucosal populations

The occurrence of mucosal bacterial populations in the gastrointestinal tracts of animals is well known, and in chickens (Lee, 1980), ruminants (Wallace et al., 1979) and termites (Breznak and Pankatz, 1977), specific microfloras have been found growing in association with epithelial surfaces. Secretory intestinal epithelia are covered in a mucus coating ranging between 100-200 µm in thickness (Pullan et al., 1994), which may be important in stabilising microbial communities growing in association with the mucosa (Savage, 1978). Due to the presence of facultative anaerobes in the mucin layer and colonic lumen, pO_2 is not an important factor affecting the growth of anaerobic species at the mucosal surface. Some reports suggest that mucosal communities in humans are generally similar to those present in the gut lumen (Nelson and Mata, 1970), with bacteroides and fusobacteria predominating. However, other organisms including eubacteria, clostridia and anaerobic Gram positive cocci are also present, either as disperse heterogeneous populations, or as microcolonies (Edmiston et al., 1982; Croucher et al., 1983).Some mucosal-associated organisms manifest unusual morphological properties and cannot be seen or cultured from lumenal contents (Lee et al., 1971). Indeed, organisms with distinct morphological characteristics have been visualised in situ on the mucosa, where scanning electron micrographs of biopsy specimens show giant helical bacteria, with lengths greater than 60 µm residing in the mucus layer (Croucher et al., 1983). Many organisms that grow in close association to the colonic mucosa are spiral shaped (Takeuchi et al., 1974), reflecting the superior efficiency of this form of locomotion in a highly viscous milieu, compared to motility mediated by bacterial flagella. Evidence suggests that the majority of bacteria associated with the intestinal epithelium inhabit the mucus layer, rather than the mucosal surface (Rozel et al., 1982). Mucin and other host secretions and sloughed epithelial cells may be important substrates for these organisms, since turnover of the colonic epithelium is rapid, with columnar absorptive cells and goblet cells having a half-life of about 6 days (Christensen, 1991).

Due to its tenuous adhesion to the underlying mucosa, and intrinsic lack of mechanical strength, the mucus layer and its associated bacterial communities are often lost during sample preparation, leaving only microorganisms that are firmly attached to the epithelial surface. Moreover, scanning electron microscopy of mucus-associated populations in the intestine can be difficult, since unless the mucin structure is stabilised, using anti-mucin antibodies for example, dehydration stages involved in sample preparation result in collapse of the gel structure (Davis *et al.*, 1977; Bayliss and Turner, 1982; Bollard *et al.*, 1986). Thus while bacterial cells and microcolonies can be readily visualised in association with fibrous-like strands of the radically condensed glycoprotein (see Fig. 1), the natural spatial relationships that exist *in vivo* are not discernible.

Mucosal populations are difficult to study in healthy individuals due to the physical inaccessibilty of the large gut, and this has severely restricted investigations of these communities. As a consequence, the metabolic and health-related significance of bacteria growing at, or near, the colonic mucosa is largely unknown.

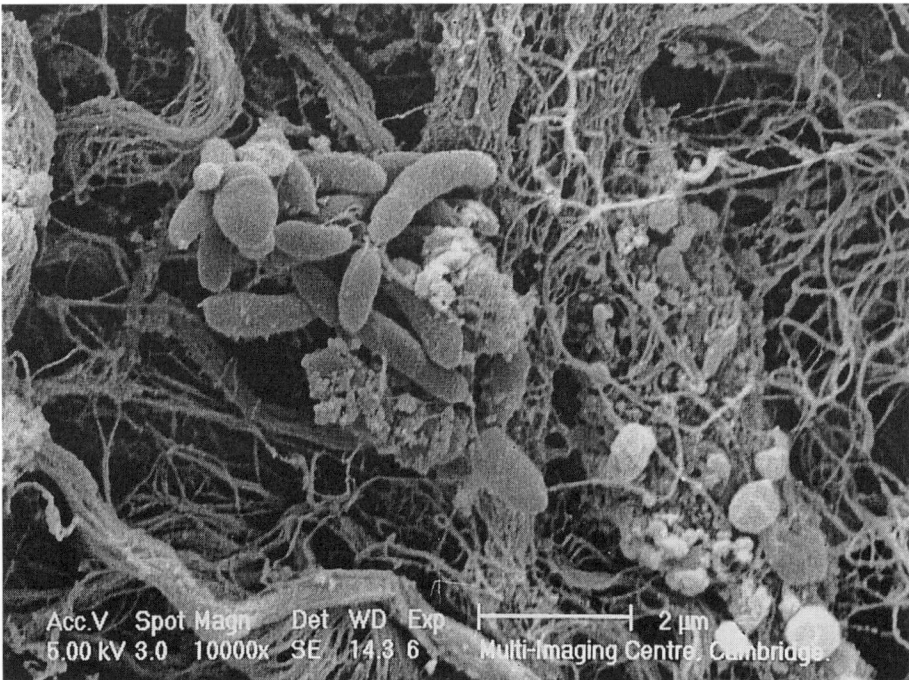

Figure 1. Scanning electron micrograph of a bacterial microcolony associated with the rectal mucosa. The mucus has hydrated during sample preparation and can be seen as dispersed fibrils in the background.

2.1   ADHESION AND COLONISATION OF THE COLONIC EPITHELIUM

The structure and composition of bacterial communities growing in association with the colonic mucosa and those existing in the mucus layer are determined by many factors such as humoral and cellular immunity, the rates of synthesis and chemical compositions of mucus, turnover rates of intestinal epithelial cells, availability of adhesion sites, lysozyme, pancreatic secretions, gut motility and colonisation resistance mediated by components of the normal colonic microflora. Innate immunity in the form of antimicrobial peptides such as defensins formed by polymorphonuclear cells and enterocytes, which are active against fungi and protozoa, as well as Gram positive and Gram negative organisms, may also be important at the mucosal surface (Mahida *et al.*, 1997).

The normal mucosal microbiota also affects mucosal and systemic immunity in the host. Intestinal epithelial cells, blood leukocytes, B and T lymphocytes, and accessory cells of the immune system are all involved (Schiffrin *et al.*, 1997). Bacterial products with immunomodulatory properties include lipoteichoic acids (LTAs), endotoxic lipopolysaccharide and peptidoglycans (Standiford *et al.*, 1994). Lipoteichoic acids of Gram positive species such as bifidobacteria possess high binding affinity for human epithelial cell membranes, and also serve as carriers for other antigens, binding them to target tissues where they can provoke an immune reaction (Op den Camp *et al.*, 1985). Maintenance of immune system homeostasis depends to some degree on cell-cell contacts, and bacteria colonising the gut mucosa play an important role in this respect.

In vitro studies with human cell lines have shown that bifidobacterial isolates of human origin adhere to enterocyte-like CACO-2 cells (Bernet *et al.*, 1993). These investigations also showed that adherent *Bifidobacterium breve*, *B. longum* and *B. infantis* variously inhibited CACO-2 cell-association and invasion by a range of intestinal pathogens including enterotoxigenic *Escherichia coli*, enteropathogenic *E. coli*, diffusely adhering *E. coli*, *Salmonella typhimurium* and *Yersinia pseudotuberculosis*. Bifidobacteria apparently attach at the apical brush border without causing epithelial damage, with cell binding being facilitated by the formation of species-specific extracellular and cell-associated protein-like adhesins by the bacteria. *Bifidobacterium infantis* and some strains of *B. breve* and *B. longum* attach strongly, while other *B. breve* and *B. longum* isolates seem to be poorly adherent.

Studies have also been made on the adherent properties of other intestinal bacteria. Although not all lactobacilli are able to attach to human intestinal epithelial cells (Kleeman and Klaenhammer, 1982), species that do colonise the gut in this way characteristically exhibit high surface hydrophobicities (Wadstrom *et al.*, 1987). Furthermore, evidence suggests that adherence of lactobacilli to epithelial cells is also mediated by secretion of a proteinaceous substance (Chauviere *et al.*, 1992). This is supported, in part, by the studies of Reid *et al.* (1993), who observed that three different strains of *L. casei* adhered to human uroepithelial and colonic epithelial cells using an extracellular protein-like adhesin. However, these authors also identified a cell wall-associated adhesin that was insensitive to protease treatment.

Protein dependent adherence of *Lactobacillus acidophilus* to CACO-2 cells prevented binding of enterotoxigenic and enteropathogenic *E. coli*, as well as *S. typhimurium* and *Yersinia pseudotuberculosis* (Bernet *et al.*, 1994). Suggested mechanisms for these effects included lactobacillus antimicrobial activity, stimulation of enterocyte production of antimicrobial substances with defensin-like characteristics, and occupation of enterocyte receptors by the lactobacilli.

Bacteria associated with the intestinal epithelium exist in close juxtaposition to immune effector cells and are likely to interact with mucosal macrophages and lymphocytes. Yoghurt lactobacilli bind to circulating peripheral blood CD4 and CD8 T lymphocytes, but not B cells (De Simone *et al.*, 1988). Lactobacilli that adhere to human intestinal epithelial cells (Kleeman and Klaenhammer, 1982) are capable of macrophage activation (Perdigon *et al.*, 1986). Lactobacilli have also been reported to degrade N-nitrosamines (Rowland and Grasso, 1975), which may be of particular significance at the mucosal surface.

Some lactobacilli, administered as probiotics, temporarily colonise the mucosal surface and displace other microorganisms. Analyses of intestinal biopsies showed that 19 test strains of lactobacilli (each 5 x 10^6 ml^{-1}) fed to healthy volunteers in 100 ml fermented oatmeal soup, colonised jejunal and rectal mucosae (Johansson *et al.*, 1993). High numbers of adherent lactobacilli were still recovered from jejunal samples up to 11 days after administration of the bacteria had ceased, while clostridial numbers decreased between 10- and 100-fold in some of the volunteers. In rectal tissue, anaerobe and enterobacterial numbers were reduced. *Lactobacillus plantarum* was the predominant adherent species, but *L. plantarum*, *L. agilis*, *L. reuteri* and *L. casei* subsp. *rhamnosus* were also present.

3 Interactions between intestinal pathogens and mucus

Pathogenic bacteria invading the body are affected by, and deal with, mucus barriers in different ways. Thus, mucus appears to have a protective role against enteric pathogens such as *Y. enterocolitica* in the colonic ecosystem, by reducing binding of the organisms to brush border membranes (Mantle *et al.*, 1989), while sulphomucins have been shown to prevent colonisation of the gastric mucosa by *Helicobacter pylori* (Piotrowski *et al.*, 1991). Conversely, the abilities of some Gram negative pathogens to colonise the mouse intestine are known to be specifically related to their abilities to adhere to mucus (Cohen *et al.*, 1986). Many motile intestinal bacteria exhibit chemotaxis (Lee *et al.*, 1997; Macfarlane *et al.*, 1997b) or possibly viscotaxis to mucin (Wilson, 1997). Virluent strains of *Serpulina hyodysenteriae* are significantly more chemotactic towards mucin than non-virulent isolates (Milner and Sellwood, 1994). A number of pathogens including campylobacters (Sylvester *et al.*, 1996) while not degrading mucus, can bind to the glycoprotein through formation of specific adhesins, as a prelude to gaining access to cell membrane receptors, but in other pathogenic species, mucus has an important nutritional function: neuraminidase for example, has been reported to be important for survival of *Bacteroides fragilis* in both *in vivo* and *in vitro* model systems during C-limited growth (Godoy *et al.*, 1993).

It has been demonstrated in our laboratory that the swarming bacterium *Clostridium septicum* synthesises a variety of mucin degrading enzymes (glycosidases, glycosulphatases), and is able to grow on purified mucins in which N-acetylglucosamine is the preferred C-source (Macfarlane *et al.,* 1997b). In these studies, dilution rate had no effect on the extent of breakdown in mucin-limited chemostats. At all dilution rates, mucin breakdown correlated with high activities of inducible cell-bound and extracellular ß-galactosidase, N-acetyl ß-glucosaminidase and neuraminidase. When mucin was replaced by glucose as C-source, synthesis of neuraminidase, hyaluronidase and glyco-sulphatase was repressed, and these enzymes were only formed by swarm cells at extremely low dilution rates. Measurements of residual mucin oligosaccharide sugars in spent culture media showed extensive utilization of N-acetylglucosamine, but neuraminic acid was not used by the bacterium, indicating that cell-associated neuraminidase is used in cell attachment or binding to mucin, rather than having a specific nutritional function.

Mucus has been shown to be chemotactic for *C. septicum* swarm cells (Wilson, 1997), and at high specific growth rates, where the planktonic populations consisted solely of short motile rods, giant hyperflagellated rods were observed to colonise surfaces of removable mucin gel cassettes that had been fitted to the chemostats. This provided evidence that swarming in *C. septicum* is in part a response to growth on mucus surfaces, and that this is an important virulence mechanism, since induction of mucinolytic enzymes by their substrate, and formation of these spreading factors by highly motile swarm cells is likely to facilitate rapid movement of the bacterium through the viscous mucus layer.

4   Studies on bacteria associated with the rectal mucosa

The benefits of using rectal biopsies to study mucosal bacterial populations are that for most of the time, the rectum is empty and the mucosa is clean, while samples are relatively easy to obtain since patients and volunteers do not need to be specially prepared, as is required when removing tissue from the proximal or distal bowel during colonoscopy.

Our studies have indicated that enterobacteria occur in low numbers on the healthy rectal mucosa. Anaerobic organisms predominate on the mucosal surface, particularly species belonging to the genera *Bacteroides* and *Bifidobacterium* (Table 1). Other investigations involving the use of colonic biopsies have also determined that bacteroides are the major anaerobes associated with mucosal surfaces in the large bowel (Poxton *et al.,* 1997). This work showed that *B. vulgatus* was the predominant organism, whereas of the eight different bacteroides species isolated in our investigation, *B. stercoris* and *B. thetaiotaomicron* were prevalent. Several different bifidobacterial isolates were obtained from rectal biopsies.

A variety of other anaerobic species were also found in our studies, some of which had unusual helical morphologies and could not be identified using standard morphological, biochemical or chemotaxonomic criteria.

Table 1. Enumeration of bacterial populations in human rectal biopsies

	Total bacteroides	Total bifidobacteria	Other anaerobic species	Total enterobacteria
Range	0 - 7.0	0 - 6.0	1.7 - 5.5	0 - 6.4
Mean	4.2 ± 1.1	3.6 ± 0.9	3.9 ± 0.7	2.4 ± 1.0

Counts are $\log_{10}$ bacteria $(cm^2$ biopsy sample$)^{-1}$, ± SEM (n = 6)

4.1 MUCIN DEGRADING ENZYMES IN BACTERIA ISOLATED FROM THE RECTAL MUCOSA

Rectal mucosal isolates were grown on mucin-containing culture media to investigate their mucinolytic activities. Table 2 shows that ß-galactosidase and N-acetyl ß-glucosaminidase synthesis predominated in the bacteroides and bifidobacteria, although the latter enzyme was not formed by any of the enterobacterial species, which produced ß-galactosidase and N-acetyl α-galactosaminidase. α-Fucosidase was formed by some bacteroides, while neuraminidase was only detected in bifidobacteria. The low content of sialomucins in the human rectum (Shamsuddin and Yang, 1995) may partly explain the lack of neuraminidase activities in these mucosal populations.

Table 2. Expression of mucin-degrading enzymes by different groups of bacteria isolated from human rectal mucosae[a]

Enzyme	Units (mg dry weight bacteria)$^{-1}$			
	Bacteroides (14)[b]	Bifidobacteria (9)	Other anaerobic species (9)	Enterobacteria (3)
ß-Galactosidase	140 ± 42	478 ± 263	36 ± 12	33 ± 3
N-Acetyl ß-glucosaminidase	150 ± 84	1336 ± 832	493 ± 435	ND
N-Acetyl α-galactosaminidase	ND[c]	ND	ND	37 ± 13
α-Fucosidase	35 ± 2	ND	ND	ND
Neuraminidase	ND	200 ± 200	ND	ND

[a]Isolates were grown on mucin-containing culture media to induce synthesis of mucinolytic enzymes.

Results show means ± SEM

[b]Number of isolates tested

[c]Not detected

Complete destruction of a complex polymer, such as mucin, is dependent on the activities of a number of different hydrolytic enzymes, that can breakdown the protein backbone and carbohydrate side chains of the macromolecule. Pure and mixed culture studies have established that in many bacteria synthesis of these enzymes, particularly ß-galactosidase, N-acetyl ß-glucosaminidase and neuraminidase (Macfarlane *et al.*, 1989; Macfarlane and Gibson, 1991; Macfarlane *et al.*, 1997b) is catabolite regulated, and is therefore dependent on local concentrations of mucin and other carbohydrates. While some colonic microorganisms can produce several different glycosidases, which in theory enables them to completely digest heterogeneous polymers (Pettipher and Latham, 1979; Degnan, 1993), the majority of experimental data point to the fact that the breakdown of mucin and other complex organic molecules is a co-operative activity.

Hoskins *et al.* (1992) observed that 60-95% of mucin oligosaccharides were degraded by pure cultures of *Ruminococcus torques* and *Bifidobacterium bifidum*, whereas other bifidobacterial species and individual bacteroides were less effective in digesting these oligomers (8-42% utilisation). However, *R. torques* and *B. bifidum* constitute a small fraction of the colonic microbiota, and it is probable that in the large intestine mucin breakdown is accomplished through the concerted actions of many different types of bacteria. Support for this comes from chemostat studies with oral microorganisms, which show that extensive mucin breakdown is a cooperative process dependent upon the synergistic activities of several disparate species (Bradshaw *et al.*, 1994).

In contrast to the depolymerisation of mucin, there is undoubtedly severe competition between colonic microorganisms for the products of oligosaccharide hydrolysis, since there are substantial populations of saccharolytic bacteria in the gut that are unable to digest the glycoprotein by themselves, and they must grow by cross-feeding on carbohydrate fragments produced by mucinolytic species. This is exemplified by data shown in Table 2, where none of the mucosal isolates were able to synthesise a range of enzymes sufficient to enable them to extensively digest mucin on their own.

5   Mucin degrading populations colonising particulate material in lumenal contents

As well as being produced by goblet cells in the large intestinal mucosa, salivary, gastric, bronchial, pancreatico-biliary and small intestinal mucins also enter the colon in effluent from the small bowel. Particulate matter, such as partly digested plant cell materials are entrapped in this viscoelastic gel which must be broken down to facilitate access of colonic microorganisms to the food residues. In early studies on gut contents obtained from human sudden death victims, Vercellotti *et al.* (1977) showed that mucins were degraded by intestinal microorganisms in all regions of the bowel. Other work has indicated that large amounts of mucus are present in lumenal contents in the mid-colon of rats (Bollard *et al.*, 1986), and this is also evident in human caecal material in Fig. 2, where bacterial cells mixed with dehydrated mucus can be seen covering food residues. While very few studies have been made on bacterial colonisation of particulate materials in the gut lumen in humans, some of the factors that are thought to affect bacterial proliferation on these substrata are shown in Table 3.

Microbiological analysis of partially digested food particles in faeces demonstrated that biofilm populations growing on the surfaces of particulate matter are members of complex multi-species consortia (Table 4). At the genus level, biofilm populations are superficially similar to those in non-adherent microbiotas, with bacteroides and bifidobacteria predominating. Interestingly however, bacilli and staphylococci, which are not considered to be autochthonous to the colonic ecosystem were unable to colonise food particles. Whilst these preliminary bacteriological observations indicate that a diverse range of microorganisms adhere to particulate surfaces in the lumen of the bowel, more detailed studies are needed to determine the genotypic relatedness of biofilm communities to non-adherent populations.

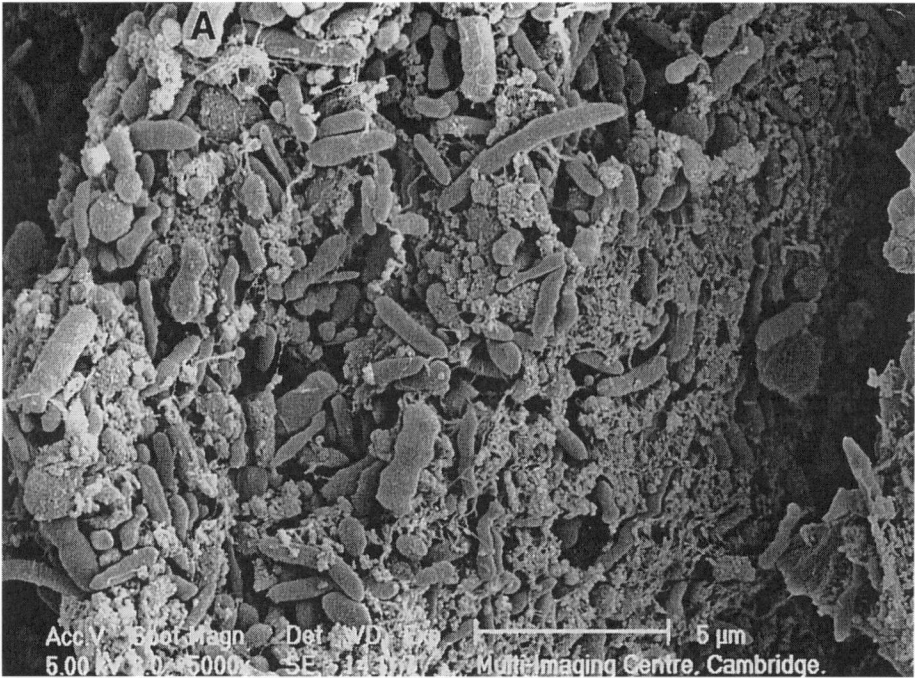

Figure 2. Scanning electron micrographs of a bacterial biofilm colonising the surface of digestive materials in human caecal contents.

5.1   HYDROLYTIC ENZYME ACTIVITIES OF LUMENAL BACTERIA GROWING ON THE SURFACES OF PARTICULATE MATERIALS IN FAECES

Enzymological measurements suggest that biofilm communities growing on the surfaces of food particles in the colon form metabolically distinct assemblages in the gut microbiota in relation to the breakdown and metabolism of mucin and other complex carbohydrates.

Table 3. Physicochemical, microbiological and host factors that affect adhesion and colonisation of surfaces in the colon

Dietary factors controlling chemical composition and physical form of substrata e.g. particle size, solubility and association of substrates with undigestible complexes and inhibitory substances

Generic and species composition of the host microbiota

Microbial competition for space and nutrients, production of antagonistic agents

Cooperative interactions between microorganisms, including synthesis of complementary hydrolytic enzyme systems involved in breakdown of complex associations of polymeric substrates, inter-species cross-feeding, catabolite regulatory mechanisms

Rates of physical destruction of substrata

Colonic transit time

Intestinal pH

Disease, drugs, antibiotic therapy

Table 4. Comparison of biofilm communities colonising food particles in faeces and non-adherent populations[a]

Bacteria	Non-adherent bacteria	Biofilm populations
Total anaerobes	10.8 ± 0.01	10.5 ± 0.16
Total facultative anaerobes	8.4 ± 0.08	7.4 ± 0.06
Bacteroides fragilis group	10.4 ± 0.04	9.5 ± 0.01
Bifidobacteria	10.4 ± 0.01	9.4 ± 0.04
Actinomycetes	10.1 ± 0.06	9.4 ± 0.06
Clostridium perfringens	9.0 ± 0.06	8.8 ± 0.06
Gram positive anaerobic cocci	10.3 ± 0.08	9.9 ± 0.02
Enterococci	7.9 ± 0.08	7.0 ± 0.33
Staphylococci	7.9 ± 0.12	ND[b]
Bacilli	7.3 ± 0.10	ND[b]

[a]Results are from (Macfarlane and Macfarlane, 1995). Counts are $\log_{10}$ bacteria (gram dry weight bacterial mass)$^{-1}$ ± SEM (n = 5)

[b]Not detected

Proteases and peptidases are involved in mucin fermentation in the large bowel, although little is known of the bacterial enzymes that are involved in digestion of the glycoprotein *in vivo*. Comparative measurements of peptidolytic enzymes in lumenal biofilm and non-adherent populations, using a range of protease inhibitors, indicated that whilst the spectrum of proteolytic/peptidolytic activity was broadly similar in both microbiotas, variations occurred in the relative importance of individual enzyme classes (Fig. 3). Thus, while serine, thiol and aspartic protease profiles were comparable, qualitative differences were evident with respect to trypsin, chymotrypsin and, to a lesser degree, metalloprotease activities. The higher trypsin and chymotrypsin in the biofilms probably resulted from adsorbed pancreatic endopeptidases, whereas lower metalloprotease activities reflected differences in bacterial enzyme expression.

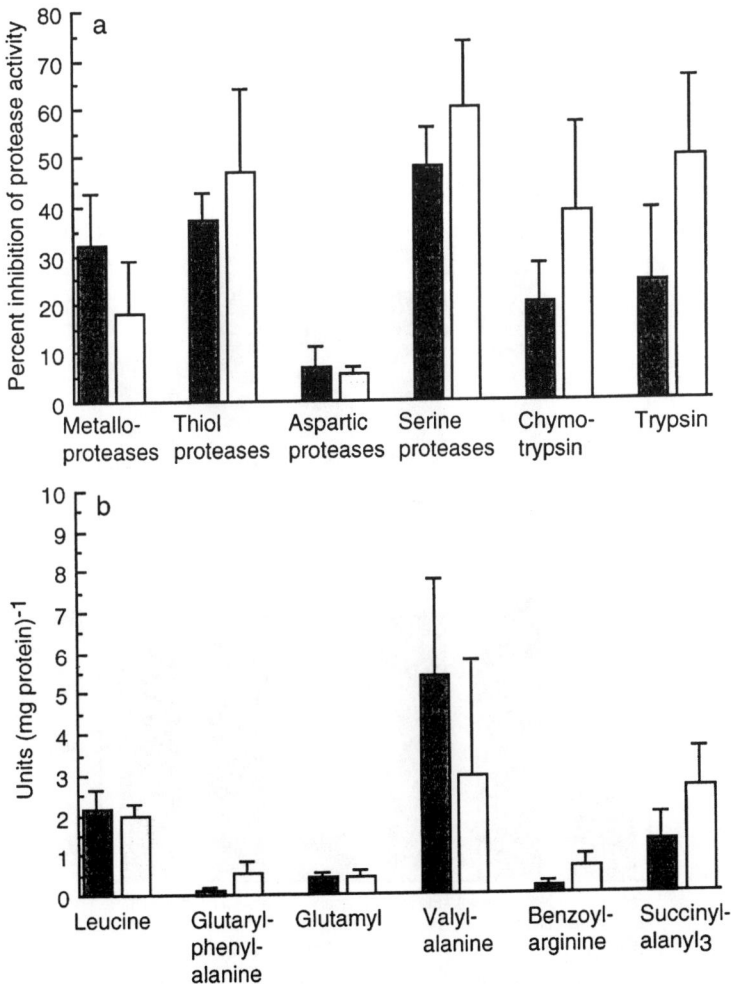

Figure 3. Bacterial protease (a) and arylamidase activities (b) associated with biofilm (open bars) and non-adherent (closed bars) populations in faecal material. Data are means of results obtained from five subjects ± SEM.

Arylamidase measurements of bacterial peptidases using p-nitroanilide substrates confirmed the higher levels of trypsin (benzoyl arginine) and chymotrypsin (glutaryl-phenylalanine) in the biofilms, and showed that this was also true for elastase (succiny-lalanyl3). Leucine and glutamyl arylamidase activities were the same in both bacterial communities, but valylalanine arylamidase, which is formed in very high levels by members of the *Bacteroides fragilis* group (Macfarlane and Macfarlane, 1991), was markedly lower in the biofilms.

Determinations of mucinolytic glycosidases (Fig. 4) showed that, with the exception of N-acetyl α-galactosaminidase, the vast majority of these enzymes were cell-associated in faecal material. Little difference was observed in bacterial expression of ß-galactosidase and N-acetyl ß-glucosaminidase in the biofilms, however, α-fucosidase and N-acetyl α-galactosaminidase activities were lower than in non-adherent populations.

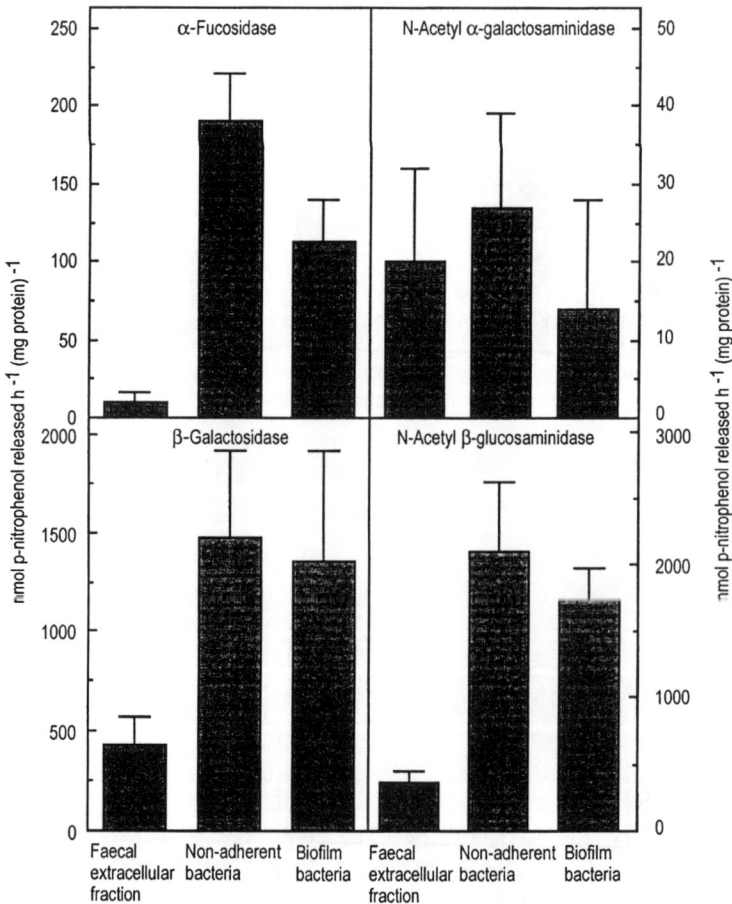

Figure 4. Mucinolytic glycosidase activities in different fractions of human faeces.
Data are means of results obtained from six subjects ± SEM.

5.3 FERMENTATIVE ACTIVITIES IN LUMENAL BIOFILMS

Intestinal bacterial fermentations are regulated by the need to maintain redox balance, principally through the reduction and oxidation of ferredoxins, flavins and pyridine nucleotides. To a large degree, this affects carbon flow through the bacteria, the energy yield obtained from the substrate, and the fermentation products formed. Synthesis of reduced products including H_2, lactate, succinate, butyrate and ethanol is used to effect redox balance during fermentation, whereas production of more oxidised substances, such as acetate, is associated with ATP generation. Conversely, formation of more reduced fermentation products results in comparatively low ATP yields.

The abilities of faecal bacteria growing in lumenal biofilms and non-adherent populations to ferment purified porcine mucin and other polymerised carbon sources was studied *in vitro*. These investigations were carried out using anaerobic short term batch culture fermentations to preclude changes in bacterial populations during the course of the experiment (Macfarlane *et al.*, 1997a). Table 5 shows that rates of fermentation acid formation from mucin were highest in the non-adherent fraction of faeces. The relative ratios of individual SCFA formed from mucin were found to be markedly different, in that acetate was the major fermentation metabolite formed by the biofilm bacteria, with less propionate and very little butyrate production. Comparison of data obtained with the other test polymers showed that adherent populations always formed proportionately more acetate than their non-adherent counterparts, and that pectin was degraded considerably less rapidly than mucin by both biofilm and non-adherent organisms.

Table 5. Comparative rates of production and molar ratios of short chain fatty acids formed from mucin and other polymerised carbon sources by biofilm communities associated with food particles and non-adherent bacterial populations[a].

Carbon source	Bacterial population	Fermentation product[b]			
		Acetate	Propionate	Butyrate	Total SCFA
Mucin	Biofilm	0.86 (82)	0.13 (12)	0.06 (6)	1.05
	Non-adherent	0.85 (62)	0.30 (21)	0.23 (17)	1.38
Xylan	Biofilm	0.92 (88)	0.07 (7)	0.05 (5)	1.04
	Non-adherent	0.83 (71)	0.15 (11)	0.21 (18)	1.19
Pectin	Biofilm	0.47 (76)	0.10 (16)	0.05 (8)	0.62
	Non-adherent	0.40 (58)	0.21 (31)	0.07 (11)	0.68

[a]Data are from (Macfarlane *et al*, 1997a)

[b]Specific rates of SCFA production are mmol h^{-1} (gram dry weight bacteria)$^{-1}$, SCFA molar ratios are given in parenthesis

6 Summary

At present we know very little about the fine structure of bacterial mucin-degrading communities in human intestinal biofilms, their ecological significance in the colonic ecosystem as a whole, or their metabolic importance to the host. However, experimental evidence shows that these microbiotas are heterogeneous assemblages that form rapidly either on the surfaces of particulate matter in the intestinal lumen, or in the mucus layer lining the mucosa. While the *in vitro* modelling studies outlined earlier provide useful comparative information on the physiological activities of adherent and non-adherent populations, the analytical methods employed were essentially destructive and did not provide useful information concerning the multicellular organisation of biofilm communities.

The current move away from culture-based procedures for studying intestinal microorganisms, and the introduction of methodologies involving measurements of ribosome abundance using quantitative hybridization techniques (Stahl and Amman, 1991) will facilitate future work on colonic biofilm communities. Quantitation of both total rRNA and that relating to specific populations will, in the future, increase our understanding of the partitioning of different bacterial communities and activities between sessile and planktonic microorganisms, while fluorescent labelling (Amman *et al.*, 1990) and whole cell hybridization combined with phase contrast, confocal and scanning electron microscopy (Wolfaardt *et al.*, 1994) will improve our understanding of the spatial organisation of these microcosms.

7   References

Amman, R.I., Krumholz, L. and Stahl, D.A. (1990) Fluorescent-oligonucleotide probing of whole cells for determinative, phylogenetic and environmental studies in microbiology. *Journal of Bacteriology* **172**, 762-70.

Anwar, H., Dasgupta, M.K. and Costerton, J.W. (1990) Testing the susceptibility of bacteria in biofilms to antibacterial agents. *Antimicrobial Agents and Chemotherapy* **34**, 2043-46.

Bayliss, C.E. and Turner, R.J. (1982) Examination of organisms associated with mucin in the colon by scanning electron microscopy. *Micron* **13**, 35-40.

Bernet, M.-F., Brassart, D., Neeser, J-R. *et al.* (1993) Adhesion of human bifidobacterial strains to cultured human intestinal epithelial cells and inhibition of enteropathogen-cell interactions. *Applied and Environmental Microbiology* **59**, 4121-28.

Bernet, M.F., Brassart, D., Neeser, J.R. *et al.* (1994) *Lactobacillus acidophilus* LA 1 binds to cultured human intestinal cell lines and inhibits cell attachment and cell invasion by enterovirulent bacteria. *Gut* **35**, 483-89.

Bollard, J.E., Vanderwee, M.A., Smith, G.W. *et al.* (1986) Location of bacteria in the mid-colon of the rat. *Applied and Environmental Microbiology* **51**, 604-08.

Bradshaw, D.J., Homer, K.A., Marsh, P.D. *et al.* (1994) Metabolic communication in oral microbial communities during growth on mucin. *Microbiology* **140**, 3407-12.

Breznak, J.A. and Pankatz, H.S. (1977) In situ morphology of the gut microbiota of wood eating termites

[*Reticulitennes flaviceps* Kollar and *Coptotermes formosanus* Shiraki]. *Applied and Environmental Microbiology* **33**, 406-26.

Chauviere, G., Coconnier M.-H, Kerneis S. *et al.* (1992) Adhesion of human *Lactobacillus acidophilus* strain LB to human enterocyte-like Caco-2 cells. *Journal of General Microbiology* **138**, 1689-96.

Christensen, J. (1991) Gross and microscopic anatomy of the large intestine, in *The Large Intestine: Physiology, Pathophysiology and Disease* (eds. S.F. Phillips, J.H. Pemberton and R.G. Shorter), Raven Press Ltd., New York, pp. 13-35.

Cohen, P.S., Wadolkowski, E.A. and Laux, D.C. (1986) Adhesion of a human fecal *Escherichia coli* strain to a 50.5 KDal glycoprotein receptor present in mouse colonic mucus. *Microecology and Therapy* **16**, 231-41.

Conrad, R., Phelps, T.J. and Zeikus, J.G. (1985) Gas metabolism evidence in support of the juxtaposition of hydrogen-producing and methanogenic bacteria in sewage sludge and lake sediments. *Applied and Environmental Microbiology* **50**, 595-601.

Croucher, S.C., Houston, A P., Bayliss, C.E. *et al.* (1983) Bacterial populations associated with different regions of the human colon wall. *Applied and Environmental Microbiology* **45**, 1025-33.

Davis, C.P., Balish, E. and Uehling, D. (1977) Bacterial microenvironments: Bacterial interaction with mucin layers in gastrointestinal tracts and bladders. *Scanning Electron Microscopy* **11**, 269-74.

Degnan, B.A. (1993) *Transport and Metabolism of Carbohydrate by Anaerobic Gut Bacteria*. Ph.D Thesis, Cambridge University, Cambridge.

De Simone, C., Grassi, P.P., Bianchi-Salvadori, B. *et al.* (1988) Adherence of specific yogurt micro-organisms to human peripheral blood lymphocytes. *Microbios* **55**, 49-57.

Edmiston, C.E. Jr., Avant, G.R. and Wilson, F.A. (1982) Anaerobic bacterial populations on normal and diseased human biopsy tissue obtained at colonoscopy. *Applied Environmental Microbiology* **43**, 1173-81.

Englyst, H.N., Hay, S. and Macfarlane, G.T. (1987) Polysaccharide breakdown by mixed populations of human faecal bacteria. *FEMS Microbiology Ecology* **95**, 163-71.

Godoy, V.G., Dallas, M.M., Russo, T.A. *et al.* (1993) A role for *Bacteroides fragilis* neuraminidase in bacterial growth in two model systems. *Infection and Immunity* **61**, 4415-26.

Hoskins, L.C., Boulding, E.T., Gerken, T.A. *et al.* (1992) Mucin glycoprotein degradation by mucin oligosac-charide-degrading strains of human faecal bacteria. Characterisation of saccharide cleavage products and their potential role in nutritional support of larger faecal bacterial populations. *Microbial Ecology in Health and Disease* **5**, 193-207.

Johansson, M.-L, Molin, G., Jeppsson, B. *et al.* (1993) Administration of different *Lactobacillus* strains in fermented oatmeal soup: In vivo colonization of human intestinal mucosa and effect on the indigenous flora. *Applied and Environmental Microbiology* **59**, 15-20.

Kleeman, E.G. and Klaenhammer, T.R. (1982) Adherence of *Lactobacillus* species to human fetal intestinal cells. *Journal of Dairy Science* **65**, 2063-69.

Kolenbrander, P.E. (1989) Surface recognition among oral bacteria: multigeneric coaggregations and their mediators. *Critical Reviews in Microbiology* **17**, 137-59.

Lee, A. (1980) Normal flora of animal intestinal surfaces, in *Adsorption of Microorganisms to Surfaces* (eds. G. Bitton and K.C. Marshall), John Wiley, New York, pp. 145-74.

Lee, S.G., Changsung, K. and Young, C.H. (1997) Successful cultivation of a potentially pathogenic coccoid organism with trophism for gastric mucin. *Infection and Immunity* **65**, 49-54.

Lee, F.D., Kraszewski, A., Gordon, J. *et al.* (1971) Intestinal spirochaetosis. *Gut* **12**, 126-33.

Macfarlane, S., McBain, A.J. and Macfarlane, G.T. (1997a) Consequences of biofilm and sessile growth in the large intestine. *Advances in Dental Research* **11**, 59-68.

Macfarlane, G.T. and Gibson, G.R. (1991) Formation of glycoprotein degrading enzymes by *Bacteroides fragilis*. *FEMS Microbiology Letters* **77**, 289-94.

Macfarlane, G.T., Hay, S. and Gibson, G.R. (1989) Influence of mucin on glycosidase, protease and arylamidase activities of human gut bacteria grown in a 3-stage continuous culture system. *Journal of Applied Bacteriology* **66**, 407-17.

Macfarlane, G.T. and Macfarlane, S. (1991) Utilization of pancreatic trypsin and chymotrypsin by proteolytic and non-proteolytic *Bacteroides fragilis*-type bacteria. *Current Microbiology* **23**, 143-48.

Macfarlane, G.T. and Macfarlane, S. (1995) Human intestinal biofilm communities, in *The Life and Death of Biofilm* (eds. J. Wimpenny, P. Handley, P. Gilbert *et al.*), Bioline, University of Wales College of Cardiff, pp. 83-89.

Macfarlane, G.T., Macfarlane, S. and Sharp, R. (1997b) Differential expression of virulence determinants in *Clostridium septicum* in relation to growth on mucin and the swarm cell cycle. *Bioscience and Microflora* **16**, 28.

Mahida, Y.R., Rose, F. and Chan, W.C. (1997) Antimicrobial peptides in the gastrointestinal tract. *Gut* **40**, 161-63.

Mantle, M., Basaraba, L., Peacock, S.C. and Gall, D.G. (1989) Binding of *Yersinia enterocolitica* to rabbit brush border membranes, mucus, and mucin. *Infection and Immunity* **57**, 3292-99.

Milner, J.A. and Sellwood, R. (1994) Chemotactic response to mucin by *Serpulina hyodysenteriae* and other porcine spirochetes: Potential role in intestinal colonization. *Infection and Immunity* **62**, 4095-99.

Mozes, N. and Rouxhet, P.G. (1992) Influence of surfaces on microbial activity, in *Biofilms-Science and Technology* (eds. L.F. Melo, T.R. Bott and B. Capdeville), Kluwer Academic Publishers, Doorderecht, pp. 125-36.

Nelson, D.P. and Mata, L.J. (1970) Bacterial flora associated with the human gastrointestinal mucosa. *Gastroenterology* **58**, 56-61.

Op den Camp, H.J.M., Oosterhof, A. and Veerkamp, J.H. (1985) Interaction of bifidobacterial lipoteichoic acid with human intestinal epithelial cells. *Infection and Immunity* **47**, 332-34.

Perdigon, G., Nader De Macios, M.E., Alvarez, S. *et al.* (1986). Effect of perorally administered lactobacilli on macrophage activation in mice. *Infection and Immunity* **53**, 404-10.

Pettipher, G.L. and Latham, M. (1979) Production of enzymes degrading plant cell walls and fermentation of cellobiose by *Ruminococcus flavifaciens*. *Journal of General Microbiology* **110**, 29-38.

Piotrowski, J., Slomiany, A., Murty, V.L.N. *et al.* (1991) Inhibition of *Helicobacter pylori* colonization by sulfated gastric mucin. *Biochemistry International* **24**, 749-56.

Poxton, I.R., Brown, R., Sawyerr, A.F. and Ferguson, A. (1997) The mucosal anaerobic Gram-negative bacteria of the human colon. *Clinical Infectious Diseases* **25**, S111-13.

Pullan, R.D., Thomas, G.A.O., Rhodes, M. *et al.* (1994) Thickness of adherent mucus gel on colonic mucosa in humans and its relevance to colitis. *Gut* **35**, 353-59.

Reid, G., Servin, A., Bruce, A.W. *et al.* (1993) Adhesion of three *Lactobacillus* strains to human urinary and intestinal epithelial cells. *Microbios* **75**, 57-65.

Rowland, I.R. and Grasso, P. (1975) Degradation of N-nitrosamines by intestinal bacteria. *Applied Microbiology* **29**, 7-12.

Rozel, K.R., Cooper, D., Lam, K. *et al.* (1982) Microbial flora of the mouse ileum mucous layer and epithelial surface. *Applied and Environmental Microbiology* **43**, 1452-63.

Salmond, G.P.C., Bycroft, B.W., Stewart, G.S.A.B. *et al.* (1995) The bacterial 'enigma': cracking the code of cell-cell communication. *Molecular Microbiology* **16**, 615-24.

Savage, D.C. (1978) Factors involved in colonization of the gut epithelial surface. *American Journal of Clinical Nutrition* **31**, S131-35.

Schiffrin, E.J., Brassart, D., Servin, A.L. *et al.* (1997) Immune modulation of blood leukocytes in humans by lactic acid bacteria: criteria for strain selection. *American Journal of Clinical Nutrition* **66**, 15S-20S.

Shamsuddin, A.K.M. and Yang, G.Y. (1995) The large intestinal mucosa, in *Gastrointestinal and Oesophageal Pathology* 2nd edn. (ed. R. Whitehead), Churchill Livingstone, Edinburgh, pp. 45-56.

Standiford, T.K., Arenberg, D.A., Danforth, J.M. *et al.* (1994) Lipoteichoic acid induces secretion of interleukin-8 from human blood monocytes: a cellular and molecular analysis. *Infection and Immunity* **62**, 119-25.

Stahl, D.A. and Amman, R.I. (1991) Development and application of nucleic acid probes in bacterial systematics, in *Sequencing and Hybridization Techniques in Bacterial Systematics* (eds. E. Stackenbrandt and M. Goodfellow), John Wiley, Chichester pp. 205-48.

Sylvester, F.A., Philpott, D., Gold, B. *et al.* (1996) Adherence to lipids and intestinal mucin by a recently recognised human pathogen, *Campylobacter upsaliensis*. *Infection and Immunity* **64**, 4060-66.

Takeuchi, A., Jervis, H.R., Nakagawa, H. *et al.* (1974) Spiral-shaped organisms on the surface colonic epithelium of the monkey and man. *American Journal of Clinical Nutrition* **27**, 1287-96.

Van Loosdrecht, M.C.M., Lyklema, J., Norde, W. *et al.* (1990) Influence of interfaces on microbial activity. *Microbiological Reviews* **54**, 75-87.

Vercellotti, J.R., Salyers, A.A., Bullard, W.S. *et al.* (1977) Breakdown of mucin and plant polysaccharides in the human colon. *Canadian Journal of Biochemistry* **55**, 1190-96.

Wadstrom, T., Andersson, K., Sydow, M. *et al.* (1987) Surface properties of lactobacilli isolated from the small intestine of pigs. *Journal of Applied Bacteriology* **62**, 513-20.

Wallace, R.J., Cheng, K-J., Dinsdale, D. *et al.* (1979) An independent microbial flora of the epithelium and its role in the microbiology of the rumen. *Nature* **279**, 424-26.

Wilson, L.M. (1997) *Physiological Studies on Swarming and Production of Virulence Determinants in Clostridium septicum*. Ph.D Thesis, Cambridge University, Cambridge.

Wolfaardt, G.M., Lawrence, J.R., Robarts, R.D. *et al.* (1994) Multicellular organisation in a degradative biofilm community. *Applied and Environmental Microbiology* **60**, 434-46.

CHAPTER 6

Probiotics

ROY FULLER
Russet House, Ryeish Green, Reading, RG7 1ES, UK

1   Introduction

The gastrointestinal microflora is a very complex collection of microorganisms living in symbiosis with the host animal. It is an essential part of the body and as Luckey (1972) pointed out, there are more microbial cells associated with the human body than there are somatic cells. It is, therefore, important that the delicate balance between microflora and the host is controlled and sustained. We must pay attention to the many features of modern existence that tend to induce changes in the composition and/or activity of the gut flora. These include:
- diet
- temperature
- humidity
- medication
- stress

Improved knowledge of the gut microflora has led to the development of probiotic preparations which attempt to respond to these problems and provide effective and natural ways of combating them. The scientific basis of the probiotic effect and practical applications have been reviewed in two recent books (Fuller, 1992, 1997).

2   Definition

The word probiotic is derived from the two Greek words meaning 'for life'. Early attempts to use the term to mean a microbial substance which stimulates the growth of another microorganism (Lilley and Stillwell, 1965) or tissue extracts which improved microbial growth (Sperti, 1971) did not gain general acceptance. Parker (1974) first used the word probiotic in the context of animal feed supplementation and defined it as: 'Organisms and substances which contribute to intestinal microbial balance'

G.R. Gibson and M.B. Roberfroid (eds.), Colonic Microbiota, Nutrition and Health, 89-99.
© 1999 *Kluwer Academic Publishers. Printed in the Netherlands.*

Fuller (1989) redefined probiotics removing the reference to 'substances' which could include antibiotics and microbial stimulants which later became known as prebiotics (Gibson and Roberfroid 1995). The revised definition is: 'A live microbial feed supplement which beneficially affects the host animal by improving its intestinal microbial balance'

This modified version stressed the need for the supplement to be composed of *viable* microorganisms. It has been suggested that this definition should be expanded to include other areas of the body such as the skin, vagina and respiratory tract. Although a case can be made for redefinition in this way, in the present context of this book, Fuller's 1989 definition is appropriate and will be used as the basis for this chapter.

Probiotics is, therefore, a beneficial effect on the host mediated through the gut microflora by ingestion of viable microorganisms. Such a definition includes not only preparations specifically designed for probiotic use but also traditional yoghurts and other fermented foods. Probiotic products currently on the market are presented in the form of powders, tablets, liquid suspensions and sprays and are intended for use by humans, farm animals and pets. Most preparations destined for human consumption are powders or tablets. They can contain one or up to seven different species of bacteria or fungi. Multistrain preparations can be regarded as broad-spectrum having the potential to be active against different adverse conditions in a variety of animal hosts.

3   Development of the probiotic concept

Although the word was not coined until 1965, the basic probiotic concept was first conceived by Metchnikoff at the beginning of this century. Up until then, fermented milks had been a common source of food but he began to give the practice a scientific basis. Metchnikoff had long believed that the complex microbial population in the colon was having an adverse effect on the host through the so-called autointoxication effect. He briefly advocated surgical removal of the colon but fortunately thought better of it and turned his attention towards modification of the activity of the colonic microflora by ingestion of soured milks. He is said to have been inspired to this approach by the discovery that Bulgarian peasants lived to a ripe old age and also consumed large quantities of soured milks. The relation between the two observations had not been established at that time but subsequent research has tended to confirm his general conclusion that the habit was having a beneficial effect on health without necessarily prolonging life.

Metchnikoff attempted to refine the supplements by isolating the lactic acid bacteria in the soured milks and using them in feeding trials. He ended up using a Gram-positive rod which he called the Bulgarian bacillus and later *Bacillus bulgaricus*. Whilst it is impossible now to be sure, it seems likely that the organism he was using is what was later known as *Lactobacillus bulgaricus* and is now called *L. delbrueckii* subsp. *bulgaricus* which together with *Streptococcus salivarius* subsp. *thermophilus* is responsible for the fermentation of milk to form yoghurt.

When Metchnikoff died in 1916, the centre of research activity moved to the United states and Rettgers and his colleagues became interested in the mechanism of the probiotic effect. Discovering that *L. delbrueckii* subsp. *bulgaricus* did not colonise the gut, they shifted their attention to intestinal isolates like *L. acidophilus* which they knew would colonise the intestine. Preparations containing *L. acidophilus* were used successfully to alleviate constipation (Rettger *et al.*, 1935). It soon became obvious that there were many other species of lactic acid bacteria in the intestine and these were subsequently incorporated into probiotic preparations. Table 1 shows lactic acid bacteria which are currently being used in probiotic preparations either singly or in combination. It should be remembered that, although a particular species may be present in several probiotic preparations, the strain may be different, producing a variable effects.

The actual composition of a probiotic preparation may not always correspond to that claimed on the product label. Maintenance of viability is the opposite of what is normally required of a pharmaceutical preparation where sterility is essential. Many preparations do not contain the stated numbers of viable cells. In a recent survey of 13 available products, Hamilton-Miller *et al.* (1996) found that only 2 matched their labelled microbiological specifications qualitatively and quantitatively. There was not only a short fall in the number of viable cells, but the organisms present were not always those stated. This sort of discrepancy, which is not checked for before trials are done, may explain some of the disappointing results seen in clinical trials.

Table 1. Examples of lactic acid bacteria used in human probiotic preparations

Lactobacilli	Bifidobacteria	Streptococci
L. delbrueckii subsp *bulgaricus*	*Bif. bifidum*	*S. salivarius* subsp *thermophilus*
L. acidophilus	*Bif. longum*	*Ent. faealis*
L. rhamnosus	*Bif. breve*	*Ent. faecium*
L. reuteri		
L. casei subsp. *casei*		
L. casei Shirota		
Lactobacillus GG		

The lack of knowledge about how probiotics have their effect has made it difficult to adopt an entirely rational approach to the selection of strains to be incorporated into a preparation. With the exception of the effect on lactose maldigestion, little is known about the way in which probiotics influence the flora and in turn produce a beneficial effect on the host animal. However, on the basis that colonisation of the gut is a prerequisite of the probiotic effect certain colonisation factors which allow survival and growth of microorganisms in the gut have been considered. Resistance to low pH and bile acids and ability to adhere to the intestinal epithelium may prove to be useful features in defining a successful probiotic microorganism.

Although adhesion of bacteria to the gut epithelium is an established colonisation factor, it does not ensure that an organism will permanently colonise the gut. Even within a single genus such as the lactobacilli, there is a constant interchange of species with one strain being replaced by another more fitted to occupy that particular niche. However some strains are more persistent than others and recently McCartney *et al.* (1996) have located in the human intestine indigenous strains of lactobacilli and bifobacteria which were detectable over a period of 12 months. Strains such as these may prove to be extremely effective probiotics, generating a response over a long period after only one dose. It remains to be seen how specific these strains are; they may require receptors which are only present in particular individuals.

Even if, in the short-term, a strain is being replaced it may be advantageous to use an adhering strain so that the period of residence in the gut is maximised. Although colonisation of the gut does appear to be a desirable characteristic, effective results have been obtained using organisms such as *Saccharomyces boulardii* which do not grow in the gut. Under these conditions, continuous administration ensures the presence in the gut of large numbers of metabolising cells and epithelial adhesion would be less important.

4   Probiotics in foods

Probiotics are commercially available for human consumption in the form of specific probiotic preparations (tablets or powders) or in foods. A large proportion of the food market is taken up by bioyoghurts (Fuller, 1993). These fermented milks are either yoghurts supplemented with probiotic strains of bifidobacteria and lactobacilli, or milks fermented solely by intestinal isolates of lactic acid bacteria. Table 2 shows the cultures used in the bioyoghurts commonly available in the UK. *Lactobacillus acidophilus* is the most commonly used bioyoghurt organism. It is often accompanied by a strain of *Bifidobacterium* such as *Bif. bifidus* or *Bif. longum*. Four of the products currently on sale fail to specify the organisms used but clearly describe them as 'special' or 'bio'.

Claims about improvement in resistance to specific diseases are avoided but general statements are made such as:

'.... long been associated with promoting good health' (Loseley)

'.... work by boosting your natural defences against harmful bacteria' (Nestlé)

'.... to help boost your body's natural resistance' (Vifit)

'Medical evidence shows that Danone Bio also helps to maintain the balance of your digestive system' (Danone)

Other product descriptions mention digestive system balance. These include:

'Medical evidence suggests that these cultures can have a beneficial effect on the digestive system where they occur naturally' (Tesco)

'These cultures are believed to aid digestion and dietary balance' (Sainsburys)

'These cultures also help to maintain the balance of the digestive system' (Onken)

Waitrose, Safeway and Marks and Spencer make no reference to beneficial physiological effects but stress the improved creamy flavour of their products. Such flavour is also highlighted on several other packs suggesting that taste is also an important selling

Table 2. Bioyoghurts currently available in the United Kingdom (partly from Fuller, 1993)

Producer	Trade Name	Cultures used as described on pack
Campina	Vifit	*Lactobacillus casei* Goldin and Gorbach
Danone	Bio	Active bifidus
Loseley	BA live	*Bifidobacterium* spp., *Lactobacillus acidophilus*, *Streptococcus thermophilus*
Marks & Spencer	Bioyoghurt	Special blend of live cultures
Nestlé	LC1	*Lactobacillus acidophilus* 1 and other cultures
Onken	Biopot	Biocultures including *Lactobacillus acidophilus*, *Bifidobacterium longum* and *Streptococcus thermophilus*
Safeway	Natural Bio	Special live cultures including *Streptococcus thermophilus, Lactobacillus acidophilus* and bifidobacteria
Sainsburys	Bioyoghurt	Special live cultures
Tescos	Healthy Eating	Special bio cultures
Waitrose	Bioyoghurt	Natural live bifids and acidophilus
Yakult	Yakult	Live *Lactobacillus casei* Shirota
Rachel's Dairy	Rachel's Organic	Bio-active cultures including *Lactobacillus bulgaricus* and *acidophilus, Streptococcus thermophilus* and bifidobacteria

point and, of course, many bioyoghurts also have fruit added which helps to sell the product without reference to its health benefits

The Yakult product is exceptional in the amount of information given on its pack. They describe the *L.casei* Shirota culture as '...beneficial bacteria that occur naturally in intestinal flora' and go on to supply some of the scientific background supporting its use: 'A well balanced intestinal flora is important to help maintain health. Levels of good bacteria can reduce as a result of bad eating habits, stress, the use of some antibiotics and age related changes. Yakult can help to maintain the all important balance... large numbers of these friendly bacteria help maintain both a favourable balance of beneficial intestinal flora and the natural rhythm of the bowel. Taken daily Yakult helps you towards a healthier and more balanced well-being'

Many bioyoghurts are described as 'live' or 'active' implying that this is in contrast to ordinary yoghurt. However, this is not a real distinction; nearly all traditional yoghurts contain viable starter bacteria with numbers depending on their age and acidity. Another selling point is the use of the term 'natural' on the pack. This green image is to some extent justified in that the aim of any probiotic preparation is to restore the 'natural' condition in the gut.

Other fermented milk products (Table 3) such as cottage cheese, cream cheese, fromage frais, sour cream and creme fraiche also contain viable organisms and although eaten primarily for taste they may have an incidental probiotic effect.

5   Nutritional effects

The nutritional content of a fermented milk is little different from that of the raw milk from which it is made (Anon, 1997). It is, therefore, a good source of protein, calcium, phosphorus, B vitamins and vitamin A. Even the lactose content may be the same. The starter organisms will hydrolyse about 25% of the lactose present but since milk is often fortified with skimmed milk powder the residual lactose concentration will be about 5% in both milk and yoghurt.

The process of fermentation will also make the proteins more available due to the proteolytic action of the starter bacteria. However, this is unlikely to significantly affect protein digestion in the intestine.

Of the B vitamins present in both foods, folic acid is significantly increased by fermentation but their value for humans is uncertain (Gurr, 1987). The most important and best defined effect of probiotics is the alleviation of the symptoms of lactose maldigestion. There is good evidence, from H_2 breath excretion tests, that lactose maldigesters can utilise lactose in yoghurt better than the same amount of lactose in milk. Gurr (1987), was of the opinion that the fermented milk was retained longer in the stomach and the slow release of the yoghurt increased contact time of the enzyme with the substrate. Others suggest an increase in lactase activity in the gut, although Lerebours *et al.*, (1989) could find no increase in the mucosal enzyme activity. However, in mice it has been shown (Garvie *et al.*, 1984) that ß- galactosidase activity increases in the contents rather than the mucosa suggesting that the increased activity is microbial, and not mucosal, in origin. Whether the extra enzyme is derived from the yoghurt starters or indigenous gut flora is uncertain but Savaiano (1989) found that yoghurt with different levels of ß-galactosidase activity was equally effective in treatment of maldigesters.

Table 3.   Classification of fermented milk products

Type of starter culture used	Products
Thermophilic :	
Classic cultures including some lactobacilli species, *e.g. Lactobacillus bulgaricus* and streptococci species, *e.g. Streptococcus thermophilus*	Yoghurt Acid buttermilk
Bio-cultures :	
More recently employed cultures including some lactobacilli species, *e.g. Lactobacillus acidophilus*, and bifidobacteria species, *e.g. Bifidobacterium bifidum*	Bio-yoghurts (active, bifidus or bifidus acidophilus [BA]) Acidophilus milk * Yakult products - Japan (some are available in the UK) Proprietary therapeutic products
Mesophilic :	
Some streptococci species, e.g. *Streptococcus lactis* subsp. *lactis* and *Streptococcus lactis* subsp. *cremoris*, lactococci, *e.g. Lactococcus lactis* and some leuconostoc species	Cultured buttermilk Cultured (sour) cream Crème fraîche Cottage cheese Cream cheese Fromage frais Quark Filmjölk - Scandinavia Ropy milks - Scandinavia
Lactic acid bacteria/yeast :	
Mesophilic bacteria are predominantly used to produce some products *e.g.* kefir, and thermophilic bacteria are mainly used to produce others *e.g.* koumiss	Kefir - Russia Koumiss (made from mare's milk) - Russia
Lactic acid bacteria/mould ripening :	
Mesophilic bacteria and mould	Viili - Finland

* Some acidophilus milk products are referred to as 'culture-containing' rather than 'cultured' because the former do not undergo a true fermentation procedures during their production *i.e.* the bacterial culture is simply added to the product.

From Anon (1997) based on Marshall and Tamine (1997) and Robinson (1991)

6   Probiotics and resistance to disease

There is good evidence that the normal gut microflora is involved in resistance to disease, especially with respect to gastrointestinal infections. This conclusion is based on the findings that:
- germfree animals are more susceptible to infection than are their conventional counterparts with a complete gut flora (Collins and Carter, 1978)
- antibiotics given by mouth increases the susceptibility of animals to infection (Freter, 1955)
- administration of enemas of faecal suspensions from a healthy adult can control antibiotic associated diarrhoea (Schwass et al., 1984).

The use of faecal suspensions has obvious disadvantages and a great deal of research has been directed towards identifying the organism(s) which are responsible for the probiotic effect. Some studies, in the chicken, suggest that a large number of different strains are required for a full protective effect but omission of lactobacilli from the complex mixture reduced the level of protection obtained (Mead and Impey, 1987). Most trials and experimental research have centred on the lactic acid bacteria. This stems from the early work of Metchnikoff using L. delbreukii subsp. bulgaricus. The lactic acid bacteria are attractive in this context because of their low pathogenic potential. They are, in FDA terminology, Generally Regarded as Safe (GRAS) and, to date, there is no record of probiotic administration being responsible for infection of the consumer. Other organisms which have been used in human trials are yeasts such as Saccharomyces boulardii. This is an isolate from fruit and is not known to be pathogenic for animals.

Much of the work done on probiotics and their effect on the gut flora has been directed towards changes in the composition of the indigenous microbiota. For example, an increase in lactic acid bacteria or decrease in Escherichia coli may be demonstrated without relating these changes to beneficial effects seen in the host animal.

Other studies have monitored the effects of probiotic supplementation by looking at changes in metabolism. Bacteria in the gut are responsible for the conversion of harmless dietary components into active toxins. For example, cycasin has different effects in germfree and conventional animals. Gut bacteria produce ß-glucosidase which hydrolyses cycasin to its aglycone which is carcinogenic, producing colon tumours in the conventional animal whereas none are seen in the germfree equivalent (Laqueur and Spatz, 1968). In germfree rats associated with a human faecal flora, ß-glucosidase activity was reduced when L. acidophilus was included in their diet (Cole et al., 1989). Lactobacillus acidophilus was also found to delay tumour formation in animals treated with the chemical carcinogen 1,2-dimethylhydrazine (Goldin and Gorbach, 1980). While these and other studies in experimental animals show that probiotics may control tumour production, there is no good evidence to suggest that a similar effect is possible in humans.

However, there are good trial results to show that probiotics can reduce the incidence of intestinal infections in humans. These trials are not always accompanied by information on the effect on the casual agent of the disease, but it must be assumed that if symptoms of the disease disappear, then the probiotic is having some antagonistic effect on the pathogen. Whether this is a direct chemical antagonism, competition for nutrients, an indirect effect *via* the immune system or activated by competition for receptors on the epithelial surface is as yet unknown.

There are good results from double blind, placebo controlled trials in a variety of different infections. McFarland *et al.* (1995) obtained a 50% reduction in antibiotic associated diarrhoea when patients were given *Sacc. boulardii.*

Isolauri *et al.* (1991) monitored the incidence of diarrhoea in children aged 4-45 months after administration of *L. casei* GG. There was a significant reduction in diarrhoea which was more pronounced when related to patients with confirmed rotavirus infection.

In another carefully controlled study (Saavedra *et al.*, 1994) children aged 5-24 months who were in hospital for non-gastrointestinal conditions were given a probiotic containing *Bifidobacterium* sp. and *S. salivarius* subsp. *thermophilus.* Over a period of 17 months there was a significant reduction in the incidence of diarrhoea in the treated group. There is also evidence, although because of the experimental design it is less reliable, that probiotics can have an antifungal effect. During chemotherapy for leukaemia, the counts of *Candida* gut increased. Treatment with a milk preparation containing *L. acidophilus* and *Bifidobacterium* sp. markedly reduced the count of *Candida* in faeces (Tomoda, 1983). There is, therefore, evidence that a variety of different probiotic preparations are having positive effects in infections caused by bacteria, viruses and fungi.

7   Future developments

Evidence for protective effects of the gut flora is incontestable. This forms the basis of the probiotic concept and development of pure culture preparations. The choice of culture presents a problem. We have no *in vitro* tests for recognition of active probiotic organisms. For the development of a more rational approach to the selection of strains it is essential that we know more about the way in which probiotics have their effect. When this is available, we will be able to detect probiotic activity in the laboratory and not have to resort to expensive and time-consuming animal feeding trials.

Knowledge of the mode of action will also allow the use of genetic manipulation techniques that enhance the production of the probiotic metabolite(s). It may also enable the development of a nonviable chemical probiotic based on active metabolites. This production can be improved by genetic manipulation without the problem of release, into the environment, of live organisms. However, the probiotic metabolite would have to be resistant to inimical factors in the gut and be delivered to the target site. The great virtue of the present probiotic approach is that it generates an active agent *in situ* at the required target site. This may be a feature of the probiotic preparations which cannot be reproduced with a non-viable chemical agent.

References

Anon (1997) Nutritional benefits of yoghurt and other fermented milk products. *National Dairy Council Topical Update*. No **8**, 1-16

Cole, C.B., Fuller, R. and Carter, S.M. (1989) Effect of probiotic supplements of *Lactobacillus acidophilus* and *Bifobacterium adolescentis* 2204 on ß-glucosidase activity in the lower gut of rats associated with a human faecal flora. *Microbial Ecology in Health and Disease* **2**, 223-25.

Collins, F.M. and Carter, P.B. (1978) Growth of salmonellae in orally infected germfree mice. *Infection and Immunity* **21**, 41-47.

Freter, R. (1955) The fatal enteric cholera infection in the guinea pig achieved by inhibition of normal enteric flora. *Journal of Infectious Diseases* **97**, 57-65.

Fuller, R. (1989) Probiotics in man and animals. *Journal of Applied Bacteriology* **66**, 365-78.

Fuller, R. (ed.) (1992) *Probiotics The Scientific Basis*. Chapman and Hall, London.

Fuller, R. (1993) Probiotic foods, Current use and future developments. *International Food Ingredients* **3**, 23-6.

Fuller, R (ed.) (1997) *Probiotics 2 Applications and Practical Aspects*. Chapman and Hall, London.

Garvie, E.K., Cole, C.B., Fuller, R. and Hewitt, D. (1984) The effect of yoghurt on some components of the gut microflora and on the metabolism of lactose in the rat. *Journal of Applied Bacteriology* **56**, 237-45.

Gibson, G.R. and Roberfroid, M.B. (1994) Dietary modulation of the human colonic microbiota: introducing the concept of prebiotics. *Journal of Nutrition* **125**, 1401-12.

Goldin, B.R. and Gorbach, S.L. (1980) Effect of *Lactobacillus acidophilus* dietary supplementation on 1, 2-dimethylhydrazine dihydrochloride-induced intestinal cancer in rats. *Journal of the National Cancer Institute* **64**, 263-5.

Gurr, M.I. (1987) Nutritional aspects of fermented milk products. *FEMS Microbiology Reviews* **46**, 337-42.

Hamilton-Miller, J.M.T., Shah, J. and Smith, C.T. (1996) "Probiotic" remedies are not what they seem. *British Medical Journal* **312**, 56.

Isolauri, E., Juntunen, M. Rautanen, T. *et al.* (1991) *Lactobacillus* strain (*Lactobacillus casei* sp strain GG) promotes recovery from acute diarrhoea in children. *Paediatrics* **88**, 90-97.

Laqueur, G.L. and Spatz, M. (1968) The toxicology of cycasin. *Cancer Research* **28**, 2262-65.

Lerebours, E., N'Djitoyap Ndam, C., Lavoine, A. *et al.* (1989) Yoghurt and fermented-then-pasteurized milk: effects of short term and long term ingestion on lactose absorption and mucosal lactase activity in lactase-deficient subjects. *American Journal of Clinical Nutrition* **49**, 823-27.

Lilley, D.M. and Stillwell, R.H. (1965) Probiotics: growth promoting factors produced by microorganisms. *Science* **147**, 747-48.

Luckey, T.D. (1972) Introduction to intestinal microecology. *American Journal of Clinical Nutrition* **25**, 1292-94.

McCartney, A.L., Wenzhi, W. and Tannock, G.W. (1996) Molecular analysis of the composition of the bifidobacterial and lactobacillus microflora of humans. *Applied and Environmental Microbiology* **62**, 4608-13.

McFarland, L.V., Surawicz, C.M., Greenberg, R.N. *et al.* (1995) Prevention of ß-lactam-associated diarrhoea by *Saccharomyces boulardii* compared with placebo. *American Journal of Gastroenterology* **90**, 439-48.

Marshall, V.W. and Tamine, A.Y. (1997) Starter cultures employed in the manufacture of biofermented milks. *International Journal of Diary Technology*. **50**, 30-41.

Mead, G.C. and Impey, C.S. (1987) The present status of the Nurmi concept for reducing carriage of food poisoning salmonellae and other pathogens in poultry. In *Elimination of Pathogenic Organisms from Meat and Poultry* (ed. F.J.M. Smulders) Elsevier, Amsterdam pp. 57-77.

Parker, R.B. (1974) Probiotics, the other half of the antibiotic story. *Animal Nutrition and Health* **29**, 4-8.

Rettger, F., Levy, M.N., Weinstein, K. and Weiss, J.E. (1935) *Lactobacillus acidophilus and its Therapeutic Application*. Yale University Press, New Haven Connecticut.

Robinson, R.K. (1991) Microorganisms of fermented milks. In *Therapeutic Properties of Fermented Milk* (ed. R.K Robinson) Elsevier Science Publishers Ltd, London pp. 23-43.

Saavedra, J.M., Bauman, N.A. Oung, I. *et al.* (1994) Feeding of *Bifidobacterium bifidum* and *Streptococcus thermophilus* to infants in hospital for prevention of diarrhoea and shedding of rotavirus. *Lancet* **344**, 1046-49.

Savaiano, D.A. (1989) In *Fermented Milks: Current Research*. Proceedings of an international congress held on December 14-16 in Paris. John Libbey Eurotext, London.

Schwass, A., Sjolin, S. Trottestam, V. and Aransson, B. (1984) Relapsing *Clostridium difficile* enterocolitis cured by rectal infusion of normal faeces. *Scandinavian Journal of Infectious Diseases* **16**, 211-15.

Sperti, G.S. (1971) *Probiotics*. Avi Publishing Col, West Point, Connecticut.

Tomada. T., Nakano, Y. and Kageyama, T. (1983) Variation of intestinal *Candida* of patients with leukemia and the effect of *Lactobacillus* administration. *Japanese Journal of Medical Mycology* **24**, 356-58.

CHAPTER 7

Prebiotics

GLENN R. GIBSON[1], ROBERT A. RASTALL[1] AND
MARCEL B. ROBERFROID[2]
*Food Microbial Sciences Unit, Department of Food Science and
Technology, The University of Reading, Reading, UK*
*Université Catholique de Louvain, Department of Pharmaceutical
Sciences, Avenue Mounier 73, B-1200 Brussels, BELGIUM*

1   Introduction

We have previously defined a **prebiotic** as 'a non-digestible food ingredient that
beneficially affects the host by selectively stimulating the growth and/or activity of one
or a limited number of bacteria in the colon' (Gibson and Roberfroid, 1995). In this context, a prebiotic is a dietary ingredient that reaches the large intestine in an intact form
and has a specific metabolism therein - directed towards advantageous rather than
adverse bacteria. This would ultimately lead to a marked change in the gut microflora
composition. The premise is based on the hypothesis that the human large gut contains
bacterial genera, and species, that are beneficial, benign and deterimental for host health.
Whilst this generalisation probably gives too simplistic a view of gut microbiology,
it is a feasible working concept. Fig. 1 gives our view of how different bacterial groups
in the colon may be categorised in this manner. Forthcoming research will no doubt
identify the realistic health values of the gut microflora, whilst other chapters in this
book have discussed some of the more useful areas of interest.

Prebiotic fermentation should be directed towards potentially health promoting
bacteria, with indigenous lactobacilli and bifidobacteria currently being the preferred
target organisms. The approach is therefore similar to that of dietary fibres, but with prebiotics acting in a much more tailored manner, in that the bacterial fermentation/growth
is selective. The most efficient prebiotics will also reduce or suppress numbers and
activities of organisms seen as pathogenic.

G.R. Gibson and M.B. Roberfroid (eds.), Colonic Microbiota, Nutrition and Health, 101-124.
© 1999 *Kluwer Academic Publishers. Printed in the Netherlands.*

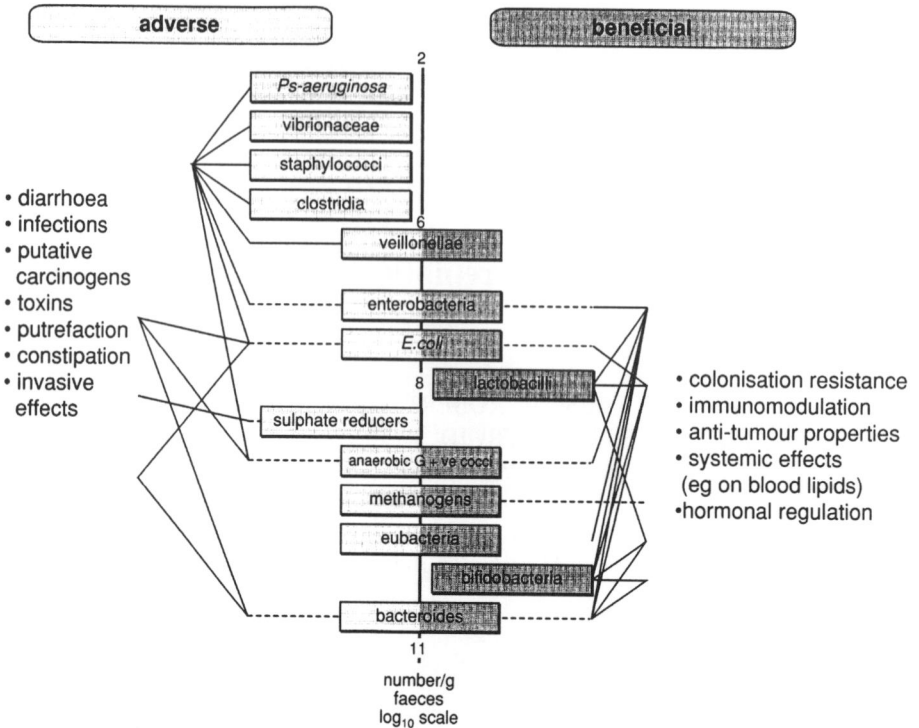

Figure 1. Generalisation of predominant groups of intestinal microorganisms into potentially harmful or beneficial categories (adapted from Gibson and Roberfroid, (1995)).

Moreover, and because of the intestinal fermentation, prebiotics may have certain physiological effects including: control of transit time and mucosal motility, regulation of epithelial cell proliferation, influence on nutrient (in particular ion) bioavailability, modulate immune activity and affect endocrine function. There may also be certain systemic influences such as carbohydrate or lipid homeostasis. For a review of these prebiotic mediated effects see Roberfroid (1996).

2   The non-digestible oligosaccharides

At present, food components which seem to exert the best prebiotic effect are the non-digestible oligosaccharides (NDO). These are oligomeric carbohydrates, the osidic bond of which is in a spatial configuration that allows resistance to hydolytic activites of intestinal digestive enzymes, but are sensitive to metabolic effects of colonic bacteria. These microorganisms can then ferment the carbohydrate to produce short chain carboxylic acids and gases, as well as increase metabolic energy, growth and proliferation

(Roberfroid, 1997; van Loo *et al.*, 1998). The carbohydrates contain mixtures of oligomers of different chain lengths and are characterised by their average degree of polymerisation (DP). As discussed recently (Cummings and Roberfroid, 1997), there is no rational physiological or chemical reason for discriminating oligosaccharides and polysaccharides on the basis of DP, and it has been recommended that alcohol precipitation be used as the practical means of delineating these two groups (Englyst and Hudson, 1996; Cummings and Roberfroid, 1997).

Non-digestible oligosacharides for which, at least some, data have been published to support a prebiotic effect are the inulin-type fructans (also called fructooligosaccharides, FOS) (Roberfroid *et al.*, 1998), soybean oligosaccharides (raffinose and stachyose) (Masai *et al.*, 1987; Wada *et al.*, 1992), galactooligosaccharides (Tanaka *et al.*, 1983; Ito *et al.*, 1990, 1993), galactosylsucrose (Yoneyama *et al.*, maltooligosaccharides (Kohmoto *et al.*, 1988), palatinose condensates (Mizutani *et al.*, 1991) and xylooligosaccharides (Imaizuna *et al.*, 1991). For a review of some effects see Roberfroid and Slavin (1998). There are many reported prebiotic oligosaccharides, principally in use in Japan. Those that have been suggested as having prebiotic potential are summarised in Table 1.

Table 1. Production of prebiotic oligosaccharides in Japan in (1995) (after Crittenden and Playne, (1996))

Oligosaccharide	Production (tonnes)
Lactulose	20 000
Galactooligosaccharides	15 000
Fructooligosaccharides	12 000
Isomaltooligosaccharides	11 000
Palatinose	5 000
Soybean oligosaccharides	2 000
Lactosucrose	1 600
Gentiooligosaccharides	400
Xylooligosaccharides	300

Currently, there are little comparative data on the relative efficiencies of these molecules, or on their selectivity at a species, or even genus, level. It is often the case that a prebiotic effect is levelled against certain NDO, or other dietary carbohydrates, without a full and careful investigation of the fermentation profile. It is critical that as many components of the gut microbiota as possible are measured during fermentation studies. These should at least include bacteroides, bifidobacteria, clostridia, Gram positive cocci, coliforms, lactobacilli, total aerobes and total anaerobes. Simple stimulation of bifidobacteria and/or lactobacilli is insufficient without determining effects on other gut microorganisms - as it is the **selectivity** of effect that determines classification as a prebiotic. Clearly, pure bacterial studies are of very limited use in this respect (*e.g.* Harteminks *et al.*, 1997), unless they are also supported by mixed culture work.

It is the effect in a **competitive** ecological environment that is key. New developments in the application of molecular biological techniques to human gut bacteriology will be of enormous value in this respect (see Chapters 9 and 10). The prebiotic effect should also be determined in vivo with human volunteers. Fig. 2 gives a scheme for identifying efficacious prebiotics.

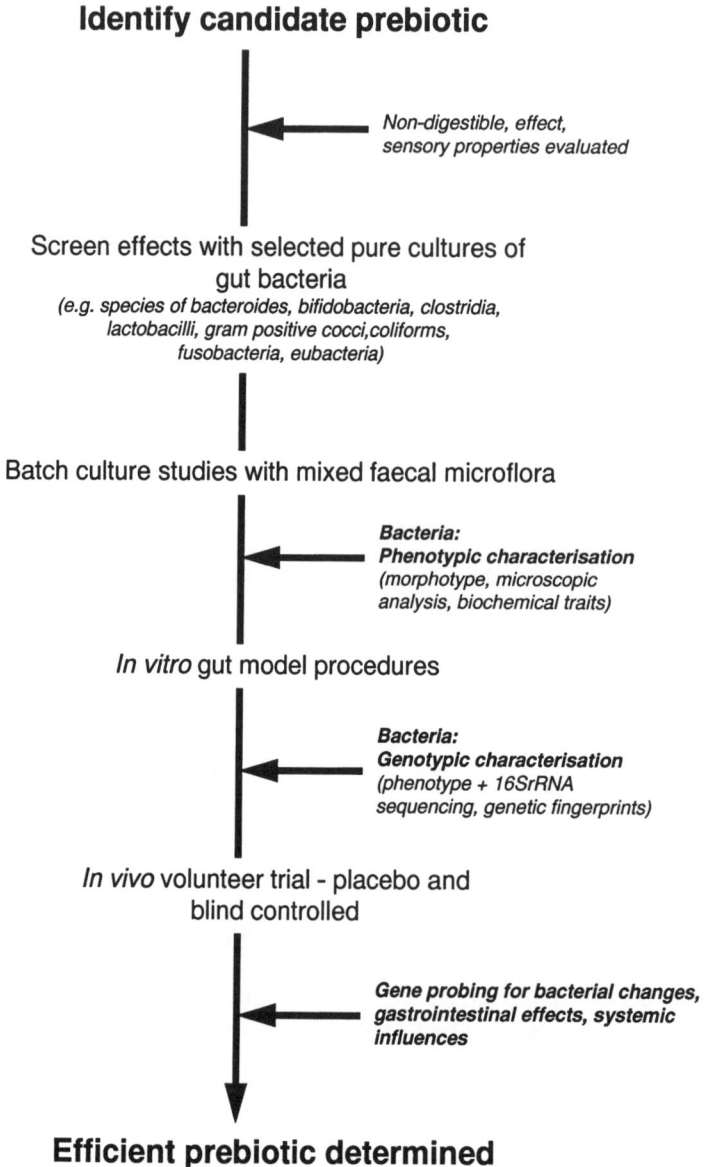

Identify candidate prebiotic

Non-digestible, effect, sensory properties evaluated

Screen effects with selected pure cultures of gut bacteria
(e.g. species of bacteroides, bifidobacteria, clostridia, lactobacilli, gram positive cocci,coliforms, fusobacteria, eubacteria)

Batch culture studies with mixed faecal microflora

Bacteria:
Phenotypic characterisation
(morphotype, microscopic analysis, biochemical traits)

In vitro gut model procedures

Bacteria:
Genotypic characterisation
(phenotype + 16SrRNA sequencing, genetic fingerprints)

In vivo volunteer trial - placebo and blind controlled

Gene probing for bacterial changes, gastrointestinal effects, systemic influences

Efficient prebiotic determined

Figure 2. Flow chart for the development of prebiotics.

Inulin-type fructans are by far the most extensively studied compounds and are clear 'market leader' prebiotics. These carbohydrates contain both $G_{py}F_n$ (α D glucopyramo-syl- [ß D fructofuranosyl]$_{n-1}$ -D fructofuranoside) and $F_{py}F_n$ (ß D fructopyranosyl-[ß D fructofuranosyl]$_{n-1}$ -D fructofuranoside) molecules, with the number of fructose units varying from 2 to more than 70 units. As food ingredients (see Chapter 19) they are available either as native inulin and high-molecular weight inulin (both abbreviated as INU), enzymatically-produced inulin hydrolysate or oligofructose (OFr), all of which occur naturally in miscellaneous edible plants such as garlic, onion, asparagus, chicory, artichoke, banana, (Van Loo *et al.*, 1995) or can be produced enzymatically (from sucrose) as synthesised compounds (SFr).

The range of glycosidic linkages found in prebiotics directed towards bifidobacteria raises questions about the biochemical basis of the effect. For FOS or inulin, bifidobac-teria are known to posses a cell-bound ß-fructofuranosidase enzyme (Pudjono *et al.*, 1993) which has a higher specificity for OFr than for sucrose (Muramatsu *et al.*, 1994). The removed fructose is then rapidly metabolised by the bacteria in the "bifidus shunt" pathway. Similarly, α-galactosidase activity has been identified in bifidobacteria (Desjardins *et al.*, 1990), potentially allowing the metabolism of soybean oligosaccha-rides (raffinose, stachyose and verbascose). This enzyme is absent from the human gastrointestinal tract.

A potentially important class of prebiotics are the galactooligosaccharides. These are manufactured from lactose by transglycosylation reactions and consist of galactosyl derivatives of lactose with ß1-3 and ß1-6 linkages. The purported prebiotic nature of galac-tooligosaccharides is explanable, in part, by the linkage-specficity of the *Bifidobacterium* ß-galactosidase; this enzyme has specificity for ß1-3 and ß1-6 linkages over ß1-4 linkages (Dumortier *et al.*, 1994).

Glucose based isomaltooligosaccharides (α1-6 linked) and gentiooligosaccharides (ß1-6 linked) are candidate prebiotics, as are xylooligosaccharides. However, specific enzymes for the degradation of these molecules have not as yet been reported.

2.1 NON DIGESTIBILITY IN THE UPPER GASTROINTESTINAL TRACT

2.1.1 *Inulin fructans*

The ß configuration of anomeric C_2, in fructose monomers, is thought to make inulin-type fructans resistant to hydrolysis by human digestive enzymes (*e.g.* α-glucosidase, maltase, isomaltase, sucrase) which are mostly specific for osidic linkages. There are both in vitro and in vivo data to support this contention.

In vitro, SFr is not hydrolysed to any significant extent by purified rat sucrase-maltase (Oku *et al.*, 1984). In contrast, yeast invertase hydrolyses GF_3 (nystose) at about 5% of the rate of sucrose, and that particular oligomer may therefore be degraded quite quickly in situ (Ziesenitz and Siebert, 1987). When incubated in the presence of homogenates of different gastrointestinal segments (duodenum, jejunum, ileum) from the rat or human small intestine, inulin-type fructans remain unchanged for up to 1 or 2 h (Hidaka *et al.*, 1986; Ziesenitz and Siebert, 1987; Nilsson *et al.*, 1988; Molis *et al.*, 1996). In addition, since stomach hydrolysis of (inulin-type) fructans is likely to be of limited physiological significance, these products proceed undigested through the upper part of the gastrointestinal tract into the colon (Nilsson *et al.,* 1988). This has been confirmed through in vivo studies with rats (Nilsson *et al.,* 1988; Nilsson and Bjork, 1988) and humans (Bach Knudsen and Hessov, 1995; Molis *et al.,* 1996; Ellegärd *et al.,* 1997).

The most convincing data are those of Bach Knudsen and Hessov (1995) and Ellegärd *et al.* (1997) who have used the ileostomy model. Both studies showed that 86-88% of the ingested dose (10, 17 or 30g) of INU and OFr were recovered in ileostomy effluent, supporting the conclusion that both were mostly undigestible in the upper gastrointestinal tract. Using an intubation technique in human volunteers, Molis *et al.* (1996) concluded that SFr was 89% unabsorbed in the small intestine. The small loss, during passage, may be due to microbial activity in the ileum, which is thought to be a much more densely populated area in ileostomists compared to persons with a conventional intestinal tract (Finegold *et al.,* 1970). Another explanation is that acid and/or enzymatic hydrolysis of the low molecular weight fructans occurred.These fructans are thought to be more sensitive to stomach and/or small intestinal hydrolysis than higher molecular weight versions (Oku *et al.,* 1984; Nilsson and Björck, 1988; Molis *et al.,* 1996).

2.1.2   *Other nondigestible oligosaccharides*
Published data on the resistance of other oligosaccharides to digestion in the upper gastrointestinal tract are less available than for the inulin fructans. Predominantly, the available evidence comes either from in vitro experiments or is based on stimulation of growth of specific gut microorganisms or faecal excretion of the oligosaccharides in animal models.

Isomaltooligosaccharides are, at least partly, hydrolysed by isomaltase in the jejunun (Dahlquist, 1964). The soybean oligosaccharides, raffinose and stachyose, are not hydrolysed to any significant extent by homogenates of rat intestine, with 90% of an ingested dose being recovered in the faeces of germ free rats (Kato *et al.,* 1991), compared to 50% in antibiotic treated rats (Yoshida *et al.,* 1969). For the galactooligosaccharides, a dose of 0.5g/kg increased breath H_2 concentration in 5 human volunteers (Tanaka *et al.,* 1983). Ito *et al.* (1990) have however concluded that the absolute amount of galactooligosaccharides reaching the colon without digestion in the small intestine could not be assessed with certainty. Studies with ileostomists would help clarify this aspect of galactooligosaccharide metabolism. For polydextrose, the available data have been derived for caloric value using radiolabelled molecules (Figdor and Rennhard, 1981, 1983; Cooley and Livesey, 1987). Wholly convincing data are still missing to demonstrate that polydextrose is totally, or partially, resistant to digestion in the upper gastrointestinal tract. Palatinose oligosaccharides, or condensates, resist in vitro hydrolysis by stomach acidity, amylase and rat intestinal homogenates (Mizutani, 1991).

3   Prebiotic oligosaccharides: New developments

It is possible to identify a range of desirable characteristics in a prebiotic (Table 2). Despite the variety of molecules reported to be prebiotic and the amount of research that has been carried out, there are incomplete data with respect to the properties identified in Table 2. These will be discussed here with respect to structural characteristics of candidate oligosaccharides.

Table 2.   Desirable properties of prebiotic oligosaccharides

Desirable attribute in prebiotic	Properties of oligosaccharides
Active at nutritionally feasible dose (it is appropriate that this is as low as possible)	Selective stimulation and proliferation of 'beneficial' bacteria *e.g.* bifidobacteria, lactobacilli, in a complex microbial ecosystem
Lack of side effects	No stimulation of gas producers, putrefactive microorganisms. pathogens, etc. in a complex microbiota
Fine control of microflora modulation	Selective fermentation in mixed culture
Persistence throughout the colon, *i.e.* towards distal areas	High molecular weight, slow fermentation
Varying viscosity	Available in different molecular weights and linkages
Good storage and processing stability	Possess 1-6 linkages and pyranosyl sugar rings
Differing sweetness	Vary monosaccharide composition
Inhibit adhesion of pathogens	Possess receptor sequences

3.1   SELECTIVE METABOLISM

The first three entries in Table 2, activity at a nutritionally feasible dose, absence of side effects and ability to achieve fine control of gut flora composition depend upon highly selective bacterial fermentation of the oligosaccharides. In order to achieve these properties, the prebiotic would ideally be metabolised by the target species, *e.g.* those of bifidobacteria and lactobacilli, and not by any other microorganism. Virtually all of the studies performed to date have characterised the gut flora at the genus level. Distiguishing individual species of *Bifidobacterium* or *Lactobacillus* in a mixed culture fermentation is extremely difficult through standard phenotypic criteria.

A very important research target is the development of convenient and reliable molecular methods of identification of bifidobacteria (see Chapters 9-11). Various health claims have been attributed towards these microorganisms. Although much promise does exist, many do not rest on a sound scientific basis. If these health benefits are to be realised, it would be surprising to find that every species of *Bifidobacterium* had equal benefits. For example, *Bifidobacterium infantis* is more adept at inhibiting pathogens, in vitro, than most other species of bifidobacteria (Gibson and Wang, 1994a). As such, it may be that preferred prebiotics will target individual species. Much more methodological development is required in this area, allowing the possibility of achieving a fine control of the gut flora to produce specific health goals. Information is needed on the relative bifidogenic activity of different prebiotics at the specific level, in mixed culture, if we are to have a practical means of achieving this aim.

3.2   ENHANCED PERSISTENCE THROUGH THE COLON

Most current prebiotics are of relatively small DP, the exception being inulin. As the oligosaccharides must be hydrolysed by cell-associated bacterial glycosidases, prior to uptake of the resultant monsaccharides, it is reasonable to assume that the longer the oligosaccharide, the further the prebiotic effect may penetrate the colon. For example, long chain inulin may exert a prebiotic effect in more distal colonic regions than the lower molecular weight FOS, which may be more quickly fermented in the saccharolytic proximal bowel. It will be of great interest to examine the persistence of prebiotics with varying chain lengths through efficient microbiological gut models. The relevance lies in the fact that most colonic disorders originate in the left (distal) side. Prebiotics that exert an effect in this region of the large intestine may therefore have added benefits.

3.3   ENHANCED TECHNOLOGICAL FUNCTIONALITY IN FOOD SYSTEMS

Oligosaccharides have a range of technological attributes in food systems (see Table 3). Varying foods may require a different set of these functional attributes. If prebiotics are to make an impact on human health, then it is highly likely that the range of foods which incorporate these ingredients will have to be expanded beyond yoghurts and 'health drinks'. Chapter 19 overviews some potential and existing food applications. For maximal effect, prebiotics with a range of physicochemical and organoleptic properties are required.

Table 3.   Some functional attributes of oligosaccharides in food systems

Functional/technological effect
Modify freezing point and moisture content
Variable stability to acid, on storage and in processing
Have bacteriostatic properties
May stabilise proteins and retain flavour and aroma
Affect colour formation
Variable sweetness, calorific value and cariogenicity
Replace fat, but retain similar "fatty" mouth feel and texture

One of the most important properties of carbohydrates in food systems is their viscosity. Prebiotics will be needed with a range of viscosities, and the rheological properties will, ideally, be controllable. This could conceivably be achieved by regulating the size of the prebiotic carbohydrate. High molecular weight oligosaccharides find application as fat replacers and bulking agents (Sester and Racette, 1992). The predominant carbohydrates in current use, however, are not prebiotic, mostly being starch-derived maltodextrins of varying size (Sester and Racette, 1992).

The organoleptic properties of prebiotic oligosaccharides must also be thoroughly investigated. Some prebiotics, notably those containing non-reducing fructose and glucose residues, have a degree of sweetness. Application in a range of different foods will demand prebiotics with varying sweet taste. In addition, some carbohydrates act synergistically with sweeteners (Wiedmann and Jager, 1997) and this aspect of prebiotic oligosaccharides will also need to be determined.

A potentially attractive approach to the incorporation of prebiotics into foods might be to develop those which specifically replace currently accepted food ingredients. For instance, it is conceivable that some bulk sweeteners, like maltodextrins, could be replaced in foods by prebiotic alternatives. Inulin has already found application as a fat replacer (Chapter 19). It is unlikely, however, that one polysaccharide such as inulin will be suitable as a fat replacer in all applications. There is, therefore, the possibility for developing other prebiotic oligosaccharides tailored for this purpose.

3.4   ANTI-ADHESIVE ACTIVITIES

It is known that certain pathogenic microorganisms attach to human cell surface monosaccharides and oligosaccharides as the first step in virulence. Many of these carbohydrate receptors are recognised (Table 4).

Table 4.   Examples of carbohydrate pathogen receptors

Galα4Gal	*Escherichia coli* (P-piliated), Vero cytotoxin
Galβ4GlcNAcβ3Gal	*Streptococcus pneumoniae*
GalNAcβ4Gal	*Pseudomonas aeruginosa, Haemophilus influenzae, Staphylococcus aureus, Klebsiella pneumoniae*
Sialic acids	*E. coli* (S-fimbriated)
Galα3Galβ4GlcNAc	*Clostridium difficile* toxin A
Fucose	*Vibrio cholerae*
GlcNAc	*E. coli, V. cholerae*
Mannose	*E. coli, K. aerogenes, Salmonella* spp. (Type 1-fimbriated)

There is potential for the manufacture of soluble oligosaccharides which mimic this interaction (Zopf and Roth, 1996). Such molecules could act as 'blocking factors' and there are several therapeutic agents based upon this approach currently in clinical trials (McAuliffe and Hindsgaul, 1997). These pharmaceutical agents are multi-valent forms of the oligosaccharide receptor. Intake of the simple receptor, however, might be expected to provide a low level of anti-adhesive activity in the gut, reducing the likelihood of colonisation with pathogenic species.

The effect would complement the colonisation resistance that is conferred by a lactic microflora (see later). Using enzymatic manufacturing techniques, the potential exists to synthesise multifunctional oligosaccharides that would posses both prebiotic and anti-adhesive properties.

4   Manufacturing technology for prebiotics

Many prebiotics, currently in use, are manufactured using enzymatic methods (Playne and Crittenden, 1996). Commonly, glycosyl transfer reactions are utilised to transfer either fructose, glucose or galactose to an acceptor sugar. Typically, cheap oligosaccharides are utilised as donors, such as sucrose, lactose and starch-derived oligomers. An example of an industrial process for the enzymatic manufacture of a trans-galactosylated oligosaccharides (TOS) is shown in Fig. 3. The product of this process is a 75% (w/w) sugar syrup with 59% (w/w) TOS containing a mixture of ß1-6, ß1-3 and ß1-4 linkages with 2-6 monomers per oligosaccharide (Ekhart and Timmermans, 1996).

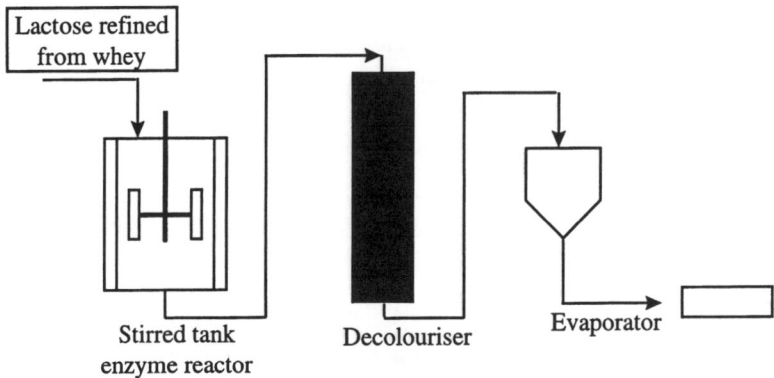

Figure 3. Manufacturing process for transgalactoligosaccharides (TOS).

The technology to perform such glycosyl transfer reactions is now well established (Rastall and Bucke, 1992). Products are generally isolated using ultrafiltration membranes and adsorption or chromatographic techniques (Playne and Crittenden, 1996).

Recent advances in the synthesis of oligosaccharides using glycosidase enzymes (Rastall *et al.*, 1992, Smith *et al.*, 1997, Suwasono and Rastall, 1996) offer the potential to manufacture a range of novel oligosaccharides. Such materials might be designed to posess desirable physicochemical, organoleptic or anti-adhesive activities. The glycosidase technology depends on the fact that the hydrolysis reaction normally catalysed by these enzymes is reversible. Under conditions of high sugar concentration, at an elevated temperature, condensation products are formed. In this manner, it is possible to build up higher oligosaccharides from simple sugar units. For example, the synthesis of hetero-oligosaccharides by glycosidase-catalysed condensation reactions may occur as follows:

$$\text{Man} + \text{Glc}\alpha1\leftrightarrow2\beta\text{Fru Man}\alpha1\rightarrow 6\text{Glc}\alpha1\leftrightarrow2\beta\text{Fru} + H_2O$$

Several glycosidases can be used in this way and many can co-condense monosaccharides with a range of 'acceptor' molecules.

This approach, to the synthesis of novel prebiotics, is currently in the research stage and requires much more development before a realistic evaluation of its commercial potential can be made. There are considerable engineering challenges to be overcome: industrial processes involving the glycosyl transfer reactions, described above, operate at around 30% (w/v) solid concentration. The glycosidase synthesis reactions require sugar concentrations of 60-70% (w/w) in order to reduce the water concentration and reverse the equilibrium. Reactor systems operating with these viscous solutions are not in common use.

A further problem lies with the downstream processing of these molecules. Separation of products from concentrated sugar solutions is difficult, although progress is being made in the application of membrane techniques. The products of these reactions are, therefore, likely to carry a price premium, and would probably need to be used in carefully targeted applications.

An alternative route to the manufacture of novel prebiotics, which has not received a great deal of attention, is the controlled hydrolysis of polysaccharides (like starches, pectin, cellulose) to give oligomeric forms. Careful manipulation of the reaction process would lead to prebiotics of varying DP. Surprisingly few food-grade polysaccharides have been tested for their prebiotic potential in this way. A potentially useful approach to this controlled hydrolysis is the use of enzyme membrane reactors. Such reactor systems are very versatile and can give high productivity (Prazeres and Cabral, 1994). These reactors are usually used for the complete saccharification of starch (Sims and Cheryan, 1992), but may be modified to allow partial hydrolysis to oligosaccharides. A membrane process potentially applicable for the manufacture of isomaltooligosaccharides is shown in Fig. 4.

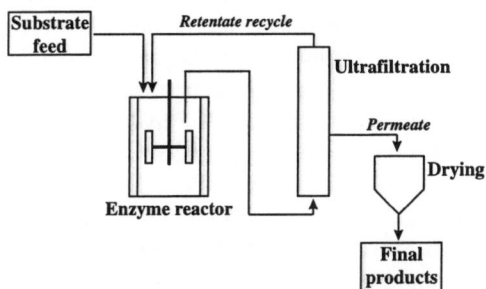

Figure 4. Enzyme-membrane reactor for controlled hydrolysis of polysaccharides.

5   Fermentation in the large intestine: The prebiotic effect

The large bowel is by far the most heavily colonised region of the human gastrointestinal tract. Through the process of fermentation, colonic bacteria, most of which are anaerobes, produce a wide variety of compounds that may affect gut, as well as systemic, physiology (see Chapters 1, 4, 12-17). Carbohydrate fermentation in the large gut produces short chain carboxylic acids (mainly acetate, propionate, butyrate and lactate) which allow the host to salvage part of the energy of resistant substrates and may play a role in regulating both cellular metabolism as well as cell division and differentiation (see Chapters 4, 8). Evidence for fermentation by bacteria colonising the large bowel may come from in vitro (both analytical and microbiological) and in vivo studies. In vitro experiments are useful to estimate the production of carboxylic acids as end products of the fermentation, in the absence of absorptive processes. However, for the authentic demonstration of a prebiotic effect, in vivo data are preferred.

5.1   IN VITRO FERMENTATION OF NON-DIGESTIBLE OLIGOSACCHA-RIDES: PRELIMINARY PREBIOTIC EFFECTS

The first line of evidence supporting the assumption that NDO are fermented by the colonic microbiota is the demonstration that these carbohydrates are utilised when incubated with either pure bacterial strains or (preferably) faecal samples in anaerobic batch cultures. Since such metabolism is known to produce an acidic environment, changes in the culture pH may be used as a very straightforward way by which to estimate fermentability. Some authors have used the decline in culture pH, during a given incubation period, to compare the degree of fermentability. However, as gut bacteria produce different acids of various strength, this type of study should only be used as an extremely preliminary guide. Such data have been reported for inulin-type fructans (Hidaka *et al.*, 1986, Wang, 1993), and they have been reviewed by Roberfroid *et al.* (1998). Quantification of INU and OFr showed that the oligosaccharides were rapidly and completely metabolised by human faecal microflora, with $G_{py}F_n$ and $F_{py}F_n$ type components disappearing from the culture media at a similar rate (Roberfroid *et al.*, 1998).

In pure culture, most species of bifidobacteria are adept at the utilisation of inulin-type fructans (Gibson and Wang, 1994b). Many other bacteria are also capable of metabolising these substrates, including *Klebsiella pneumoniae, Staphylococcus aureus* and *epidermidis, Enterococcus faecalis* and *faecium, Bacteroides vulgatus, thetaiotaomicron, ovatus,* and *fragilis, Lactobacillus acidophilus* and *Clostridium* spp. (mainly *Cl. butyricum*) (Roberfroid *et al.*, 1998). In mixed batch and chemostat culture studies, it has been demonstrated that both inulin and its hydrolysate selectively stimulated the growth of the bifidobacteria which, at the end of the incubation period, become numerically predominant (Wang and Gibson, 1993; Gibson and Wang, 1994c).

In vitro data are also available for other NDO. In pure culture, soybean oligosaccharides (raffinose and stachyose) were fermented by various species of *Bifidobacterium*, except *Bifidobacterium bifidum*, and to a lesser extent by some strains of lactobacilli and bacteroides (Masai *et al.*, 1987). In one sample of human faecal flora, soybean oligosaccharides were shown to double the number of total viable bacteria and also increase numbers and the relative proportion of bifidobacteria (Saito *et al.*, 1992). However, as other bacteria were not enumerated in this experiment it is not possible to adequately ascertain the prebiotic effect.

The galactooligosaccharides are readily fermentable by bifidobacteria as well as some, but not all, strains of bacteroides, lactobacilli and enterobacteria. They are not thought to be metabolised by eubacteria, fusobacteria, clostridia and most strains of streptococci, although this has not been determined in vivo (Tanaka *et al.*, 1983). Palatinose oligosaccharides or condensates, are metabolised by most pure cultures of bifidobacterial species (Mizutani, 1991).

For a true prebiotic effect to be demonstrated, we propose that candidate carbohydrates be scrutinised according to the scheme given in Fig. 2.

5.2   IN VIVO FERMENTATION: THE PREBIOTIC EFFECT

For the inulin-type fructans, galactooligosaccharides and xylooligosaccharides, a selective stimulation of faecal bifidobacteria has been demonstrated in efficient animal model systems (Campbell *et al.*, 1997; Rowland and Tanaka, 1993). Moreover, NDO have been used in human volunteer studies with the aim to confirm, in vivo, the prebiotic effect.

Gibson *et al.* (1995) carried out a volunteer trial with adult subjects on strictly controlled diets supplemented with 15g/d of either OFr or INU. Sucrose was used as the control and faecal samples were processed, in a blind manner, within 30 min of passage. Importantly, the study used follow-up characterisation techniques (phenotypic) to fully identify the microflora that developed during the feeding regimes. These studies showed that the intake of OFr or INU significantly modified the composition of the faecal microbiota by stimulating the growth of bifidobacteria which, after 2 weeks of the feeding period, became by far the most numerically predominant bacterial group. The data from this study are summarised in Fig. 5. In addition, the feeding of OFr significantly reduced the count of bacteroides, fusobacteria and clostridial populations. These effects lasted for as long as the prebiotic was consumed. However, at the end of a 2 week follow up on the control diet, the faecal flora of all volunteers still had higher bifidobacterial numbers than when recruitment started. Similar human studies in adult European, Japanese, and North American populations have also been reported for these fructans, at various doses (from 4-40g/d), (Hidaka *et al.*, 1986; Mitsuoka *et al.*, 1987; Hidaka *et al.*, 1991; Williams *et al.*, 1994; Bouhnik *et al.*, 1996; Buddington *et al.*, 1996; Kleesen *et al.*, 1997).

For other NDO, in vivo studies have also been performed with doses ranging from 3g up to 15g/d, given for 1, 2 or 3 week periods. For soybean oligosaccharides, a dose of 10g given twice daily for 3 weeks, significantly increased the number of bifidobacteria whilst slightly decreasing clostridial counts (Masai *et al.*, 1987). A dose of 3g/d not only increased bifidobacteria but also bacteroides and eubacteria (Wada *et al.*, 1992). For the galactooligosaccharides, Tanaka *et al.* (1983) and Ito *et al.* (1990, 1993) have reported an increase both in bifidobacteria and lactobacilli for doses ranging from 3 to 10g/d. Rowland and Tanaka (1993) have confirmed these data for germ-free rats inoculated with human faecal bacteria. A daily dose of galactosylsucrose (5 or 10g/d) similarly stimulated the growth of bifidobacteria after 1 and 2 weeks of ingestion (Yoneyama *et al.*, 1992). An isomaltooligosaccharide dose of 13.5g/d for 2 weeks significantly increased bifidobacteria both in adult and elderly volunteers (Kohmoto *et al.*, 1988). Palatinose condensates may also stimulate the growth of bifidobacteria (Mizutani *et al.*, 1991).

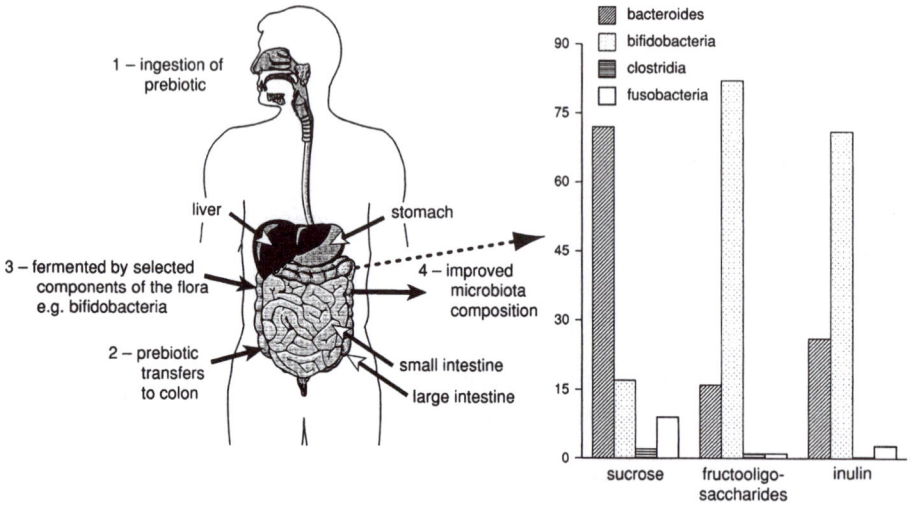

Figure 5. The prebiotic concept, showing bar diagrams of how oligofructose and inulin (15 g/d) can influence the composition of the human large gut microbiota in comparison to sucrose (data from Gibson *et al.* (1995)).

Care must be taken when interpreting data from in vivo, as well as in vitro, studies. Problems with the latter have been alluded to earlier, in that:

1) It is important that pure culture studies are interpreted in a realistic context - in that they do not reflect any competitive interactions
2) As high a microbial diversity as is possible should be determined in response to prebiotic fermentation
3) Follow-up characterisation of colonies that develop on ostensibly 'selective' agars is critical. It is worthwhile mentioning here that fully reliable selective growth media do not even exist!

For in vivo studies, these criteria can be even more exposed. Again, follow up characterisation of the bacteria is essential, as is the use of fresh specimens if a conventional microbiological approach towards identification is taken. It is not feasible to enumerate the prebiotic target organisms (lactobacilli, bifidobacteria) in isolation. The lowest effective dose ought to be advocated and studies should be placebo-controlled in a blinded assay procedure.

6   Health consequences of prebiotic intake

In functional food development, there are certain criteria that ought to met. One strategy for the implementation of foods with added functionality is given in Fig. 6. For prebiotics, a **functional** claim may be levelled against NDO, or other carbohydrates, that alter the gut flora composition. However, the real interest in this area lies in potential **health** outcomes of prebiotic intake. Whilst it is fair to say these have not been definitively proven in humans, there are a number of areas that offer much current promise.

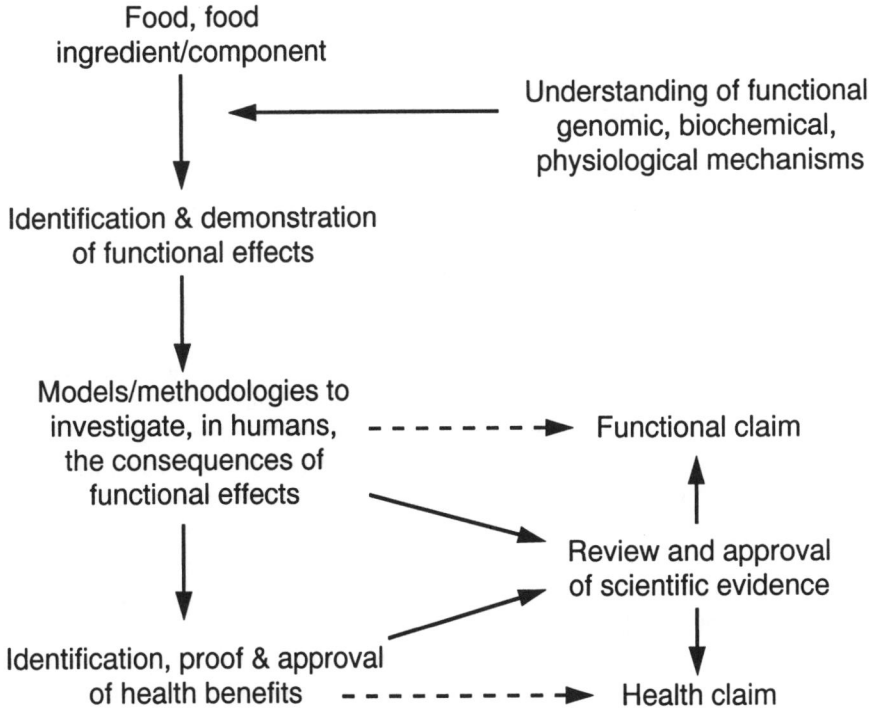

Figure 6. Strategy for the development of 'functional' and 'health' foods.

6.1   MICROBIOLOGICAL

In terms of health and medicinal burden, as well as economic consequences, the symptoms of acute gastroenteritis, mostly elaborated by foodborne pathogens, pose an immense burden. The gastrointestinal ecosystem can become contaminated by bacteria (*e.g.* campylobacters, salmonellae, listeria, *Escherchia coli*), viruses (*e.g.* rotavirus) and parasites (*e.g. Giardia* spp.). This is especially significant in certain vulnerable population groups such the the elderly or infirm, infants and hospitalised patients - where antibiotic associated diarrhoea is a major problem (*e.g.* that caused by the proliferation of *Clostridium difficile* in the gut). In the United Kingdom alone, reported campylobacter infections are as high as 50000 cases per annum (the real incidence may be up to 10 fold greater). Recent outbreaks of *E. coli* 0157, as well as ongoing problems with BSE, ensure that food safety has never had such a high consumer profile as is the current case.

Great effort is being expended on cleaning up the microbiology of the food chain, *i.e.* from 'farm to fork' or from 'plough to plate.' Better diagnostics, detection of pathogens, improved food processing and handling, efficient tracking of virulence traits and sensitive monitoring of the possibility for cross-genetic transfer, will all contribute towards improved food hygiene. However, it is beyond the fork, or plate, that transient pathogens exert their debilitating effects. Moreover, those pathogens that are normally held in a suppressed state, such as *Clostridium difficile*, seem to problematic only when the gut flora is compromised. It is therefore implicit that a more prophylactic approach towards gut infection can be of much value. In this context, some promising effects have been gained through probiotic usage (Gibson *et al.*, 1997). For prebiotics, it is clear that certain NDO have the capacity to influence the gut flora composition. One important avenue for further research is the preventative aspects for prebiotics against gut pathogens.

A gut flora which is predominated by bifidobacteria and/or lactobacilli may have a number of relevant protective factors. Firstly, metabolic end products such as acids excreted by these microorganisms may lower the gut pH, in a microniche, to levels below those at which pathogens are able to effectively compete. Another factor is the competitive effects towards colonisation sites and available nutrients.

Many lactic species are both able to excrete natural anti-microbial agents which can have a broad spectrum of activity (*e.g.* lactocins, helveticins, lactacins, curvacins, nisin, bifidocin). For the bifidobacteria, our studies have indicated that some species are able to exert antimicrobial effects on various Gram positive and Gram negative intestinal pathogens (Gibson and Wang, 1994a). As mentioned earlier, oligosaccharides themselves may act as anti-infective agent through the occupation of pathogen colonisation / receptor sites (Zopf and Roth, 1996).

The plant derived carbohydrate cellobiose is able to attenuate virulence in *Listeria monocytogenes* through direct regulatory effects on listeriolysin and phospholipase C (Park and Kroll, 1993). In the presence of cellobiose (*e.g.* in soils) the organism is not pathogenic. This is not the case in the human body however, where free cellobiose is absent. It may be that further incorporation of this disaccharide to foods susceptible to listeria infection could therefore reduce its virulence. This aspect of prebiotics, where the activities of pathogens are directly affected, warrants much closer consideration.

Immune regulatory mechanisms, mediated by probiotic bacteria, may often have positive effects against pathogenic microorganisms (see Chapter 17), whilst the anti-infective nature of prebiotics was discussed earlier.

In essence, there are a number of feasible mechanisms where attaining the correct "microflora balance" may help resist the effects exerted by pathogens.

6.2 EFFECTS ON MINERAL ABSORPTION

Certain dietary fibres have been regularly touted as causing an impairment in the small intestinal absorption of minerals due to their binding/sequestering abilities. However, the minerals affected are not absorbed in the small intestine but reach the colon where they may be released from the carbohydrate matrix and absorbed. Moreover, a high concentration of short-chain carboxylic acids, resulting from the colonic fermentation of carbohydrate may facilitate the colonic absorption of minerals, particularly Ca^{2+} and Mg^{2+}.

Independent of any binding effect, NDO may improve mineral absorption and balance, because of an osmotic effect which transfers water into the large bowel, thus allowing the minerals to become more soluble. In addition to being well fermented, they can cause acidification of colonic contents and consequently raise the concentration of ionised minerals, thus favouring passive diffusion.

It has been shown that inulin-based prebiotics do not impair mineral absorption in the small intestine (Ellegärd *et al.*, 1997). These studies demonstrated that the amount of Ca, Mg and Fe ions recovered in ileostomates over a 3 day period was not significantly modified after supplementing the diet with 17 g/d of these fructans. Using growing rats, it has been reported that the fructans enhanced Ca^{2+} and Mg^{2+} absorption as well as iron and zinc balance without having any significant effect on Cu^{2+} bioavailability (Demigne *et al.*, 1989; Delzenne *et al.*, 1995; Ohta *et al.*, 1995).

More recent human studies have been performed which confirm the positive effect of FOS on the absorption and balance of dietary calcium but not of iron, magnesium or zinc. Nine male men (21.5 ± 2.5 years) who had an intake of about 850 mg calcium/d received a dietary supplement of 40 g/d inulin. This caused a significant increase in the apparent absorption (± 12%) and balance (+ 100 mg/d) of calcium without any change in urinary excretion (Coudray *et al.*, 1997). In another study, twelve 15-18 year old volunteers consumed 16.8 g OFr/d and their calcium balance, as measured by the double stable isotope technique, showed an 11% increase (p = 0.09) with no effect on urinary excretion (van den Heuvel and Schaafsma - personal communication).

6.3 SYSTEMIC EFFECTS ON BLOOD LIPIDS

The effects of inulin-type prebiotics on triglyceridaemia have been studied both in human subjects and in animals. In rats, a decrease in serum triglyceridaemia (both in the fed and fasted state) has consistently been reported in several studies (Fiordaliso *et al.*, 1995; Kok *et al.*, 1996a,b), whereas in healthy humans, only fasting triglycerides have been measured and are not usually modified (Rumessen *et al.*, 1990; Luo *et al.*, 1996; Pedersen *et al.*, 1997). No data have hitherto been published reporting studies performed in hypertriglyceridemic patients. Data concerning the effects of prebiotics on cholesterolaemia and/or lipoproteinaemia are not well documented.

The feeding of rats a diet supplemented with OFr (10% w/v in the diet) significantly lowered serum triglycerides and phospholipids, but did not modify free fatty acids concentration in the serum (Fiordalisco *et al.*, 1995). Hypotriglyceridaemia is mostly due to a decrease in the concentration in plasma of very low density lipoproteins (VLDL). This effect is likely to result from a decrease in the hepatic synthesis of triglycerides rather than a higher catabolism of triglyceride-rich lipoproteins (for more details see Chapter 13).

Preliminary data have been reported in slightly hypercholesterolemic human volunteers which have indicated that inulin (18g/d for 3 weeks) may lower both total and LDL serum cholesterol (Davidson *et al.,* 1998).

6.4   REDUCTION IN DISEASE RISK

For prebiotics, 'disease risk reduction claims' which are based on presently available scientific information are tentative. This still needs further research to be supported and validated, but may include:
a) prophylactic effects on the causes of acute gastroenteritis
b) relief from constipation due to faecal bulking and, possibly, effects on intestinal motility
c) inhibition of diarrhoea, particularly in association with intestinal infection
d) reduction of the risk of osteoporosis if prebiotics can improve the bioavailability of Ca and if this effect is followed by a more physiological change in peak bone density and bone mass
e) reduction of the risk of atheroschlerotic cardiovasular disease associated with dyslipidaemia, especially hypertriglyceridaemia, and insulin resistance
f) reduction of the risk of obesity and possibly non-insulin dependent diabetes

A further area for important consideration in the context of disease risk reduction by prebiotics is gut cancer. The colon is the second most popular site for carcinogenesis in humans. Experimental data have been recently been published which demonstrate that the feeding of rats with inulin significantly reduced the incidence of the so-called 'aberrant crypt foci' induced by colonic carcinogens like azoxymethane or dimethylhydrazine (Reddy *et al.,* 1997; Rowland *et al.,* 1998; also see Chapters 8, 16). These effects, in experimental animals, correspond only to a first step in terms of the health bonuses for chronic disease, but do warrant further evaluation in relevant human studies.

7   Conclusions

Dietary carbohydrates are a large family of miscellaneous compounds with different physiological effects and nutritional properties (Cummings and Roberfroid, 1997). In this context, the NDO have been shown to be of particular interest because of their prebiotic properties. It is important however, that this area of research moves beyond functional food science and towards the validation of health advantages. This chapter has briefly highlighted some of the beneficial properties that can be associated with prebiotics and identified how the functionality can be enhanced.

In order to justify health claims, most of these data will have to be confirmed in humans in relevant nutritional studies focusing on well validated endpoints, and that rely on sound mechanistic hypotheses. If any one of these postulated outcomes is definitively proven, the consequences are large - because such dietary effects are very straightforward to carry out and are 'consumer friendly' in terms of product development.

In our opinion, it is most realistic to envisage that careful microflora management, through diet, holds most promise in the reduction of risk of acute gastroenteritis. In this case, the mechanisms are proven in vitro and it is easy to envisage that fortification of a lactic microflora can offer more than one prophylactic measure in vivo. However, improving the microflora composition, modulation of the metabolism of triglycerides, modulation of insulinaemia, improved bioavailability of dietary calcium, and negative modulation of colon carcinogenesis all appear to be promising areas for further research.

A combination of probiotics and prebiotics, in what has been called a synbiotic (Gibson and Roberfroid, 1995) can be defined as 'a mixture of pro- and prebiotics which beneficially affects the host by improving survival and implantation of live microbial dietary supplements in the gastrointestinal tract' is a possibility whereby probiotic survival may be improved after ingestion. It is evident that probiotics, when used in the correct context, can have certain health advantages (see Chapters 6 and 17). In particular, results against the symptoms of lactose malabsorption, food allergy and rotavirus infection are compelling. Use of efficient probiotic strains in combination with selective substrate(s) ought to effect improved survival of the live microbial strain much more. The synergistic effects of combining the anticarcinogeneic effects of inulin and bifidobacteria in experimental animals have already been demonstrated (Rowland *et al.,* 1998).

For prebiotics, it is clear that adjusting the composition and activities of the colonic microflora in such a way that the 'health promoting' activity is optimised remains the goal. Our original definition implied that this may occur through the stimulation of one or a limited number of colonic bacteria. This definition is still appropriate, because selectivity of the fermentation in a complex microbial ecosystem, like the colon, is a key prerequisite. To move towards authentic health promoting values however, it may be more realistic that the integrated microbial ecosystem as a whole should be considered. For example, a reduction in known detrimental species and/or attenuation of virulence and/or inhibition of pathogen binding through prebiotic action may be equally as feasible as a stimulation in 'beneficial' genera. Moreover, new prebiotic developments may result in efficacious molecules that can encompass more than one biological activity. In this case, prebiotics could simply be viewed as 'food components which improve the health promoting activities of the gut microflora'.

In line with our original definition (Gibson and Roberfroid, 1995), the demonstration of a prebiotic effect requires an extensive qualitative and quantitative **in vivo** analysis of the colonic flora. This is critical to prove a significant modification of the composition - either beneficial or detrimental. As discussed in other chapters of this book, new molecular methodologies are now being extensively applied to human gut microbiology and promise the degree of reliability required for detecting subtle changes in microflora composition. Whatever the definition/jargon used for microflora modulation through diet, these are essential developments that provide extremely useful tools for use in well controlled, large scale, volunteer studies. Such trials should also exploit the use of efficient clinical markers/outcome.

References

Bach Knudsen, K.E. and Hessov I. (1995) Recovery of inulin from Jerusalem artichoke (*Helianthus tuberosus* L.) in the small intestine of man. *British Journal of Nutrition* **74**, 101-13.

Bouhnik, Y., Flourie, B., Riottot, M., Bisetti, N., Gailing, M., Guibert, A., Bornet, F. and Rambaud, J.C. (1996) Effects of fructooligosaccharide ingestion on fecal bifidobacteria and selected metabolic indexes of colon carcinogenesis in humans. *Nutrition and Cancer* **26**, 21-29.

Buddington, R.K., Williams, C.H., Chen, C.S. and Witherly, S.A. (1996) Dietary supplementation of Neosugar alters the faecal flora and decreases activities of some reductive enzymes in human subjects. *American Journal of Clinical Nutrition* **63**, 709-16.

Campbell, J.M., Fahey, G.C. and Wolf, B.W. (1997) Selected indigestible oligosaccharides affect large bowel mass, cecal and fecal short-chain fatty acids, pH and microflora in rats. *Journal of Nutrition* **127**, 130-36.

Cooley, S. and Livesey, G. (1987) The metabolisable energy value of polydextrose in mixed diet fed to rats. *British Journal of Nutrition* **57**, 235-43.

Coudray, C., Ballange, J., Castiglia-Delhaut, C., Remesey, C., Vermorel, M. and Demigne, C. (1997) Effect of soluble and partly soluble dietary fibre supplementation on absorption and balance of calcium, magnesium, iron and zinc in healthy young men. *European Journal of Clinical Nutrition* **51**, 375-80.

Crittenden, R.G. and Playne M.J. (1996) Production, properties and applications of food-grade oligosaccharides. *Trends in Food Science and Technology* **7**, 353-61.

Cummings, J.H. and Roberfroid M.B. (1997) A new look at dietary carbohydrate: chemistry, physiology and health. *European Journal of Clinical Nutrition* **51**, 417-23.

Dahlquist, A. (1964) Method for assay of intestinal disaccharidases. *Analytical Biochemistry* **7**, 18-25.

Davidson, M.H., Maki, K.C., Synecki, C., Torri, S.A. and Drennan, K. (1998) Evaluation of the influence of dietary inulin on serm lipids in adults with hypercholesterolemia. *Nutrition Research* **18**, 503-17.

Delzenne, N., Aertssens, J., Verplaetse, H., Roccaro, M. and Roberfroid, M.B. (1995). Effect of fermentable fructo-oligosaccharides on mineral, nitrogen and energy digestive balance in rats. *Life Science* **57**, 1579-87.

Demigne, C., Levrat, A.M. and Remesy, C. (1989) Effects of feeding fermentable carbohydrates on caecal concentration of minerals and their fluxes between the caecum and blood plasma in the rat. *Journal of Nutrition* **119**, 1625-30.

Desjardins, M.L., Roy, D. and Goulet J. (1990) Growth of Bifidobacteria and their enzyme profiles. *Journal of Dairy Science* **73**, 299-307.

Dumortier, V., Brassart, C. and Bouquelet, S. (1994) Purification and properties of a ß-galactosidase from *Bifidobacterium bifidum* exhibiting a transgalactosylation reaction. *Biotechnology and Applied Biochemistry* **19**, 341-54.

Ekhart, P.F. and Timmermans, E. (1996) Techniques for the production of transglycosylated oligosaccharides. *Bulletin of the International Dairy Federation* **313**, 59-64

Ellegärd, L., Andersson, H. and Bosaeus, I. (1997) Inulin and oligofructose do not influence the absorption of cholesterol, or the excretion of cholesterol, Ca, Mg, Zn, Fe or bile acids but increases energy excretion in ileostomy subjects. *European Journal of Clinical Nutrition* **51**, 1-5.

Englyst, H.N. and Hudson, G.J. (1996) The classification and measurement of dietary carbohydrates. *Food Chemistry* **57**, 15-21.

Figdor, S.K. and Rennhard, J.R. (1981) Caloric utilisation and deposition of [^{14}C]-polydextrose in the rat. *Journal of Agriculture and Food Chemistry* **29**, 1181-89.

Figdor, S.K. and Rennhard, J.R. (1983) Caloric utilisation and deposition of [^{14}C]-polydextrose in man. *Journal of Agriculture and Food Chemistry* **31**, 389-93.

Finegold, S.M., Sutter, V.L., Boyle, J.D. and Shimada, K. (1970) The normal flora of ileostomy and transverse colostomy effluents. *Journal of Infectious Diseases* **122**, 376-81.

Fiordaliso, M.F., Kok, N., Desager, J.P., Goethals, F., Deboyser, D., Roberfroid, M.B. and Delzenne, N. (1995) Dietary oligofructose lowers triglycerides, phospholipids and cholesterol in serum and very low density lipoproteins of rats. *Lipids* **30**, 163-67.

Gibson, G.R. and Roberfroid, M.B. (1995) Dietary modulation of the human colonic microbiota: Introducing the concept of prebiotics. *Journal of Nutrition* **125**, 1401-12.

Gibson, G.R. and Wang, X. (1994a) Regulatory effects of bifidobacteria on the growth of other colonic bacteria. *Journal of Applied Bacteriology* **77**, 412-20.

Gibson, G.R. and Wang, X. (1994b) Enrichment of bifidobacteria from human gut contents by oligofructose using continuous culture. *FEMS Microbiology Ecology* **118**, 121-28.

Gibson, G.R. and Wang, X. (1994c) Bifidogenic properties of different types of fructo-oligosaccharides. *Food Microbiology* **11**, 491-98

Gibson, G.R., Beatty, E.B., Wang, X. and Cummings, J.H. (1995) Selective stimulation of bifidobacteria in the human colon by oligofructose and inulin. *Gastroenterology* **108**, 975-82.

Gibson, G.R., Saavedra, J.M., Macfarlane, S. and Macfarlane G.T. (1997) Gastrointestinal Microbial Disease, in *Probiotics 2: Application and Practical Aspects* (ed. R. Fuller) Chapman and Hall, London, pp. 10-39.

Harteminks, R., Van Laere, K.N.J. and Rombouts, F.M. (1997) Growth of enterobacteria on fructo-oligosaccharides. *Journal of Applied Bacteriology* **83**, 367-74.

Hidaka, T., Eida, T., Takizawa, T., Tokunga, T and Tashiro, Y. (1986) Effects of fructooligosaccharides on intestinal flora and human health. *Bifidobacteria Microflora* **5**, 37-50.

Hidaka, T., Tashiro, Y. and Eida, T. (1991) Proliferation of bifidobacteria by oligosaccharides and their useful effect on human health. *Bifidobacteria Microflora* **10**, 65-79.

Imaizumi, K., Nakatsu, Y., Sato, M., Sedamawati, Y. and Sugano, M. (1991) Effects of xylooligosaccharides on blood glucose, serum and liver lipids and short chain fatty acids in diabetic rats. *Agricultural Biological Chemistry* **55**, 199-205.

Ito, M., Deguchi, Y., Miyamori, A., Kikuchi, H., Matsumoto, K., Kobayashi, Y., Yajima, T. and Kan, T. (1990) Effect of administration of galacto-oliosaccharides on the human faecal microflora, stool weight and abdominal sensation. *Microbial Ecology in Health and Disease* **3**, 285-92.

Ito, M., Kimura, M., Deguchi, Y., Miyamori-Watabe, A., Yajima, T. and Kan, T. (1993) Effets of transgalactosylated disaccharides on the human intestinal flora and their metabolism. *Journal of Nutritional Science and Vitaminology* **39**, 279-88.

Kato, Y., Ikeda, N., Iwanami, T., Ozaki, A. and Ohmura, K. (1991) Change of soybean oligosaccharides in the digestive tract. *Nippon Shokuryo Gakkaishi* **44**, 29-35.

Kleesen, B., Sykura, B., Zunft, H.J. and Blaut, M. (1997) Effect of inulin and lactose on faecal microflora, microbial activity and bowel habit in elderly constipated patients. *American Journal of Clinical Nutrition* **65**, 1397-1402.

Kohmoto, T., Kukui, F., Takaku, H., Machida, Y., Arai, M. and Mitsuoka, T. (1988) Effect of isomaltooligosaccharides on human fecal flora. *Bifidobacteria Microflora* **7**, 61-69.

Kok, M., Roberfroid, M. and Delzenne, N. (1996a) Involvement of lipogenesis in the lower VLDL secretion induced by oligofructose in rats. *British Journal of Nutrition* **76**, 881-90.

Kok, M., Roberfroid, M. and Delzenne, N. (1996b) Dietary oligofructose modifies the impact of fructose on hepatic triacylglycerol metabolism. *Metabolism* **45**, 1547-50.

Luo, J., Rizkala, S., Alamowitch, C., Boussairi, A., Blayo, A., Barry, J.L., Laffitte, A., Guyon, F., Bornet, F.R.J. and Slama, G. (1996) Chronic consumption of short-chain fructooligosaccharides by healthy subjects decreased basal glucose production but had no effect on insulin-stimulated glucose metabolism. *American Journal of Clinical Nutrition* **63**, 939-45.

Masai, T., Wada, K. and Hayakawa, K. (1987) Effects of soybean oligosaccharides on human intestinal flora and metabolic activities. *Japanese Journal of Bacteriology* **42**, 313-29.

McAuliffe, J.C. and Hindsgaul, O. (1997) Carbohydrate drugs - an ongoing challenge. *Chemistry and Industry* 170-75.

Mitsuoka, T., Hidaka, H. and Eida, T. (1987) Effect of fructo-oligosaccharides on intestinal flora. *Die Nahrung* **31**, 426-36.

Mizutani, T. (1991) Properties and use of palatinose condensates. *New Food Industry* **33**, 9-16.

Molis, C., Flourie, B., Ouane, F., Gailing, M.F., Lartigue, S., Guibert, A., Bornet, F. and Galmiche, J.P. (1996) Digestion, excretion and energy value of fructooligosaccharides in healthy humans. *American Journal of Clinical Nutrition* **64**, 324-28.

Muramatsu, K., Onodera, S., Kikuchi, M. and Shiomi, N. (1994) Substrate specificity and subsite affinities of ß-fructofuranosidase from *Bifidobacterium adolescentis* G1. *Bioscience, Biotechnology and Biochemistry*, **58**, 1642-45.

Nilsson, U. and Björck, I. (1988) Availability of cereal fructans and inulin in the rat intestinal tract. *Journal of Nutrition* **118**, 1482-86.

Nilsson, U., Oste, R., Jagerstad, M. and Birkhed, D. (1988) Cereal fructans: in vitro and in vivo studies on availability in rats and humans. *Journal of Nutrition* **118**, 1325-30.

Ohta, A., Ohtsuki, M., Baba, S., Adachi, T., Sakata, T. and Sakaguchi, E. (1995) Calcium and magnesium absorption from the colon and the rectum are increased in rats fed fructooligosaccharides. *Journal of Nutrition* **125**, 2417-24.

Oku, T., Tokunga, T. and Hosoya, H. (1984) Nondigestibility of a new sweetener 'Neosugar' in the rat. *Journal of Nutrition* **114**, 1574-81.

Park, S.F. and Kroll, R.G. (1993) Expression of listeriolysin and phosphatidylinositol-specific phospholipase C is repressed by the plant-derived molecule cellobiose in *Listeria monocytogenes*. *Molecular Microbiology* **8**, 653-61.

Pedersen, A., Sanstrom, B. and Van Amelsvoort, J.M.M. (1997) The effect of ingestion of inulin on blood lipids and gastrointestinal symptoms in healthy females. *British Journal of Nutrition* **78**, 215-22.

Playne, M.J. and Crittenden, R. (1996) Commercially available oligosaccharides. *Bulletin of the International Dairy Federation* **313**, 10-22.

Prazeres, D.M.F. and Cabral, J.M.S. (1994) Enzymatic membrane reactors and their applications. *Enzyme and Microbial Technology* **16**, 738-50.

Pudjono, G., Barwald, G. and Amanu, S. (1993) Activity of inulinase of some strains of *Bifidobacterium* and their effects on the consumption of foods containing inulin and other fructans, in *Inulin and Inulin Containing Crops*, (ed. A Fuchs), Elsevier, Amsterdam, pp. 373-79.

Rastall, R.A. and Bucke, C. (1992) Enzymatic synthesis of oligosaccharides. *Biotechnology and Genetic Engineering Reviews* **10**, 253-81.

Rastall, R.A., Rees, N.H., Wait, R., Adlard, M.W. and Bucke, C. (1992) α-Mannosidase-catalysed synthesis of novel manno-, xylose-, and heteromanno-oligosaccharides: a comparison of kinetically and thermodynamically mediated approaches. *Enzyme and Microbial Technology* **14**, 53-57.

Reddy, B.S., Hamid, R. and Rao, C.V. (1997) Effect of dietary oligofructose and inulin on colonic preneoplastic abberant crypt foci inhibition. *Carcinogenesis* **18**, 1371-74.

Roberfroid, M.B. (1996) Functional effects of food components on the gastrointestinal system. *Nutrition Reviews* **54**, S38-S42.

Roberfroid, M.B. (1997) Health benefits of non-digestible oligosaccharides, in *Dietary Fiber in Health and Disease*. (eds. Kritchevsky, D. and Bonfield, C.) Plenum Press, New York, pp. 211-19

Roberfroid, M.B. and Slavin, J. (1998) Resistant oligosaccharides, in *Health Effects and Applications of Complex Carbohydrates* (eds. Lineback, D. and Dreher, M.) ILSI Press, Washington - in press.

Roberfroid, M.B., Van Loo, J.A.E. and Gibson, G.R. (1998) The bifidogenic nature of inulin and its hydrolysis products. *Journal of Nutrition* **128**, 11-19.

Rowland, I.R., Rumney, C.J., Coutts, J.T. and Lievense, L.C. (1998) Effect of *Bifidobacterium longum* and inulin on gut bacterial metabolism and carcinogen-induced abberant crypt foci in rats. *Carcinogenesis* **19**, 281-85.

Rowland, I.R. and Tanaka, R. (1993) The effects of transgalactosylated oligosaccharides on gut flora metabolism in rats associated with human faecal microflora. *Journal of Applied Bacteriology* **74**, 667-74.

Rumessen, J.J., Bode, S., Hamberg, O. and Gudmand-Hoyer, E. (1990) Fructans of Jerusalem artichokes: intestinal transport, absorption, fermentation and influence on blood glucose, insulin, and C-peptide response in healthy subjects. *American Journal of Clinical Nutrition* **52**, 675-81.

Saito, Y., Takono, T. and Rowland, I.R. (1992) Effects of soybean oligosaccharides on the human gut microflora in in vitro culture. *Microbial Ecology in Health and Disease* **5**, 105-10.

Sester, C.S. and Racette, W.L. (1992) Macromolecule replacers in food products. *Critical Reviews in Food Science and Nutrition* **32**, 275-97.

Sims, K.A. and Cheryan, M. (1992) Hydrolysis of liquified corn starch in a membrane reactor. *Biotechnology and Bioengineering* **39**, 960-67.

Smith, N.K., Gilmour, S.G. and Rastall, R.A. (1997) Statistical optimisation of enzymatic synthesis of derivatives of trehalose and sucrose. *Enzyme Microbial Technology* **21**, 349-54.

Suwasono, S. and Rastall, R.A. (1996) A highly regioselective synthesis of mannobiose and mannotriose by reverse hydrolysis using specific α1,2-mannosidase from *Aspergillus phoenicis*. *Biotechnology Letters* **18**, 851-56.

Tanaka, R., Takayama, H., Morotomi, M., Kuroshima, T., Ueyama, S., Matsumoto, K., Kuroda, A. and Mutai, M. (1983) Effects of administration of TOS and *Bifidobacterium breve* 4006 on the human fecal flora. *Bifidobacteria Microflora* **2**, 17-24.

Van Loo, J.A.E., Coussement, P., De Leenheer, L., Hoebregs, H. and Smits, G. (1995) On the presence of inulin and oligofructose as natural ingredients in the western diet. *CRC Critical Reviews in Food Science and Nutrition* **35**, 525-52.

Van Loo, J.A.E., Cummings, J.H., Delzenne, N., Englyst, H.N., Franck, A., Hopkins, M., Kok, N., Macfarlane, G.T., Newton, D., Quigley, M.E., Roberfroid, M.B., van Vliet, T. and van den Heuvel, E. (1998) Functional food properties of non-digestible carbohydrates: a consensus report from the 'ENDO' project (DGX11 AIR11-CT 94-1095). *British Journal of Nutrition* - submitted.

Wada, K., Wayabe, J., Mizutani, J., Tomoda, M., Suzuki, and Saitoh, Y. (1992) Effects of soybean oligosaccharides in a beverage on human fecal flora and metabolites. *Nippon Nogeikagaku Kaishi* **68**, 127-35.

Wang, X. (1993) Comparative aspects of carbohydrate fermentation by colonic bacteria. Ph.D. Thesis University of Cambridge, UK.

Wang, X. and Gibson, G.R. (1993) Effects of the *in vitro* fermentation of oligofructose and inulin by bacteria growing in the human large intestine. *Journal Applied Bacteriology* **75**, 373-80.

Wiedmann, M. and Jager, M. (1997) Synergistic sweeteners. *Food Ingredients and Analysis International* 51-56.

Williams, C.H., Witherly, S.A. and Buddington, R.K. (1994) Influence of dietary Neosugar on selected bacterial groups of the human faecal microbiota. *Microbial Ecology in Health and Disease* **7**, 91-97.

Yoneyama, M., Mandai, T., Aga, H., Fuji, K., Sakai, S. and Katayama, Y. (1992) Effects of 4G-D-galacto-sylsucrose (lactosucrose) intake on intestinal flora in healthy humans. *Nippon Eijo Shokuryo Gakkaishi (Journal of Japanese Society for Nutition and Food Science)* **45**, 101-07.

Yoshida, A, Umai, A., Kurata, Y. and Kawamura, S. (1969) Utilization of soybean oligosaccharides by the intact rat. *Eiyo to Shpkuryo* **22**, 262-65.

Ziesenitz, S. and Siebert, G. (1987) In vitro assessment of nystose as sugar substitute. *Journal of Nutrition* **117**, 846-51.

Zopf, D. and Roth, S. (1996) Oligosaccharide anti-infective agents. *The Lancet* **347**, 1017-21.

CHAPTER 8

Dietary Fibre and Non-Digestible Oligosaccharides

JOANNE SLAVIN
*Department of Food Science and Nutrition, University of Minnesota, 1334
Eckles Avenue, St. Paul, MN 55108 USA*

1   Introduction

Dietary fibre has long been realised to be an important component of many foods. The term was not coined until 1953, but the anti-constipating effects of high-fibre foods have been long appreciated. In 430 BC, Hippocrates described the laxative effects of coarse wheat in comparison with refined wheat (McCance and Widdowsen, 1955). Graham, of graham cracker fame, denounced the harmful effects of refined carbohydrate foods during the 19th century, and the Kellogg and Post cereals owe their start to interest in increasing the fibre content of the diet. In the 1920's, J.H. Kellogg published extensively on the attributes of bran, claiming that it increased stool weight, promoted laxation, and prevented disease. Denis Burkitt is usually credited with popularising, in the 1970's, the idea that dietary fibre may protect against the development of Western diseases. Despite more than 20 years of research in dietary fibre, there is still disagreement on what fibre is and how it can be measured.

Oligosaccharides are constituents of many commonly consumed foods. Foods high in oligosaccharides include onions, chicory, Jerusalem artichoke, asparagus, globe artichoke, leek, garlic, banana, and wheat (Roberfroid, 1993). Thus, oligosaccharides have been dietary staples since antiquity. Oligosaccharides have received much less attention than other carbohydrates, including simple sugars or dietary fibre. Oligosaccharides were thought to have only one interesting physiological effect, a propensity for gas formation. More recently however, interest in oligosaccharides has increased because of their functional properties. These include sweetening ability, fat replacement, and enhancement of a 'healthy' gastrointestinal tract. Furthermore, oligosaccharides are metabolised in a similar manner to soluble dietary fibre and should also provide positive effects, such as a lowering of blood lipids and improving glucose control. Both dietary fibre and oligosaccharides will be described in this chapter including their definition, measurement, and physiological effects.

G.R. Gibson and M.B. Roberfroid (eds.), Colonic Microbiota, Nutrition and Health, 125-147.

2   Definition of dietary fibre

Dietary fibre has been defined in many ways. The most abundant compounds identified as fibre are in the plant cell wall; others are part of the intracellular cement; and/or secreted by the plant in response to injury (Scheeman, 1986). Thus, dietary fibre cannot be solely equated to the plant cell wall. Dietary fibre is plant cellular material resistant to digestion by the endogenous enzymes of humans. As such, starch that resists pancreatic enzyme action and passes to the colon would be considered dietary fibre using this definition. Also, according to the physiological definition, the disaccharide lactulose would be considered a liquid fibre, as would lactose escaping digestion in lactase-deficient individuals. In an attempt to circumvent this, dietary fibre has been defined as the sum of polysaccharides and lignin not digested by the endogenous secretions of the human gastrointestinal tract (Trowell, 1974).

The term dietary fibre is not accurate because many of its components are not fibrous. Gums and mucilages, for example, are classified as dietary fibre because they are not digested by mammalian enzymes or secretions. Only one component of fibre, namely cellulose, is truly fibrous; yet dietary fibre is the accepted nomenclature when describing roughage or residue in the human diet.

Research shows that not all dietary fibre is created equally and attempts have been made to describe individual components that may help explain physiological effectiveness. Dietary fibre can be divided into the following three major fractions:

a)   *Structural polysaccharides*
   - associated with the cell wall
   - includes noncellulose polysaccharides (hemicellulose and some pectins) and cellulose

b)   *Structural nonpolysaccharides*
   - predominantly lignin

c)   *Nonstructural polysaccharides*
   - includes gums and mucilages secreted by cells
   - includes polysaccharides from algae and seaweed

For simplicity, dietary fibre can be divided into noncellulosic polysaccharides, cellulose, and lignin. Human foodstuffs contain mainly noncellulosic polysaccharides, some cellulose, and a little lignin. The average content of noncellulosic polysaccharides, cellulose and lignin for common foodstuffs is about 70%, 20% and 10% respectively (Slavin, 1987).

Generally, fruits and vegetables tend to have more cellulose than cereals. Lignin is highest in fruits with edible seeds (*e.g.* strawberries) or in mature vegetables such as carrots or other root vegetables. The dietary fibre composition of a plant depends upon plant species, maturity and component (*i.e.* leaf, root or stem).

The human diet contains, in addition to polysaccharides and lignin, plant-derived materials similar to fibre that resist digestion in the small bowel. These include cutin, waxes, small amounts of protein and lipids, and phenolic compounds. Non enzymatic browning products and nonhydrolysable starch, are often indigestible and may also be considered as dietary fibre. Fibres have also been classified by solubility in water, since water-soluble and water-insoluble types have distinct physiological effects.

The solubility of polysaccharides is analytically defined as soluble in 80% ethanol, corresponding to chain lengths of at least 10 monomers. Branching of monosaccharides does affect solubility in ethanol. Undigestible oligosaccharides, such as the raffinose family or fructans, are neither included in the dietary fibre concept nor determined as dietary fibre with any of the current analytical methods. This is currently being discussed, since these oligosaccharides have dietary fibre-like physiological effects (Sunvold *et al.,* 1995).

Fermentability of dietary fibre is another property that has been linked to physiological effects. It is difficult to measure the fermentation of fibre since this is determined by breakdown in the gut and no accepted in vitro methods exist by which to quantitatively measure fermentability. Generally, highly-soluble fibres such as oat bran, guar gum, and pectin will be extensively fermented, whilst insoluble fibres such as cellulose are not well fermented. Hundreds of fibre products are now on the market and their fermentability is unknown. This is especially true when fibres are processed which may enhance or limit fermentability. For example, "oat fibre" may be mostly insoluble dietary fibre if it is extracted from oat hulls, rather than oat bran.

It is well accepted that carbohydrate fermentation in the colon produces short chain fatty acids (SCFA) that have a role in maintaining the gut integrity (Cummings and Macfarlane, 1997). Unfortunately, it is difficult to measure in vivo fermentation. In vitro methods of determining fermentability of fibres have been developed. Eleven fibre-rich substrates were subjected to incubation with faecal bacteria and SCFA production was measured (Sunvold *et al.,* 1995). Citrus pectin was 83% fermentable while oat fibre was about 6% fermented. Short chain fatty acid production was correlated to fermentability in the in vitro system. In a similar study (McBurney and Thompson, 1989), the most rapid fermentation rate was with pectin followed by psyllium gum, tragacanth gum, guar gum, soy fibre and finally cellulose.

3   Definition of oligosaccharides

Oligosaccharides are carbohydrates with a low (2-20) degree of polymerisation (DP) and molecular weight (Yun, 1996). This includes a wide range of substances that vary greatly in chemical composition and physiological effects. Standard oligosaccharide categories do not exist so it is difficult to compare the many market products. Some general categories of oligosaccharides include fructooligosaccharides (FOS), inulo-oligosaccharides, glucooligosaccharides (Djouzi *et al.,* 1995), galactooligosaccharides, isomaltooligosaccharides, polydextrose, soy oligosaccharides, and xylooligosaccharides. As interest in oligosaccharides has increased so has the supply. Since large chain polysaccharides can be hydrolysed to oligosaccharides, the number of oligosaccharides that may enter the marketplace is great and creates confusion. This is especially the case as chemical and physiological standards have not been agreed.

Obvious chemical differences among commercially available oligosaccharides include chain length, monosaccharide composition, degree of branching, and purity. Analytical measurements include HPLC, but standard solutions for many fractions are not available (Yun, 1996). Also, oligosaccharides prepared from naturally occurring foods such as chicory, Jerusalem artichoke, and soy will contain a range of oligosac-charides of various chain lengths and chemical compositions. Inulin contains up to 60 monosaccharide units and would not therefore meet the chemical definition of an oligosaccharide.

3.1   OTHER PROPERTIES OF OLIGOSACCHARIDES

Oligosaccharides are readily water soluble and exhibit some sweetness. Fructooligo-saccharides (FOS) have been described as 0.4-0.6 times as sweet as sucrose (Roberfroid, 1993). Sweetness decreases with longer chain length. Inulin with a DP higher than 20 does not have a sweet taste. Depending upon chain length and composition, oligosac-charides may contribute functional benefits to the diet, such as water binding, gelling and fat replacement value. Since oligosaccharides are not digested and absorbed in the small intestine, they have no calorific value in the traditional sense. It is thought that oligosaccharides have a calorific contribution of about 1.5 kcal/g due to colonic fermentation, which is similar to that of soluble dietary fibre. A study with radiolabeled FOS supports this value (Hosoya *et al.*, 1988).

4   Preparation of fructooligosaccharides

Fructooligosaccharides can be prepared by the hydrolysis of inulin or by an enzymatic reaction with sucrose (Spiegel *et al.,* 1994). Fructooligosaccharides and inulin occur naturally in many common food products, including banana, rye, garlic, onions, and dandelion (Van Loo *et al.,* 1995). Chicory roots and Jerusalem artichoke are particularly rich sources of FOS and inulin. The DP varies greatly among inulin sources with globe artichoke and dahlia inulin with a high DP, but onion and Jerusalem being lower (Van Loo *et al.,* 1995).

Yun (1996) has suggested that FOS is a common name only for fructose oligomers that are mainly composed of kestose (GF2), nystose (GF3), and 1-F-fructofuranosyl nystose (GF4), in which fructosyl units are bound at the ß-2-1 position of sucrose (GF). This type of FOS, known commercially as Neosugar, NutraFlora and Actilight, has been determined to be 95% pure FOS. Yun (1996) noted that many authors have included FOS with fructan, glucofructosan, and inulin-type oligosaccharides.

Fructooligosaccharides have been described as inulin-type oligosaccharides of D-fructose attached by ß-2-1 linkages that carry a D-glucosyl residue at the end of the chain. They constitute a series of homologous oligosaccharides, derived from sucrose, represented by the formula Gfn. Fructooligosaccharides can be produced by enzymatic conversion of sucrose by different procedures (Yun, 1996). By one method, FOS is produced by the action of a fungal (*Aspergillus niger*) ß-fructofuranosidase on sucrose (Spiegal *et al.,* 1994). The product is a mixture of GF2, GF3, GF4, sucrose, glucose, and fructose. Fructooligosaccharides manufactured by this procedure are no different to those found naturally occurring in foods. This product was originally called Neosugar.

Fructooligosaccharides can also be isolated from food components naturally high in FOS. For example, chicory is chemically extracted and different fractions are prepared. Generally, inulin is defined as DP from 2-60+ (Raftiline) while oligofructose is usually DP < 8 (Raftilose). Other manufacturers of inulin describe their product as having an average DP of 9. Commercially available FOS manufactured by enzymatic procedures has a DP of 4 or less. Inulin products are stable both to heat and relatively low pH, are mildly sweet and dissolve readily. Shorter chain inulin hydrolysis products are sweeter than longer versions.

5   Methods to measure dietary fibre contents

By accepting a physiological definition for dietary fibre, measurement becomes problematic. How do you measure dietary fibre except by running it though humans and measuring the residual? Historically, values have been given as crude fibre, a method that can seriously underestimate the total content of food, recovering only 50% to 80% of the cellulose, 10% to 50% of the lignin, and 20% of the hemicellulose (Slavin, 1987). Dietary fibre values are usually 3 to 5 times higher than crude values, but no correction factors can be used, because the relationship between crude and dietary fibre varies depending upon the various chemical components. Bran flakes, for example, contain six times more dietary than crude fibre, yet strawberries contain only 1.6 times more.

Despite the enormous scientific interest in dietary fibre, few reliable values exist in the literature for the contents of human foods (Lanza and Butrum, 1986; Fredstrom *et al.*, 1991). Unfortunately, the values that are available are generated with different methods and cannot be directly compared. Nutrient databases contain some dietary fibre values but there are also many that are missing. The difficulties in devising an accurate method for assessing total dietary fibre can be appreciated if one considers its very diverse nature. A simple, reproducible method to remove protein, fat, and soluble sugars and starch from food, whilst retaining both water-soluble and insoluble components of dietary fibre is difficult from an analytical standpoint.

Nutritional labels in the United States now contain values for total dietary fibre and these values are generated by the Association of Official Analytical Chemists (AOAC) method. Techniques to measure soluble and insoluble dietary fibre are available. Values for dietary fibre vary greatly, depending on the laboratory doing the analysis, sample issues, etc. The largest compilation of published dietary fibre values has been generated with a modified Theander method. As such, these cannot be directly compared with the AOAC method (Marlett, 1992). Furthermore, when the dietary fibre content and composition of different forms of fruits was measured by a modified Theander method and the AOAC method, the two sets were significantly different (Marlett and Vollendorf, 1994). A discussion on the difficulty of measuring dietary fibre in foods high in soluble fibre, *i.e.* psyllium-containing cereals, or other viscous fibres has been published (Lee *et al.*, 1995).

6   Measurement of oligosaccharides in foods

Research on oligosaccharides has been hampered by lack of good analytical procedures. Currently, there is an effort to include FOS and inulin in the soluble dietary fibre fraction of foods. A collaborative study of the AOAC found an ion exchange chromatographic procedure appropriate for measurement of fructans in foods and food products (Hoebregs, 1997). The inulin content of foods has also been measured by the AOAC procedure (Quemener *et al.*, 1997). Selected FOS in foods and feeds have been measured by ion chromatography (Campbell *et al.*, 1997). It is expected that FOS and inulin will be included as soluble dietary fibre once the methodology for their measurement reaches final approval within AOAC.

7   Fibre sources and intake

Dietary fibre is found only in plant products - fruits, vegetables, nuts, and grains. Refining decreases the fibre content of grain. The most concentrated sources of dietary fibre are the bran layers of grains; products like wheat bran. Because of their high water content, fruits and vegetables provide less dietary fibre than drier grains and cereals per gram of ingested material. The effect of cooking on the fibre content of foods is unclear. Cooking can cause browning reactions that increase the apparent fibre content of the food - since these products are analysed as lignin. Cooking also drives out water that would have the effect of increasing the percentage dietary fibre in the product.

Commonly consumed foods are low in dietary fibre. Marlett and Cheung (1997) reported that 3 out of 4 of commonly consumed foods contain 2 g of dietary fibre or less per serving. Generally, the fibre content of foods was quite consistent and the content could be estimated by multiplying fruit or vegetable servings by 1.5 g dietary fibre, refined grains by 1 g, and whole grains by 2.5 g. More concentrated fibre sources, including legumes and high fibre cereals, add additional dietary fibre to the diet.

Dietary fibre intakes continue to be less than recommended levels in the United States population, which generally fall in the range of 20-35 g/d (US Department of Health and Human Services, 1990). Others have recommended dietary fibre intakes based on kcal intake, - 10-13 g of dietary fibre per 1000 kcal (Pilch, 1987). Usual intakes of dietary fibre in the USA average only 14-15 g/d (Marlett and Slavin, 1997), so few people attain the recommended levels. An obvious reason is that most popular foods consumed do not have a high content, for example, most common servings of grains, fruits, and vegetables contain between 1-3 g of dietary fibre (Slavin, 1987). Thus, for the recommended levels of dietary fibre intake, one would need to consume at least 10 or more servings of the fibre-containing foods per day - assuming the average food contained 2 g of dietary fibre. If consumers were eating according to the USDA Food Guide Pyramid and choosing cereals, whole grains and intact fruits and vegetables, it would be possible to obtain the recommended levels of dietary fibre. Despite the efforts of nutrition educators, dietary fibre intakes average only half of the recommended amount and do not appear to be increasing.

8 Oligosaccharide intake

Van Loo *et al.* (1995) have estimated the daily intakes of inulin and oligofructose for the United States and Western Europe. They estimated that an average North American consumes between 0.014 and 0.054 g inulin or oligofructose per day per kg of body weight. This corresponds to an intake of 1 to 4 g by a 165 pound person. Wheat, onion, and banana are the most important sources of inulin and oligofructose in the North American diet. Western Europe intake is estimated at 3.2 to 11.3 g for a 75 kg person. Consumption will be higher with intake of French onion soup or other concentrated inulin sources such as leeks and garlic.

9 Physiological effects of dietary fibre

Dietary fibre affects the digestive tract from mouth to anus and has other important physiological functions. Effects of dietary fibre on the intestinal tract are influenced by the type of fibre ingested, physical state of the subject, previous diet and other dietary components. Thus, confusion in the field of dietary fibre is the result of many variables that are not controlled in research studies.

The fate of dietary fibre in the human digestive tract is summarised in Fig 1. Digestible carbohydrates, including starches, are hydrolysed by enzymes and broken down to component sugars ready for absorption. Although it was generally believed that this process was complete, new research suggests that between 3 to 20% of starch is not completely digested and absorbed in the small intestine. That fraction escaping into the large intestine is termed "resistant starch" and probably functions similar to dietary fibre in the gut.

When carbohydrates reach the caecum, they are fermented with a resultant products of SCFA, and gases, including CO_2, H_2 and CH_4. These gases may escape through lungs or gastrointestinal tract as flatus. This active fermentation process is difficult to study in vivo. Dietary fibre fermentation has been estimated by measuring the food content and comparing it to that remaining in faeces. Researchers have also estimated H_2 and CH_4 production as a means to determine fibre fermentation. Short chain fatty acid production has been measured by having subjects swallow dialysis bags and collecting these bags in freshly passed faeces. None of these methods is ideal and little accurate information is available on the fate of dietary fibre in the body.

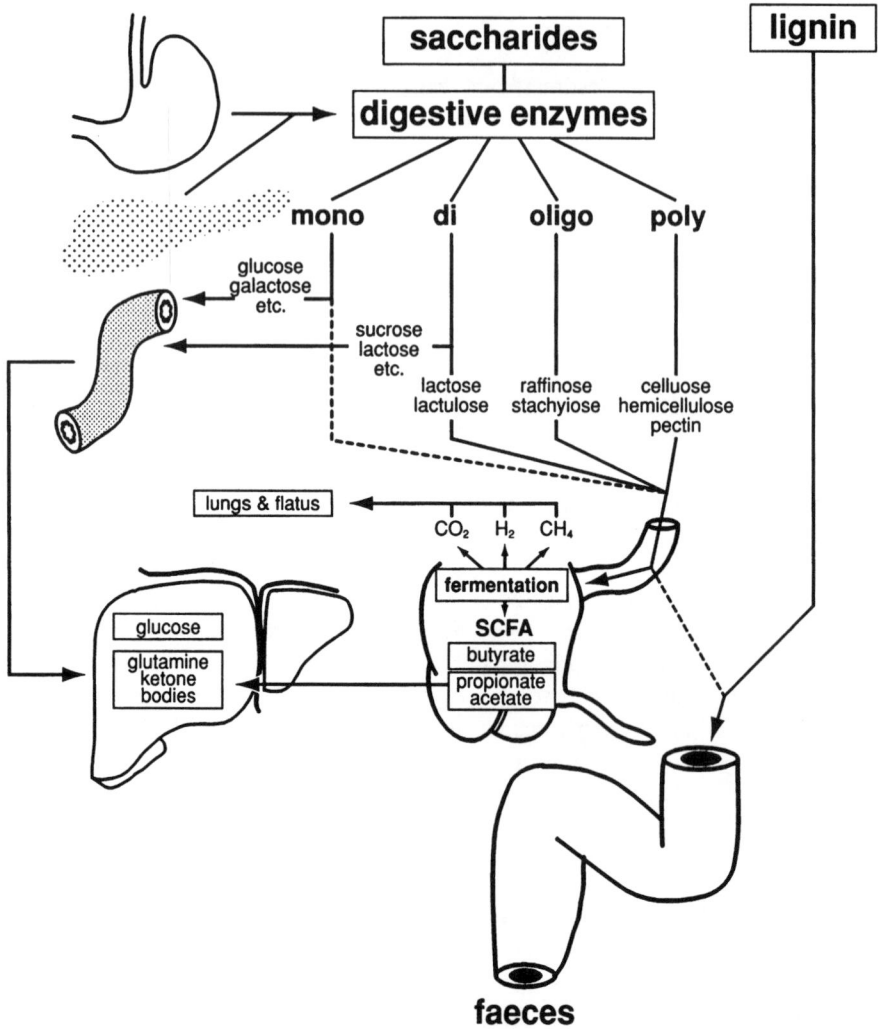

Figure 1. Summary of the fate and metabolic breakdown of dietary polysaccharides.

9.1   EFFECT OF FIBRE ON THE GASTROINTESTINAL TRACT

9.1.1 *Mouth and stomach*

In the mouth, fibre stimulates the flow of saliva, primarily by increasing food volume. When dietary fibre reaches the stomach, it may dilute the contents and perhaps prolong storage. Pectin and guar gum generally increase the gastric emptying time, while other fibres have no effect. Viscosity of the fibre source may be an important variable in gastric emptying.

9.1.2 *Small intestine*

Dietary fibre is thought to dilute the contents of the small intestine. The most viscous dietary fibres may delay absorption of carbohydrates and fats in the small gut. This is one potential mechanism which explains why viscous fibres (soluble) are usually more effective than insoluble fibres in lowering serum cholesterol and moderating glucose response. Changes in the absorption of macronutrients, seen with fibre feeding, are of little practical significance in the USA, but may be significant in countries where food supplies are scarce.

Fibre may affect pancreatic enzyme activity, although the data are confusing. In vitro, insoluble fibre can reduce the activities of amylase, lipase, and trypsin, while pectin had no effect. It is generally accepted that dietary fibre in practical doses has little effect on mineral absorption in the small intestine.

Fibres also may lower serum cholesterol by binding bile acids. If bile acids are bound to fibre, they may not be reabsorbed in the small intestine but will be lost in faeces. Since the body can resynthesise bile acids, it is unlikely that such binding is a primary mechanism for the effect of fibre on serum cholesterol. Even if soluble fibres can effectively bind bile acids and therefore lower serum cholesterol, there is concern that such therapy would be unwise. In the colon, bacteria convert primary bile acids into smaller, secondary bile acids that may increase colon cancer risk. Insoluble fibres may dilute the concentration of secondary bile acids in the colon may therefore be protective against large bowel cancer.

Dietary fibre also affects small intestine morphology and epithelial cell regeneration. The intestinal villi of vegetarians are broad and leaf-shaped while jejunal villi in humans consuming usual American diets are fingerlike and regular. The level and type of dietary fibre consumed may alter the structure of the small intestine.

9.1.3 *Large intestine*

Most research on dietary fibre is directed towards its metabolism in the large intestine. More than 75% of dietary fibre in an average American diet is broken down in the large intestine resulting in the production of gas and SCFA (*e.g.* butyrate, propionate and acetate). Propionate and acetate are thought to be metabolised in colonic epithelial cells or peripheral tissues. Butyrate may regulate colonic cell proliferation and serve as an energy source for colonic cells. Propionic acid is transported to the liver, and some research suggests that it may suppress cholesterol synthesis. This may be another potential explanation for how soluble dietary fibre lowers serum lipids. Faecal excretion of SCFA may not reflect colonic levels, and other methods to quantitate these metabolites in the large intestine have not been well studied. Production of SCFA will lower the colonic pH, which may be relevant for the prevention of certain gastrointestinal diseases, including colon cancer. Dietary fibre has been shown to inhibit colonic cell proliferation and influence its morphology, both of which may be related to the protective role of fibre in colon cancer.

According to calculations by Cummings (1982), if approximately 20 g of fibre is fermented in the colon each day, approximately 200 mM of SCFA should be produced, of which 62% will be acetate, 25% propionate, and 16% butyrate. Colonic absorption of SCFA is concentration dependent, with no evidence of a saturable process. The mechanism by which SCFA cross the colonic mucosa is thought to be passive diffusion of the unionised acid into the mucosal cell. Short chain fatty acids are respiratory fuels for the colonic mucosa. In isolated human colonocytes, butyrate is actively metabolised to both CO_2 and ketone bodies, which accounts for about 80% of the oxygen consumption of colonocytes. Butyrate is almost completely consumed by the colonic mucosa, while acetate and propionate enter the portal circulation, extending the effects of dietary fibre beyond the intestinal tract.

Butyrate may be an important protective agent in colonic carcinogenesis (Valazquez *et al.,* 1996). Trophic effects on normal colonocytes in vitro and in vivo are induced by butyrate. In contrast, butyrate arrests the growth of neoplastic colonocytes and inhibits the preneoplastic hyperproliferation induced by some tumour promoters in vitro. Butyrate induces differentiation of colon cancer cell lines and regulates the expression of molecules involved in colonocyte growth and adhesion.

Our laboratory has been unable to correlate apparent fibre digestibility to SCFA production, but did find that propionate and butyrate concentrations were greater from wheat bran than vegetable fibre. This supports the requirement to carefully choose the dietary fibre source if SCFA production is the desired endpoint (Fredstrom *et al.,* 1994). Kapadia *et al.* (1995) compared bowel function and SCFA production in healthy subjects consuming diets containing either no fibre or 15 g of soy oligosaccharide fibre or oat fibre, or soy polysaccharide. The soy oligosaccharide fibre was associated with higher production of butyrate than the other sources. Compared to a fibre-free polymeric enteral diet, the daily consumption of an enteral diet supplemented with 30 g of total dietary fibre derived from a poorly fermentable oat fibre, a highly fermentable soy oligosaccharide fibre, or a moderately fermentable soy polysaccharide fibre had little impact on bowel function measures.

The relationship between famine, fibre, fatty acids, and failed colonic absorption has been reviewed by Roediger (1994). Another review by Scheppach and Bartram (1993) concludes that data from clinical studies with dietary fibre have been disappointing. However, the model that fibre is fermented, thus generating SCFA, which serve as nutrients for colonic mucosal cells is correct. In vivo studies have been too short, measurements are semi-quantitative and the methods (such as measuring dietary fibre and SCFA), are not well-developed. In vivo, researchers are dependent on faeces as a source of SCFA or must have subjects swallow dialysis bags which are retrieved in faeces. It is not clear that results from in vitro fermentation studies have direct application to the in vivo setting (McBurney and Thompson, 1991).

It is generally accepted that dietary fibre is the treatment of choice for many bowel disorders. Increased dietary fibre has been recommended for constipation (Read *et al.,* 1995; Hillemeier, 1995) and diverticular disease (Ozick *et al.,* 1994). Most of this therapy must be individualised. According to Ozick *et al.* (1994), treatment of diverticular disease depends on severity and clinical presentation. A high-fibre diet is recommended if there is no acute inflammation.

10   Physiological effects of oligosaccharides

Oligosaccharides have many interesting physiological effects which has increased scientific interest. Because oligosaccharides are handled by the human body in a similar manner to dietary fibres, they may share some of their desirable properties (Tomomatsu, 1994). These effects are diverse and include an improvement in the gut environment, reduced serum cholesterol and triglycerides, and normalisation of blood glucose levels. Furthermore, since oligosaccharides can function as sugar or fat replacers, they have potential in the food industry for calorific reduction. Finally, oligosaccharides have received interest because of their role as prebiotics.

Prebiotics are defined as "non-digestible food ingredients that beneficially affect the host by selectively promoting the growth and /or activity of one or a limited number of health-promoting bacteria in the colon, thus improving host health" (Gibson and Roberfroid, 1995). All oligosaccharides not digested in the upper gut and are extensively fermented in the colon could potentially serve this role. However, the degree of bacterial growth is different with each oligosaccharide. An extensive fermentation will induce luminal or systemic effects which may be of benefit to the host.

The gastrointestinal effects of oligosaccharide ingestion, including information on gut environment, stool weight, transit time, faecal pH, SCFA and colonic microflora are relevant and will be discussed in this review. Available data on the potential positive effects of oligosaccharide ingestion on cholesterol lowering and glucose control will also be reviewed. Although an attempt is made to include all oligosaccharides, most research has been carried out with FOS. Often the product fed to subjects is not well defined which makes data interpretation difficult.

10.1   GASTROINTESTINAL EFFECTS OF FRUCTOOLIGOSACCHARIDES

When carbohydrate is not capable of being digested and absorbed in the upper gastrointestinal tract, it enters the caecum and colon where it can be acted upon by the large gut microflora. It is difficult to estimate how much carbohydrate is fermented and the rate of this utilisation since the products are quickly absorbed and metabolised. Most likely, shorter chain carbohydrates that are highly purified would be preferred substrates for the gut microflora, compared to complicated, less refined fibres such as wheat bran.

An estimate of the colonic breakdown of carbohydrates can be determined by measuring breath H_2. Malabsorption of fermentable substrates, such as carbohydrates, results in H_2 production by the colonic flora. This gas is absorbed and excreted in expired air and breath excretion can serve as a semi-quantitative indicator of its production rate in the colon. Breath H_2 studies in humans have found that FOS is fermentable, resulting in amounts in breath roughly similar to that excreted after ingestion of an identical load of lactulose, a non-digestible disaccharide (Stone-Dorshow and Levitt, 1987). Oligosaccharide ingestion from Jerusalem artichoke (Rumessen *et al.*, 1990) and chicory (Gibson *et al.*, 1995) has been associated with increased breath H_2 excretion, suggesting that these oligosaccharides were not digested in the small intestine, but fermented by the gut microflora.

A study with oligosaccharides extracted from chicory (Raftilose P95) measured the fate of two different doses (5 and 15 g/d) compared to glucose in the intestine of healthy men (Alles *et al.,* 1996). The highest dose significantly increased 24h excretion of breath H_2. Excretion after the ingestion of 5 g Raftilose was higher than in controls, but did not reach statistical significance.

Molis *et al.,* (1996) examined the fate of an oligosaccharide (Actilight) in the human gastrointestinal tract of 6 healthy volunteers. After an equilibration phase, 20.1 g Actilight/day was given in three identical postprandial doses. Distal ileal output of Actilight and constituent components were determined by intestinal aspiration after a single meal, and the amounts of Actilight excreted in stools and urine were also measured. Most of the ingested dose, was not absorbed in the small intestine, and none was excreted in stools, indicating that the portion reaching the colon was completely fermented by colonic flora. A small fraction of ingested Actilight was recovered in urine. The mean estimated energy value of Actilight was 9.5 kJ/g.

Recovery of inulin in the small intestine of humans can also be studied with the ileostomy model. Bach Knudsen *et al.,* (1995) have found that the recovery of inulin from Jerusalem artichoke was 87% when it was fed to seven subjects with well-established ileostomies. This same value was found when subjects consumed either a low (10g/d) or a high (30 g/d) inulin diet. The authors suggest that any small loss of inulin during passage through the small intestine was due to hydrolysis by either acid, enzymes or microbial degradation in the distal small intestine.

Generally, soluble dietary fibres that are extensively fermented have less effect on stool size and transit time than more insoluble fibres. Gibson *et al.,* (1995) found that both Raftilose and Raftiline significantly increased faecal wet and dry matter, nitrogen and energy excretion. They found little change in faecal SCFA with Raftilose and Raftiline feeding. Bouhnik *et al.,* (1996) found that 12.5 g of Actilight had no significant effect on faecal pH, the activities of the faecal enzyme nitroreductase, azoreductase, and ß-glucuronidase, and the faecal concentration of bile acids and neutral sterols. Actilight does increase SCFA production in vitro when added to human faecal innocula (Hoebregs, 1997).

Alles *et al.,* (1996) also found that neither 5 nor 15 g/d Raftilose extracted from chicory had any effects on faecal composition. Raftilose was not recovered in the faeces and little change was detected in stool weight, faecal SCFA and pH. The authors found this surprising, but noted that the subjects were already on a fairly high fibre intake (40 g/d) so even 15 g of oligosaccharide per day may not have had any significant effect.

Inulin has been shown to have a desirable laxative effect in elderly constipated persons when compared to lactose (Kleessen *et al.,* 1997). The oligosaccharide was well tolerated as part of an enteral diet with no undesirable effects on serum chemistry (Garletb *et al.,* 1996). More complaints of flatus and slight distention were reported with the FOS-containing diets, but these symptoms were mild and subsided as the study progressed.

10.2   PREBIOTIC EFFECTS OF FRUCTOOLIGOSACCHARIDES

Fructooligosaccharides have consistently been found to significantly alter the human faecal flora. McKellar *et al.,* (1993) reported that *Bifidobacterium* spp. grew well on Neosugar. Wang and Gibson (1993) found that in a batch culture system, Raftilose and Raftiline exerted a preferential stimulatory effect on numbers of the possible health-promoting bifidobacteria, while maintaining populations of potential pathogens (*Escherichia coli*, clostridia) at relatively low levels. Similar effects were found in a continuous culture system which should more closely represent the actual human gut conditions (Gibson and Wang, 1994).

When FOS (Nutraflora) was fed to healthy volunteers, anaerobe densities were increased (Buddington *et al.,* 1996). Also, the densities of bifidobacteria were higher after FOS. Supplementation of an uncontrolled diet with 4 g/d FOS caused bifidobacteria to increase from 1.3 % to 6.8 % of the total bacterial flora, although there was significant individual variation in response. Bouhnik *et al.,* (1996) found that Actilight increased faecal bifidobacterial counts and ß-fructosidase activity.

Mitsuoka *et al.,* (1987) fed 8 g/d of FOS (Neosugar) to 23 elderly patients for 2 weeks. Stools were collected and bacteriologically examined before and during the test. During ingestion of FOS, numbers of bifidobacteria in the stools increased about 10 times compared to before ingestion. Moreover, the frequency of occurrence of bifidobacteria increased from 87% to 100% of the patients. These data were not statistically significant, but suggested that FOS did affect the microflora. Garleb *et al.,* (1996) reported that FOS was a bifidogenic factor in low residue polymeric formulas.

Little research has been published either in vivo and in vitro on differences among oligosaccharides and their prebiotic effects. In other words, do certain oligosaccharides have more potent effects on the microflora? In the Gibson *et al.,* (1995) in vivo study, both Raftilose and Raftiline significantly increased bifidobacteria whereas bacteroides, clostridia, and fusobacteria decreased. Total bacterial counts were unchanged. Raftilose and Raftiline are obtained from chicory and differed only in their chain-lengths. Additional studies are needed to compared the prebiotic effects of the oligosaccharides on the market. Just because the existing FOS studies show support for the prebiotic effect does not mean that we can generalise this effect to all oligosaccharides.

10.3   LIPID LOWERING EFFECTS OF FRUCTOOLIGOSACCHARIDES

Since soluble dietary fibres lower serum lipids, it has been hypothesised that FOS can have a similar effect. The proposed mechanism is that fermentation increases SCFA production which has been associated with lipid lowering. Little research has been conducted in this respect. Yamashita *et al.,* (1984) reported that daily intakes of 8 g of FOS (Neosugar) for 14 days significantly reduced serum cholesterol levels by 19 mg/d/L and LDL-cholesterol levels by 17 mg/d/L in diabetic subjects compared to a sucrose control. A limitation of this study is the lack of a cross-over design where subjects served as their own control.

When FOS (Neosugar) was given to healthy subjects for 12 days, no changes in serum cholesterol or serum triglycerides were found (Stone-Dorshaw and Levitt, 1987). Again, this study was not designed to test the effect of a substrate on cholesterol lowering because the human subjects did not have elevated serum cholesterols and the feeding periods were short. In a recent study (Pedersen *et al.,* 1997), 14 g inulin was added to a low-fat spread in a trial of 64 young, healthy women. Inulin was consumed for 4 weeks and serum lipids were measured. No significant difference was found with the inulin ingestion. Inulin consumption did cause a significant increase in gastrointestinal discomfort that did not appear to lessen as the study progressed.

10.4   GLUCOSE MODULATING EFFECTS OF FRUCTOOLIGOSACCHARIDES

Yamashita *et al.,* (1984), gave daily intakes of 8 g FOS (Neosugar) and significantly reduced mean fasting blood glucose levels by 15 mg/d. In contrast, when 12 healthy volunteers received either 20 g Actilight or sucrose for 4 weeks in a double-blind crossover design, Actilight did not modify fasting plasma glucose and insulin concentrations (Luo *et al.,* 1996). Actilight consumption decreased basal hepatic glucose production but had no detectable effect on insulin-stimulated glucose metabolism in healthy subjects. The authors suggested that the colonic fermentation pattern of undigestible carbohydrates may be relevant to predicting their metabolic effects.

10.5   OTHER POTENTIAL METABOLIC EFFECTS OF FRUCTOOLIGOSAC-CHARIDES

Roberfroid (1995) has suggested that intake of inulin may improve absorption of minerals such as iron by increasing the amount in the ferrous state as compared to the ferric state. This hypothesis has not been extensively tested in humans but is of interest because of marginal intakes of minerals such as calcium, iron, and zinc in the population. Coudray *et al.,* (1997) found that inulin improved calcium balance in healthy young men. Apparent absorption and balance of Mg, Fe, and Zn were not significantly altered by the ingestion of inulin.

Another interesting effect of FOS could be on immune function. It is established that a healthy gut microflora is critical for protecting against the unwanted establishment of pathogens (Elmer *et al.,* 1996). Increased levels of certain bacteria in the gut have been suggested as causative agents in colonic disorders such antibiotic-associated colitis, inflammatory bowel diseases, and colorectal cancer. Overgrowth of pathogenic bacteria such clostridia and *E. coli* are major factors in gut disorders. Prebiotics such as FOS may play a role in prevention or treatment. Soy fibre, another prebiotic, has been shown to delay disease onset and prolong survival in experimental *Clostridium difficile* ileocecitis (Frankel *et al.,* 1994).

11 Epidemiological evidence for a protective effect of dietary fibre and oligosaccharide consumption on chronic disease

Epidemiological studies that relate diet and disease generally rely on food frequency questionnaires or records to estimate intake. These need to be translated into nutrient values and depend on databases for this calculation. Dietary fibre values have been added to these data bases. Values for oligosaccharides have not. Therefore, most published data is on dietary fibre, not oligosaccharides. Thus, this review will be limited to dietary fibre/disease relationships.

11.1 LARGE BOWEL CANCER

Extensive evidence supports the theory that dietary fibre may be protective against large bowel cancer. Correlation studies that compare colorectal cancer incidence or mortality rates among countries with estimates of national dietary fibre consumption suggest that intake may be protective (Bingham, 1990). Case-control studies are considered stronger than population-based correlation studies, since individual exposure to dietary variables can be related to individual outcome. Studies reviewed by both Bingham (1990) and Willett (1989) support the protective role of dietary fibre in colorectal cancer.

Data collected from 20 populations in 12 countries showed that the average stool weight varied from 72 to 470 g/d and was inversely related to colon cancer risk (Cummings *et al.*, 1992). When results from 13 case-control studies of colorectal cancer rates and dietary practices were pooled, the authors concluded that the data provided substantive evidence that intake of fibre-rich foods was inversely related to risks of both colon and rectal cancer (Howe *et al.*, 1992). The authors estimated that risk of colorectal cancer in the U.S. population could be reduced by about 31% through an average increase in fibre intake from food sources of about 13 g/d.

Alberts *et al.*, (1990) studied the effects of wheat bran fibre (an additional 13.5 g/d as wheat bran cereal) on rectal epithelial cell proliferation in patients with resection for colorectal cancers. They found that the fibre cereal inhibited DNA synthesis and rectal mucosal cell proliferation in this high-risk group, which they argued should be associated with reduced cancer risk. They suggested that this might be used as a chemopreventive agent for colorectal cancers.

In a randomised trial of intake of fat, fibre and beta carotene to prevent colorectal adenomas, patients on the combined intervention of low fat and added wheat bran had no large adenomas at both 24 and 48 months, a statistically significant finding (MacLennan *et al.*, 1995). In an animal model, wheat bran and soluble fibre psyllium had a synergistic effect for protection against colon cancer (Alabaster *et al.*, 1996). Greenwald *et al.*, (1995) suggested that dietary fibre may prevent colon cancer by lowering colorectal mucosal proliferation.

11.2 BREAST CANCER

Limited epidemiological evidence has been published on fibre intake and risk of human breast cancer. Since the fat and fibre content of the diet are generally inversely related, it is difficult to separate independent effects of these nutrients and most research has focused on the fat and breast cancer hypothesis. As reviewed by Rose (1992), international comparisons show an inverse correlation between breast cancer death rates and the consumption of fibre-rich foods. An interesting exception to the high-fat diet hypothesis in breast cancer, is seen in Finland where intake of both fat and fibre is high and the breast cancer mortality rate is considerably lower than in the USA or other Western countries consuming high-fat diets. Rose (1992) suggested that a high level of fibre in the rural Finnish diet may modify breast cancer risk associated with a high-fat diet.

A meta-analysis of 12 case-control studies of dietary factors and risk of breast cancer found that a high dietary fibre intake was associated with reduced breast cancer risk (Howe et al., 1990). Baghurst and Rohan (1994) found that highest risk reductions in breast cancer were associated with intake of dietary fibres that had high densities of mannose from insoluble non-starch polysaccharides (NSP) and of glucose from soluble NSP. Thus the type of dietary fibre, as well as other properties such as particle size, chemical composition, food matrix, hydration, fermentability, etc. are important factors in determining protection. Dietary fibre intake has also been linked to lowered risk of benign proliferative epithelial disorders of the breast (Baghurst and Rohan, 1995). Not all studies find a relationship between dietary fibre intake and breast cancer incidence however, including a prospective cohort study reported by Willett et al., (1992).

Considerable evidence suggests that both breast and colon cancers are hormone-mediated diseases (Potter, 1992). Case-control studies have found elevated levels of circulating oestrogens in breast cancer patients (Rose, 1992). Interest in the diet-hormone relationship is increasing since it is generally accepted that oestrogens play a key role in promoting the development and subsequent progression of breast cancer (Rose, 1993).

Few studies have examined the effects of dietary fibre on hormone metabolism, while fat content of the diet was held constant. Rose et al., (1991) reported that when wheat bran was added to the normal diet of premenopausal women, it significantly reduced serum oestrogen concentrations, while neither corn bran nor oat bran had any effect. Dietary fibre intake was increased from about 15 to 30 g/d in this study, an increase similar to that recommended by the National Cancer Institute. Goldin et al., (1994) reported that high fibre, low fat diet significantly decreased serum concentrations of estrone, estrone sulphate, testosterone, and sex hormone binding globulin in premenopausal women. Dietary fibre also caused a lengthening of the menstrual cycle by 0.72 day and a lengthening of the follicular phase by 0.85 day, changes thought to lower the overall risk of developing breast cancer. Bagga et al., (1995) also found that an increased intake of dietary fibre significantly decreased serum estradiol and estrone, which may lower breast cancer risk.

11.3   DIETARY FIBRE AND OTHER DISEASES

Dietary fibre may lower the risk of coronary heart disease by decreasing serum lipids, lowering blood pressure, improving glucose metabolism, and aiding in weight maintenance (Geil and Anderson, 1992). Soluble fibres appear most effective in lowering serum cholesterol. A meta-analysis of 10 trials evaluating the lipid lowering effects of oats supported the notion that they had hypo-cholesterolemic effects, independent of other dietary changes (Ripsin *et al.,* 1992). Lesser cholesterol-lowering effects are found with vegetables, fruits, barley, and rice bran. The most effective fibre sources for lowering serum cholesterol include psyllium, pectin, guar, and soy polysaccharide. Large epidemiological studies find protection against cardiovascular disease with vegetable, fruit, and cereal fibre (Rimm *et al.,* 1996).

Soluble fibres have a greater potential to alter serum lipid levels than insoluble forms. In the recent review by Glore *et al.* (1994), a significant reduction in serum cholesterol by soluble fibre was found in 68 of the 77 (88%) human studies reviewed. Often, soluble fibres also decrease LDL-cholesterol while maintaining HDL-cholesterol (Basu *et al.,* 1995). Furthermore, soluble fibres lower serum cholesterol even when the main dietary modifiers of blood lipids, saturated fats and cholesterol, are greatly reduced in the diet (Jenkins *et al.,* 1993). Multiple mechanisms appear to be involved in the hypocholesterolemic response, and those for cholesterol lowering may vary considerably among different dietary fibre sources.

Some clinical research has suggested that dietary fibre may have a role in improving glycemic control in individuals with diabetes. Dietary fibre, especially soluble fibre, may delay glucose absorption and reduce insulin requirements in both insulin-dependent and non-insulin-dependent diabetes mellitus. Obese persons with diabetes often respond to high-fibre diets with weight loss and decreases in insulin requirements (Guevin *et al.,* 1996).

Guevin *et al.,* (1996) compared the postprandial glucose, insulin, and lipid responses when noninsulin-dependent diabetic subjects consumed 4 different fibre diets. The diets contained two levels of total dietary fibre (10 g vs. 20 g) and two soluble: insoluble fibre ratios (1: 4 vs. 2: 3). The incremental area under the curve for glucose and insulin was lowered after consuming 20 g compared to 10 g fibre, but was not affected by the soluble: insoluble ratio. The authors concluded that the proportion of soluble to insoluble fibre in cereal and fruit would not necessarily predict effects of fibre on glycemic response, while the overall quantity of fibre did appear to affect postprandial glucose metabolism.

High fibre diets tend to be lower in fat and, thus calories, and should be appropriate for weight control. High fibre cereals eaten at breakfast time lowered caloric intake when subjects were served a buffet lunch compared to not consuming a high fibre breakfast (Levine *et al.,* 1989). Thus, high fibre diets may affect satiety, but whether that translates into weight loss is not known. Despite the theoretical reasons whereby high-fibre diets should aid weight loss, data from clinical trials are inconsistent and long-term clinical trials of dietary fibre effectiveness for weight management have not been conducted.

12   Potential negative effects of dietary fibre

Negative effects of dietary fibre may include a lower absorption of vitamins, minerals, proteins, and kcalories. It is unlikely that healthy adults who consume fibre amounts within the recommended range will have problems with nutrient absorption. However, there is concern that dietary fibre recommendations of 25 g/d are not appropriate for children and the elderly, since little research has been conducted in these populations.

Generally, dietary fibre in recommended amounts is thought to normalise transit time and should help when either constipation or diarrhoea is present. However, case histories have reported diarrhoea when excessive amounts of dietary fibre are consumed (Cooper and Tracey, 1989; Saibil, 1989). As such, it is difficult to individualise intakes based on bowel function measurement. Thus, stool consistency cannot be used as an accuarate bench mark for appropriate intake.

Dietary fibre is fermented in the gut resulting in gas production which may be related to complaints of distention or flatulence. In a trial in our laboratory (Slavin *et al.*, 1985), subjects who were on an enteral diet, with and without soy polysaccharide, reported no significant differences in subjective measurements of bowel function among diets containing 0, 30 and 60 g of soy polysaccharide per day. Patil *et al.*, (1985) found that either fibre-free or fibre-containing diets were equally tolerated. Generally, the gas and bloating sometimes associated with fibre consumption subsides as the body adapts to increased fibre. Fibre intake should be accompanied by elevated fluid intake and should be gradually increased to allow the gastrointestinal tract an adaptation period to the additional fibre intake. Intestinal gas formation is also a common complaint in research trials with oligosaccharides (Stone-Dorshow and Levitt, 1987; Pederson *et al.*, 1997). Since these studies tend to be short-term, it is not known how long it takes the gastrointestinal tract to adapt appropriately.

Other potential concerns are the effects of dietary fibre on micronutrient absorption, especially minerals. Heymsfield *et al.*, (1988) found reduced rates of apparent absorption for phosphorus, magnesium, and zinc when 12.4 grams of soy polysaccharide was added to a fibre-free elemental diet. Taper *et al.*, (1988) found that 40 g of soy fibre caused a negative balance for copper and iron, whereas zinc, calcium, and magnesium remained positive. Zinc, copper and magnesium absorption was lower on a high fibre, high phytate conventional food diet (Knudson *et al.*, 1988). Since the phytate content of most enteral diets should be minimal, binding effects ought not to be a problem. When mineral intake is above recommended levels, the effects of fibre on mineral absorption appear to be insignificant (Hosig *et al.*, 1996).

Even less information is available on the effects of dietary fibre on vitamin absorption. Shinnick *et al.*, (1989) found that 15 g/d soy fibre added to a fibre-free formula apparently reduced folate absorption. In this study, the dietary folate content was double the recommended daily allowance.

13   Conclusions

In Western populations, consumption of complex carbohydrates such as oligosaccharides and dietary fibre has decreased and may be associated with increased risk of chronic disorders like heart disease and cancer. Despite many years of research and nutritional education, dietary fibre intakes are not increasing. Since dietary fibre and oligosaccharides share many positive physiological properties, it may be more feasible to increase intake of oligosaccharides than dietary fibre. This is especially true since isolated oligosaccharides from chicory and other natural products are now being incorporated into processed foods.

Our nutrition message must continue to be increased consumption of foods high in complex carbohydrates, including resistant starch, oligosaccharides, and dietary fibre. As many consumers depend on processed foods as the mainstay of their diet, efforts should be made to increase levels in popular foods so as to assist consumers in obtaining recommended levels of unavailable carbohydrate.

References

Alabaster, O., Tang, Z. and Shivapurkar, N. (1996) Dietary fibre and the chemopreventive modelation of colon carcinogenesis. *Mutation Research* **350** (1), 185-97.

Alberts, D.S., Einspahr, J. and Rees-McGee, S. (1990) Effects of dietary wheat bran fibre on rectal epithelial cell proliferation in patients with resection for colorectal cancers. *Journal of Clinical Investigation* **82** (15), 1280-85.

Alles, M.S., Hautvast, J.G.A., Nagengast, F.M., Hartemink, R., Lan Laere, K.M.J. and Jansen, J.B.M.J. (1996) Fate of fructo-oligosaccharides in the human intestine. *British Journal of Nutrition* **76**, 221-21.

Bach Knudsen, K.E. and Hessov, I. (1995) Recovery of inulin from Jerusalem artichoke (Helianthus Tuberosus L.) in the small intestine of man. *British Journal of Nutrition* **74**, 101-13.

Bagga, D., Ashley, J.M., Geffrey, S.P., Wang, H., Barnard, J., Korenman, S. and Heber D. (1995) Effects of very low fat, high fibre diet on serum hormones and menstrual function. *Cancer* **76**, 2491-96.

Baghurst, P.A. and Rohan, T.E. (1995) Dietary fibre and risk of benign proliferative epithelial disorders of the breast. *International Journal of Cancer* **63**, 481-85.

Baghurst, P.A. and Rohan, T.E. (1994) High-fibre diets and reduced risk of breast cancer. *International Journal of Cancer* **56** (2), 173-76.

Basu, T.K. and Ooraikul, B. (1995) Lipid lowering effects of dietary fibre. *Journal of Clinical Biochemical Nutrition* **18** (1), 1-9.

Bingham, S.A. (1990) Mechanisms and experimental and epidemiological evidence relating dietary fibre (non-starch polysaccharides) and starch to protection against large bowel cancer. *Proceedings of Nutrition Society* **49** (2), 153-71.

Bouhnik, Y., Flourie, B., Riottot, M., Bisetti, N., Gailing, M., Guibert, A., Bornet, F. and Rambaud, J. (1996) Effects of fructo-oligosaccharides ingestion on fecal bifidobacteria and selected metabolic indexes of colon carcinogenesis in healthy humans. *Nutrition Cancer* **26**, 21-29.

Buddington, R.K., Williams, C.H., Chen, S. and Witherly, S.A. (1996) Dietary supplement of neosugar alters the fecal flora and decreases activities of some reductive enzymes in human subjects. *American Journal of Clinical Nutrition* **63**, 709-16.

Campbell, J.M., Bauer, L.L., Fahey, G.C., Hogarth, A.J.C.L., Wolf, B.W. and Hunter, D.E. (1997) Selected fructooligosaccharide (1-kestose, nystose, and 1(f)-beta fructofuranosylnystose) composition of foods and feeds. *Journal of Agriculture Food Chemistry* **45**, 3076-82.

Cooper, S.G. and Tracey, E.J. (1989) Small bowel obstruction caused by oat bran bezoar. *New England Journal of Medicine* **320** (7), 1148-49.

Coudray, C., Ballanger, J., Castigliadelavaud, C., Remsy, C., Vermorel, M. and Rayssignuier, Y. (1997) Effect of soluble or partly soluble dietary fibres supplementation on absorption and balance of calcium, magnesium, iron and zinc in healthy young men. *European Journal of Clinical Nutrition* **51**, 375-80.

Cummings, J.H. (1982) Consequences of the metabolism of fibre in the human large intestine. In Vahouny, G., Kritchevsky, D. (eds.): Dietary Fibre in Health and Disease, New York, Plenum Press.

Cummings, J.H., Bingham S.A., Heaton K.W. and Eastwood M.A. (1992) Fecal weight, colon cancer risk and dietary intake of nonstarch polysaccharides (dietary fibre). *Gastroenterology* **103** (6), 1783-89.

Cummings, J.H. and Macfarlane, G.T. (1997) Colonic microflora: Nutrition and health. *Nutrition* **13**, 476-78.

Djouzi, Z., Andrieuz, C., Pelenc, V., Somarriba, S., Popot, F., Paul, F., Monsan, P. and Szylit (1995) Degradation and fermentation of alpha-gluco-oligosaccharides by bacterial strains from human colon: in vitro and in vivo studies in gnotobiotic rats. *Journal of Applied Bacteriology* **79**, 117-27.

Elmer, G.W., Surawica, C.M. and McFarland, L.V. (1996) Biotherapeutic agents. A neglected modality for the treatment and prevention of selected intestinal and vaginal infections. *JAMA* **275**, 870-76.

Frankel, W.L., Choi, D.M., Z..hang, W., Rother, J.A., Don, S.H., Alfonso, J.J., Lee, J.H., Klurfeld, D.L. and Rombeau, J.L. (1994) Soy fibre delays disease onset and prolongs survival in experimental *Clostridium difficile* ileocecitis. *Journal of Parenteral and Enternal Nutrition* **18**, 55-61.

Fredstrom, S.B., Baglien, K.S., Lampe, J.W. and Slavin, J.L. (1991) Determination of the fibre content of enteral feedings. *JPEN* **15** (4), 450-53.

Fredstrom, S.B., Lampe, J.W., Jung, H.J. and Slavin, J.L. (1994) Apparent fibre digestibility and fecal short-chain fatty acid concentrations with ingestion of two types of dietary fibre. *JPEN* **18** (1), 14-19.

Garleb, K.A., Snook, J.T., Marcon, M.J., Wolf, B.W. and Johnson, W.A. (1996) Effect of fructooligosaccharide containing enteral formulas on subjective tolerance factors, serum chemistry profiles, and faecal bifidobacteria in healthy adult male subjects. *Microbial Ecology in Health and Disease* **9**, 279-85.

Geil, P.B. and Anderson, J.W. (1992) Health benefits of dietary fibre. *Medicine Exercise Nutrition Health* **1** (5), 257-71.

Gibson, G.R., Beatty, E.R., Wang, X. and Cummings, J.H. (1995) Selective stimulation of bifidobacteria in the human colon by oligofructose and inulin. *Gastroenterology* **108**, 975-82.

Gibson, G.R. and Roberfroid, M.B. (1995) Dietary modulation of the human colonic microbiota: Introducing the concept of prebiotics. *Journal of Nutrition* **125**, 1401-12.

Gibson, G.R. and Wang X. (1994) Enrichment of bifidobacteria from human gut contents by oligofructose using continuous culture. *FEMS Microbiology Letters* **118**, 121-28.

Glore, S.R., Van Treeck, D., Knehans, A.W. and Guild M. (1994) Soluble fibre and serum lipids: a literature review. *Journal of American Dietary Association* **94** (4), 425-36.

Goldin, B.R., Woods, M.N.L., Spiegelman, D., Longscope, C., Morrill-LaBrode, A., Dwyer, J.T., Gualtier, L.J., Hertzmark, E. and Gorbach, S.L. (1994) The effect of dietary fat and fibre on serum estrogen concentrations in premenopausal women under controlled dietary conditions. *Cancer* **74** (3 Suppl.), 1125-31.

Greenwald, P., Kelloff, G.J., Boone, CW and Mcdonald, S.S. (1995) Genetic and cellular changes in colorectal cancer - proposed targets of chemopreventive agents. *Cancer Epidemiological Biomarkers* **4** (7), 691-702.

Guevin, N., Jacques, H., Nadeau, A. and Galibois, I. (1996) Postprandial glucose, insulin, and lipid responses to four meals containing unpurified dietary fibre in noninsulin-dependent diabetes mellitus (NIDDM), hypertriglyceridemic subjects. *Journal of American College Nutrition* 15 (4), 389-96.

Heymsfield, S.B., Roongspisuthipong, C., Evert, M., Casper, K, Heller, P. and Akrabawi, S.S. (1988) Fibre supplementation of enteral formulas: Effects on the bioavailability of major nutrients and gastrointestinal tolerance. *JPEN* 12 (3), 265-73.

Hillemeier, C. (1995) An overview of the effects of dietary fibre on gastrointestinal transit. *Pediatrics* 96, 997-99.

Hoebregs, H. (1997) Fructans in foods and food products, ion-exchange chromatographic method - collaborative study. *Journal AOAC International* 80, 1029-37.

Hosig, K.G., Shinnick, F.L., Johnson, M.D., Story, J.A. and Marlett, J.A. (1996) Comparison of large bowel function and calcium balance during soft wheat bran and oat bran consumption. *Cereal Chemistry* 73 (3), 392-98.

Hosoya, N., Dhorranintra, B. and Hidaka, H. (1988) Utilization of [U-14C] fructooligosaccharides in man as energy resources. *Journal of Clinical Biochemical Nutrition* 5, 67-74.

Howe, G.R., Benito, E., Castelleto, R., *et al.* (1992) Dietary intake of fibre and decreased risk of cancers of the colon and rectum: Evidence from the combined analysis of 13 case-control studies. *Journal of Clinical Investigation* 84 (24), 1887-96.

Howe, G.R., Hirohata, T. and Hislop, T.G. (1990) Dietary factors and risk of breast cancer: combined analysis of 12 case-control studies. *Journal of Clinical Investigation* 82 (7), 561-69.

Jenkins, D.J.A., Wolever, T.M.S., Rao, A.V., Hegele, R.A., Mitchell, St, Ransom, T.P.P., Boctor, D.L., Spadafora, PJ, Jenkins, A.L., Mehling, C., Katzman, L., Connelly, P.W., Story, J.A., Furumoto, E.J., Corey, P. and Wursch, P. (1993) Effect on blood lipids of very high intakes of fibre in diets low in saturated fat and cholesterol. *New England Journal of Medicine* 329 (1), 21-26.

Kapadia, S.A., Raimundo, A.H., Grimble, G.K., Aimer, P. and Silk, D.B.A. (1995) Influence of three different fibre-supplemented enteral diets on bowel function and short-chain fatty acid production. *JPEN* 19 (1), 63-68.

Klessen, B., Sykura, B., Zunft, H. and Blaut, M. (1997) Effects of inulin and lactose on fecal microflora, microbial activity, and bowel habit in elderly constipated persons. *American Journal of Clinical Nutrition* 65, 1397-1402.

Knudsen, E., Sandstrom, B. and Solgaard, P (1996) Zinc, copper and magnesium absorption from a fibre-rich diet. *Journal Trace Elements Medicine Biology* 10, 68-76.

Lanza, E. and Butrum, R., 1986. A critical review of food fibre analysis and data. *Journal of American Dietary Association* 86 (6), 732-43.

Lee, S.C., Rogriguez, F., Storey, M., Farmakalidis, E. and Prosky, L, (1995) Determination of soluble and insoluble dietary fibre in psyllium-containing cereal products. *Journal AOAC International* 78 (3), 724.

Levine, A.S., Tallman, J.R., Grace, M.K., Parker, S.A., Billington, C.J. and Levitt M.D. (1989) Effect of breakfast cereals on short-term food intake. *American Journal of Clinical Nutrition* 50 (6), 1303-7.

Luo, J., Rizkala, S., Alamowitch, C., Boussairi, A, Blayo, A., Barry, J., Laffitte, A., Guyon, F., Bornet, F.R.J. and Slama, G. (1996) Chronic consumption of short-chain fructooligosaccharides by healthy subjects decreased basal glucose production but had no effect on insulin-stimulated glucose metabolism. *American Journal of Clinical Nutrition* 63, 939-45.

MacLennan, R., Macrae, F., Bain, C., Battistutta, D., Chapuis, P., Gratten, H., Lambert, J., Newland, R.C., Ngu, M., Russell, A., Ward, M. and Wahlqvist, M.L. (1995) Randomized trial of intake of fat, fibre, and beta carotene to prevent colorectal adenomas. *Journal National Cancer Institute* 87, 1760-66.

Marlett, J.A. and Cheung, T. (1997) Database and quick methods of assessing typical dietary fibre intakes using data for 228 commonly consumed foods. *Journal of American Dietary Association* 97, 1139-51.

Marlett, J.A. and Slavin, J.L. (1997) Position of the American Dietetic Association: Health implications of dietary fibre. *Journal of American Dietary Association* **97**, 1157-59.

Marlett, J.A. and Vollendorf, N.W. (1994) Dietary fibre content and composition of different forms of fruits. *Food Chemistry* **51**, 39-44.

Marlett, J.A. (1992) Content and composition of dietary fibre in 117 frequently consumed foods. *Journal of American Dietary Association* **92** (2), 175-86.

McBurney, M.I. and Thompson, L.U. (1991) Dietary fibre and total enteral nutrition: fermentative assessment of five fibre supplements. *JPEN* **15** (3), 267-70.

McBurney, M.I. and Thompson, L.U. (1989) In vitro fermentabilities of purified fibre supplements. *Journal Food Science* **54** (2), 347-50.

McCance, R.A. and Widdowson, E.M. (1955) Old thought and new work on breads white and brown. *Lancet* **2**, 205-10.

McKellar, R.C., Modler, H.W. and Mullin, J. (1993) Characterization of growth and inulinase production by *Bifidobacterium* spp. on fructooligosaccharides. *Bifidobacteria Microflora* **12**, 75-86.

Mitsuoka, T., Hidaka, H. and Eida, T. (1987) Effect of fructo-oligosaccharides on intestinal microflora. *Die Nahrung* **31**, 5-6, 426-36.

Molis, C., Flourie, B, Ouarne, F., Gailing, M., Lartigue, S, Guibert, A., Bornet, F. and Galmiche, J. (1996) Digestion, excretion, and energy value of fructooligosaccharides in healthy humans. *American Journal of Clinical Nutrition* **64**, 324-28.

Ozick, L.A., Salazar, C.O. and Donelson, S.S. (1994) Pathogenesis, diagnosis, and treatment of dieverticular disease of the colon. *Gastroenterology* **2** (4), 299-10.

Patil, D.H., Grimble, G.K., Keohane, P., *et al.* (1985) Do fibre-containing enteral diets have advantages over existing low residue diets? *Clinical Nutrition* **4**, 67.

Pedersen, A., Snadstrom, B. and Vanamelsvoort, J.M.M. (1997) The effect of ingestion of inulin on blood lipids and gastrointestinal symptoms in healthy females. *British Journal of Nutrition* **78**, 215-22.

Pilch, S. (1987) Physiological Effects and Health Consequences of Dietary Fibre, Bethesda, M.D. Life Sciences Research Office, Federation of American Societies for Experimental Biology.

Potter, J.D. (1992) Reconciling the epidemiology, physiology, and molecular biology of colon cancer. *JAMA* **268** (12), 1573-77.

Prosky, L. (1997) Dietary fibre and complex carbohdrates. *Journal AOAC International* **80**, 138-40.

Quemener, B., Thibault, J.F. and Coussement, P. (1997) Integration of inulin determination in the AOAC method for measurement of total dietary fibre. *International Journal of Biological Macromolecules* **21**, 175-78.

Read, N.W., Celik, A.F. and Katsinelos, P. (1995) Constipation and incontinence in the elderly. *Journal of Clinical Gastroenterology* **20** (1), 61-70.

Rimm, E.B., Ascherio, A., Giovannucci, E., Spiegelman, D., Stampfer, M.J. and Willett, W.C. (1996) Vegetable, fruit, and cereal fibre intake and risk of coronary heart disease among men. *JAMA* **275**(6), 447-51.

Ripsin, C.M., Keenan, J.J., Jacobs, D.R., Elmer, P.J., Welch, R.R., Van Horn L., Liu K., Turnbull, W.H., Thye, F.W., Kestin, M. (1992) Oat products and lipid lowering: a meta-analysis. *JAMA* **267** (24), 3317-25.

Roberfroid, M. (1993) Dietary fibre, inulin, and oligofructose: a review comparing their physiological effects. *Critical Reviews in Food Science and Nutrition* **33**, 103-48.

Roberfroid, M.R. (1995) Chicory fructooligosaccharides: A colonic food with prebiotic activity. The World of Ingredients pp. 42-44.

Roediger, W.E. (1994) Famine, fibre, fatty acids, and failed colonic absorption: Does fibre fermentation ameliorate diarrhea? *JPEN* **18** (1), 4-8.

Rose, D.P., Goldman, M., Connolly, J.M., Strong, L.E. (1991) High-fibre diet reduces serum estrogen concentrations in premenopausal women. *American Journal of Clinical Nutrition* **54** (3), 520-25.

Rose, D.P. (1993) Diet, hormones, and cancer. *Annual Review Public Health* **14**, 1-17.

Rose, D.P. (1992) Dietary fibre, phytoestrogens, and breast cancer. *Nutrition* **8** (1), 47-51.

Rumessen, J.J., Bode, S, Hamberg, O. and Gudmand-Hoyer, E. (1990) Fructans of Jerusalem artichokes: intestinal transport, absorption, fermentation, and influence on blood glucose, insulin, and C-peptide response in healthy subjects. *American Journal of Clinical Nutrition* **52**, 675-81.

Saibil, F. (1989) Diarrhea due to fibre overload. *New England Journal of Medicine* **320** (9), 599.

Scheppach, W.M. and Bartram, H.P. (1993) Experimental evidence for and clinical implications of fibre and artificial enteral nutrition. *Nutrition* **9** (5), 399-405.

Schneeman, B. (1986) Dietary fibre: Physical and chemical properties, methods of analysis, and physiological effects. *Food Technology* **40** (2), 104-10.

Shinnick, F.L., Hess, R.L., Fischer, M.H. and Marlett, J. (1989) Apparent nutrient absorption and upper gastrointestinal transit with fiber-containing enteral feedings. *American Journal of Clinical Nutrition* **49** (3), 471-75.

Slavin, J.L., Nelson, N.L., McNamara, E.A. and Cashmere, K. (1985) Bowel function of healthy men consuming liquid diets with and without dietary fibre. *JPEN* **9** (3), 317-21.

Slavin, J.L. (1987) Dietary fibre: Classification, chemical analyses, and food sources. *Journal of American Dietary Association* **87** (9), 1164-71.

Spiegel, J.E., Rose, R., Karabell, P., Frankos, V.H., Schmitt, D.F. (1994) Safety and benefits of fructooligosaccharides as food ingredients. *Food Technology* January, 85-89.

Stone-Dorshow, T. and Levitt, M.D. (1987) Gaseous response to ingestion of a poorly absorbed fructooligosaccharide sweetener. *American Journal of Clinical Nutrition* **46**, 61-65.

Sunvold, G.D., Titgemeyer, E.C., Bourquin, L.D., Fahey, G.C. and Garleb, K.A. (1995) Alteration of the fibre and lipid components of defined-formula diet: effects on stool characteristics, nutrient digestibility, mineral balance, and energy metabolism in humans. *American Journal of Clinical Nutrition* **62**, 1251-60.

Taper, L.J., Milam, R.S., McCallister, M.S., Bowen, P.E. and Thye, F.W. (1988) Mineral retention in young men consuming soy-fibre-augmented liquid-formula diets. *American Journal of Clinical Nutrition* **48** (2), 305-11.

Tomomatsu, H. (1994) Health effects of oligosaccharides. *Food Technology* October, 61-65.

Trowell, H. (1974) Definitions of fibre. Lancet 1(856), 503.

US Department of Health and Human Services. Healthy People 2000, Public Health Service, (1990).

Valazquez, O.C., Lederer, H.M. and Rombeau, J.L. (1996) Butyrate and the colonocyte - implications for neoplasia. *Digestive Disease Science* **14** (4), 727-39.

Van Loo, J., Coussement, P., De Leenheer, L., Hoebregs and H., Smits, G. (1995) On the presence of inulin and oligofructose as natural ingredients in the Western diet. *Critical Reviews in Food Science and Nutrition* **35**, 525-52.

Wang, X. and Gibson, G.R. (1993) Effects of the in vitro fermentation of oligofructose and inulin by bacteria growing in the human large intestine. *Journal of Applied Bacteriology* **75**, 373-80.

Willett, W. (1989) The search for the causes of breast and colon cancer. *Nature* **338** (6214), 389-94.

Willett, W.C., Hunter, D.J., Stampfer, M.J., Colditz, G., Manson, J.E., Spiegelman, D., Rosner, B., Hennekens, C.H. and Speizer, F.E. (1992) Dietary fat and fibre in relation to risk of breast cancer. An 8-year follow-up. JAMA **268** (15), 2037-44.

Yamashita, K, Kawai, K and Itakura, M. (1984) Effect of fructo-oligosaccharides on blood glucose and serum lipids in diabetic subjects. *Nutrition Research* **4**, 961-66.

Yun, J.W. (1996) Fructooligosaccharides - occurrence, preparation, and application. *Enzyme Microbial Technology* **19**, 107-17.

CHAPTER 9

Taxonomy and Systematics
of Predominant Gut Anaerobes

PAUL A. LAWSON
Food Microbial Sciences Unit, Department of Food Science and Technology, The University of Reading, Reading, UK

1   Introduction

Historically, microorganisms have been identified using morphological and phenotypic criteria. As a consequence, classification frameworks have been constructed with limited and often fragmentary information. A major obstacle for microbiologists in the identification and classification of organisms is the ability to apply individual tests universally across all bacterial groups. For example, biochemical tests incorporating carbohydrates have been of great use in the affiliation of isolates to a particular taxon, but are of little value in the identification of anaerobic, assacharolytic organisms which use alternative biochemical pathways to derive their energy requirements (Busse *et al.*, 1996; Vandamme *et al.*, 1996). The outcome was that many taxa were created using a few poorly defined criteria and have been used as a repository for organisms that could not be classified with certainty.

Classification schemes are never static with new information bearing on the taxonomy of bacteria continually being generated by researchers. Each generation of microbiologists relies on the techniques available to them at that point in time. The result has been that microbiology was not constructed on a "natural" phylogenetically valid system of classification. Because each organism is the product of its history, a knowledge of phylogenetic relationships is essential for understanding the nature of the organism. Although put forward as a conceptual idea (Haeckel, 1866; van Neil, 1946), it was not until the 1950's that molecular sequences were used to determine evolutionary relationships. Zuckerkandl and Paulings's seminal article "Molecules as documents of evolutionary history" provided a major impetus in the study of molecular evolution (Zuckerkandl and Pauling, 1965). Nevertheless, it was not until the end of the 1970's that rRNA sequences were shown to provide a key to prokaryotic phylogeny. The pioneering work

G.R. Gibson and M.B. Roberfroid (eds.), Colonic Microbiota, Nutrition and Health, 149-166.
© 1999 *Kluwer Academic Publishers. Printed in the Netherlands.*

and vision of Woese (1975, 1987) was accepted and taken up by others (*e.g.* Lane *et al.*, 1985, Pace *et al.*,1985). The rRNA molecule is universally distributed, functionally homologous across all prokaryotes and has regions that are conserved - thus allowing sequences to be aligned for comparison and thereby fulfilling the criteria of an excellent chronometer (Woese, 1987). Unlike physiological and phenotypic traits, the genetic material is not affected by culture media or growth conditions. Typically, the number of differences in nucleotide sequences in a particular gene that are counted, the greater the difference between genes, the more evolutionary separated are the pairs of organisms. The evolutionary distances, taken as a fraction of sequence differences between pairs in a collection of sequences, is used to construct phylogenetic trees (Felsenstein, 1982; Fitch and Margoliash, 1967; Jukes and Cantor, 1969). Because of the slow rate of evolutionary change of small-subunit rRNA gene sequences, as well as ease of extraction and manipulation they have become the molecule of choice, and form the "gold standard" in construction of phylogenetic classification schemes. 16S rRNA gene sequence analysis has not only facilitated new insights into the phylogenetic interrelationships of organisms but has provided molecular systematists with an immensely powerful means for characterising new diversity within any given microbiological environment.

The molecular approach has had major repercussions not only in taxonomy but also, for example, microbial ecology. Investigations by Pace *et al.* (1985) have demonstrated that microorganisms could be (phylogenetically) identified directly in their niches-through a combination of rRNA gene cloning and sequencing. The design and use of rRNA-directed "phylogenetic stains" or "phylotypes" has given new insights into microbial communities. These advances in molecular tools have encouraged their use in a wide range of biological niches (Delong 1992; Liesack and Stackebrandt, 1992; Pace *et al.*, 1985, Pace, 1996) including the gastrointestinal tract of man and animals (Dore, 1995; Stahl *et al.*, 1988 Wilson and Blitchington, 1996).

The purpose of this article is to review and, where appropriate, use examples in the context of identification, taxonomy and classification of organisms found in the human gastrointestinal tract. A comprehensive review of the phylogeny of the enormous diversity of the gut is beyond the scope of this article. However, groups thought to have significant biological relevance are discussed, namely the bacteroides, clostridia and lactic acid bacteria.

2   Strategies based on rRNA sequences for characterising microbial communities

The human colon contains in excess of 10^{11} bacterial cells per gram of contents belonging to over 400 different species (Finegold, 1983; Moore and Holdeman, 1974). Despite the large numbers of published articles on the identification, classification and microecology of this habitat, almost all the information is derived from culture-based methodologies which have many limitations (Amann, *et al.*,1990; Ward *et al.*, 1990). Studies in the 1970s and 1980s estimated that, even with the use of selective media and strictly anaerobic techniques, numbers ranged between 10-60% of human faecal organisms recovered when compared to direct microscopic counts (Drasar and Borrow, 1985; Holdeman *et al.*, 1976). More recent investigations have shown that these figures could be greatly over estimated and that we may not culture the vast majority (> 99%) of naturally occurring microbes using standard techniques (Amann *et al.*, 1995; Pace, 1996; Ward *et al.*, 1990).

Data amassed from 16S rRNA gene sequences have been derived from native rRNA or, more recently, from rRNA genes amplified from the DNA directly using the polymerase chain reaction (PCR). In the majority of cases the starting genetic material has been derived from cultured cells - a tedious and time consuming process. Advances in molecular biology have removed the need to firstly culture the organism. Total DNA can be directly extracted from the environment containing microbial populations (Fig. 1). The DNA can then be analysed by using established cloning and sequencing strategies (Maniatis, 1982) to investigate the microorganisms present (Delong, 1992; Hugenholtz and Pace, 1996 Pace, 1996, Pace *et al.*, 1985). Attempting to simplify this, a second strategy is employed which involves PCR to amplify rRNA genes selectively from the environment and sequenced accordingly. Because rRNA is highly conserved a further method is to synthesise primers universal primers that anneal to all three phylogenetic domains (Archaea, Bacteria and Eucarya). These primers are then used to amplify rDNAs of all organisms present before individual types are separated from the mixture and then sequenced. Although PCR-amplified cloning analysis of microbial communities provides a convent and rapid alternative to standard cloning techniques, caution is advised. PCR artefacts, such as mistakes in the amplification and the possible production of chimera sequences (a sequence which raisers from the combination of two or more genes) may arise. Sufficient quality controls must be in place to detect these erroneous sequences which could lead to major incorrect phylogenetic conclusions. Dot- or colony- blotting, using specific probes may be used to rapidly identify targets for sequencing or other analysis (Britschgi and Giovannoni, 1991; Liesack and Stackebrandt, 1992). Screening of clones in libraries can also be facilitated by restriction fragment length polymorphisms (RFLP) (Britschgi and Giovannoni, 1991) and single-nucleotide sequencing (Reysenbach *et al.*, 1994). Non sequencing and cloning techniques are now being developed and applied to the screening process. Denaturing gradient gel electrophoresis (DGGE) separates different rRNA genes on the basis of their G+C content (Kowalchuk, 1997), RFLP analysis is then used to produce patterns which may be specific for different community members. In addition Amplified Fragment Length Polymorphism (AFLP) analysis is now recognised as a tremendously powerful tool for the identification of bacterial species down to the strain level (Vos *et al.*, 1995). These, and other, techniques as applied to gut microbiology are discussed in Chapter 10.

When used in conjunction with sequence information studies of natural microbial communities performed to date have uncovered novel 'phylotypes', which often represent major new lineages (Amann *et al.*, 1995; Delong, 1992; Hugenholtz and Pace, 1996; Pace *et al.*, 1985). Many of the sequences obtained have belonged to organisms termed non-culturable. A more precise term would be 'not-yet culturable', as it may be that the artificial media and growth conditions have not yet been developed that will sustain growth of such organisms. An unexpected result of these studies is in the development of such media. It is now possible to phylogenetically place an unknown isolate on the 'tree of life', from this it is reasonable to assume that the unknown organism and its phylogenetic neighbours share a common ancestry and may have retained physiological similarities (including enzyme pathways used to derive energy). Using this as a base line and extrapolating back to the nearest living relative, optimum culture media and environmental conditions could be refined to allow the growth of the organism in the laboratory.

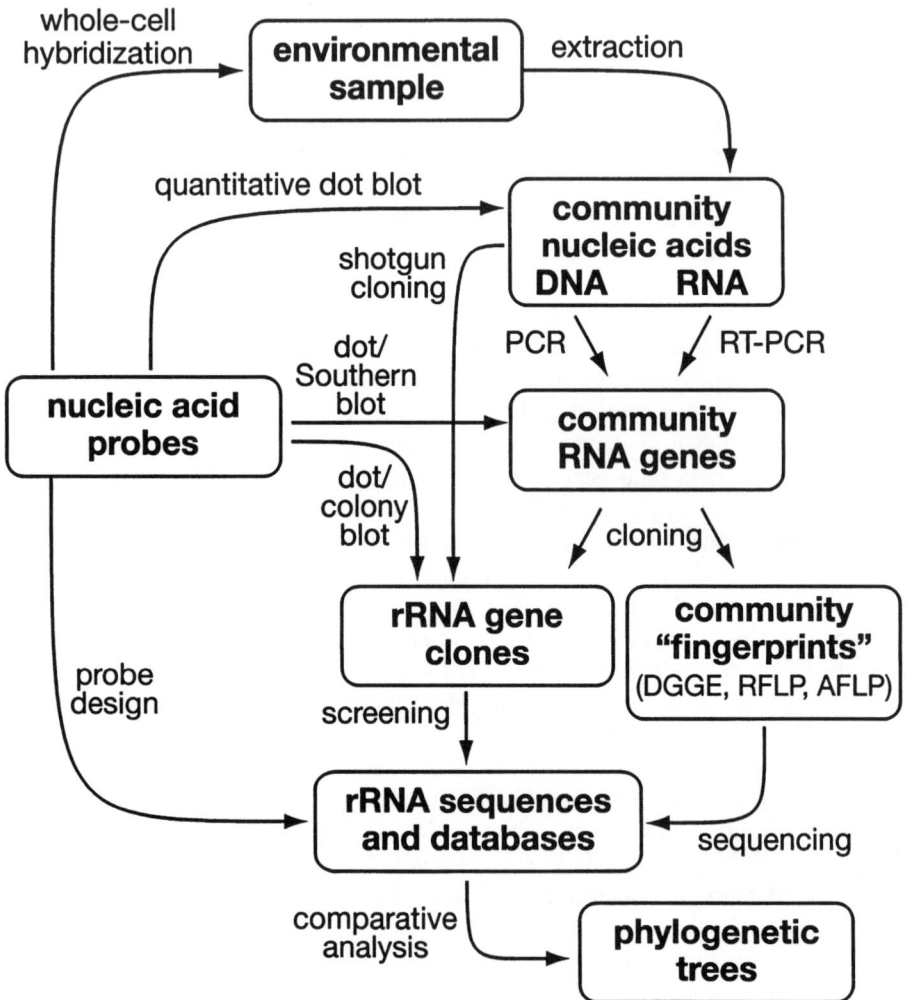

Figure 1. Strategies based on rRNA sequences for characterising microbial communities without the prior need for cultivation of organisms. Adapted from Hugenholtz and Pace (1996).

3   16S rRNA gene analysis and its application to the taxonomy of the colonic microbiota

Most studies on the bacterial microflora of the large intestine have centred on the recovery of bacterial isolates from faeces. Those few studies, which have been made on the inner contents of the large intestine, suggest that the microflora, at least distally, is quantitatively similar to that of faecal material (Drasar and Roberts, 1990). Obligate anaerobic bacteria predominate and some 30-40 presently described species make up 99% of the bacterial mass. Most of the bacteria growing in the intestine are strict anaerobes including *Bacteroides, Bifidobacterium, Clostridium, Enterobacteriacae* and members of the lactic acid group of bacteria (Finegold *et al.*, 1983; Holdeman *et al.*, 1976 Moore and Holdeman, 1974.).

This chapter will discuss the genera *Bacteroides, Clostridium* and organisms known as the Lactic Acid Bacteria (LAB). Each group is found in large numbers in the human gut and is thought to have a major nutritional function in the colonic microbial ecology. Moreover, the application of molecular, and in particular rRNA based methodologies have led to major advances in the taxonomy and classification of these microorganisms.

3.1   BACTEROIDES AND THEIR RELATIVES

Bacteroides are the numerically most important genus of colonic bacteria in humans and animals (Holdeman *et al.*, 1972). Their taxonomy has undergone significant change in recent years. The genus as described in Bergey's *Manual of Systematic Bacteriology* comprises species that are obligately anaerobic, gram-negative, non-sporulating, pleomorphic rods. In addition, G+C contents of the DNA of the bacteroides were shown to range from 28 to 61 mol% (Holdeman *et al.*, 1984). Because of these fairly loose criteria, the genus as been used as a repository for organisms that could not be classified with confidence. What arose was a very large genus comprising an ill-defined association of organisms of marked heterogeneity. Although it was universally accepted that the taxonomy of this group has been unsatisfactory, and despite overwhelming phenotypic and phylogenetic evidence, workers continued to describe new species which showed little resemblance to *B. fragilis* and related species, *e.g. B. forsythus* (Tanner *et al.*, 1986), *B. galacturonicus* and *B. pectinophilus* (Jensen and Canale-Parola, 1986). Attempts to better classify this bacterial group have been made on the basis of physiological characteristics (Holdeman *et al.*, 1984), electrophoretic patterns of dehydrogenases (Shah and Williams, 1982), cellular fatty acid and sugar composition (Brondz *et al.*, 1989; Brondz *et al.*, 1991; Mayberry *et al.*, 1982), lipid analysis (Miyagawa *et al.*, 1982), serology (Lambe, 1974), and bacteriophage typing (Booth *et al.*, 1979). The diversity of this group was further demonstrated in studies using 16S rRNA oligonucleotide cataloging (Paster *et al.*, 1985), rRNA-DNA hybridization (Johnson and Harich, 1986), DNA homology (Johnson, 1978; Van Steenbergen *et al.*, 1982). On the evidence of such heterogeneity, Shah and Collins (1989) proposed that the genus *Bacteroides* be divided into (i) *Bacteroides* sensu stricto, consisting of saccharolytic, nonpigmenting species such as the type species, *B. fragilis*, and its relatives (Shah and Collins, 1989); (ii) *Prevotella*, consisting of moderately saccharolytic, bile-sensitive, predominately oral species (Shah and Collins, 1990); and (iii) *Porphyromonas*, generally consisting of asaccharolytic, black pigmenting species (Shah and Collins, 1988). In addition, a number of predominately monospecific genera have been described for some species which did not conform to the description of these three genera, some of which are found amongst the colonic microflora (Table 1). Phylogenetic relationships of the aforementioned genera were studied by Paster *et al.*, (1994) and the groupings were found to be robust, confirming that the three genera were distinct, yet closely, related. The phylogenetic interrelationships of species that are found in the human gastrointestinal tract are shown in Fig. 2.

Table 1. *Bacteroides* species isolated from the human gastrointestinal tract which are listed in *Bergey's Manual of Systematic Bacteriology* and have been reclassified

Species or group	Taxonomic status	Reference
B. fragilis and related species	*Bacteroides* (emended definition)	Shah and Collins (1989)
B. ruminicola }	*Prevotella*	Shah and Collins (1990)
B. oralis }		
B. melaninogenica }		
B. asaccharolyticus	*Porphyromonas*	Shah and Collins (1988)
B. succinogenes	*Fibrobacter*	Montgomery *et al.* (1988)
B. multiacidus	*Mitsuokella*	Shah and Collins (1982)
B. microfusus	*Rikenella*	Collins *et al.* (1985)
B. amylophilus	*Ruminobacter*	Stackebrandt and Hippe (1986)
B. praeacutus	*Tissierella*	Collins and Shah (1986)

Bacteroides and relatives form a subgroup of the Cytophaga-flavobacter-bacteroides (CFB) phylum (Gherna and Woese, 1992) and form three major phylogenetic clusters containing most of the species of *Bacteroides*, *Prevotella* and *Porphyromonas*. *Bacteroides fragilis* is the type strain of *Bacteroides* sensu stricto. Other human gastrointestinal species of this group include *B. eggerthii, B. ovatus, B. thetaiotamicron, B. uniformis* and *B. vulgatus*. The 16S rRNA sequence similarities demonstrated within this group are between 91-97%, the lower values being obtained with *B. vulgatus* which is more distantly related to this group. *Bacteroides melaninogenica, B. oralis* and *B. ruminicola* (approx. 88-89% sequence similarity) have now been placed in the genus *Prevotella* that contains over 30 species with an average interspecies sequence similarity of 91%. Presently, the only representative of the genus *Porphyromonas* reported to be found in the gastrointestinal tract out of 13 currently described species is *P. asaccharolytica* (the type strain). In addition to the former *Bacteroides* spp. which have been reclassified with some confidence (Table 1), the taxonomic status of a number of species remains to be clarified. *Bacteroides distasonis, B. putredinis, B. splanchnicus* and *B. ureolyticus* clearly belong in the CFB subgroup, but because of their phylogenetic depths it is uncertain whether these species should be considered to be a members of currently recognised genera or constitute the nuclei of several new taxa.

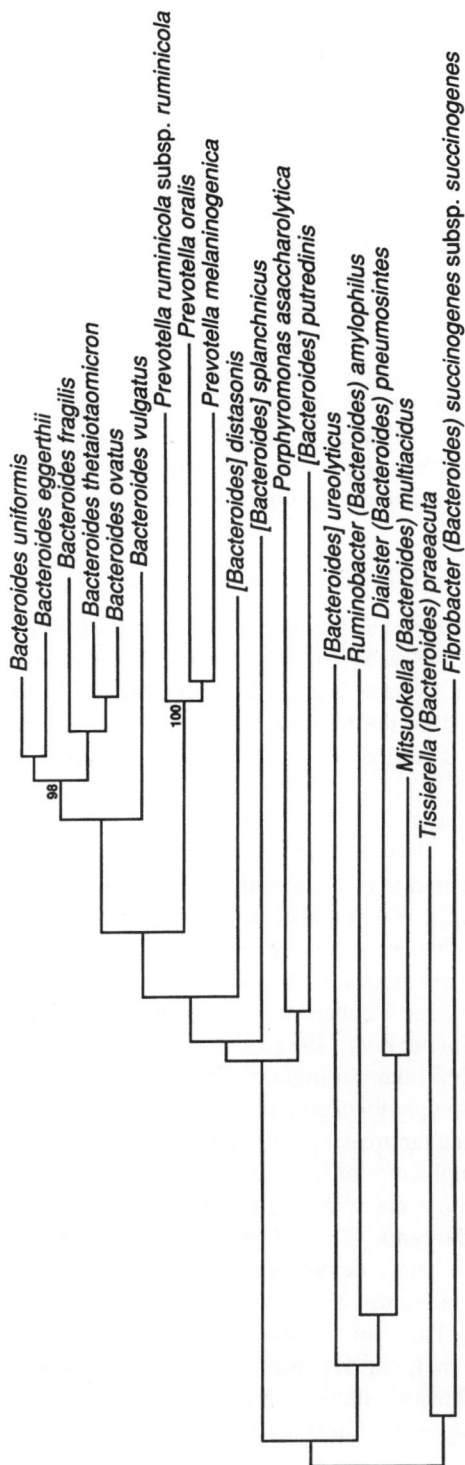

Figure 2. Unrooted phylogenetic tree for the bacteroides subgroup of the CFB phylum. Species represented in square brackets are presently awaiting reclassification. Bootstrap values, expressed as a percentage of 500 replications, are given at branching points.

Bacteroides putridinis shares a sequence similarity value of 89% with *Rikenella* (Paster *et al.*, 1994) and may subsequently be placed in this genus. Further investigation using DNA-DNA hybridisation studies and phenotypic criteria will be used to clarify these loosely associated organisms.

It is immediately apparent that the "bacteroides" have undergone major revisions and is still is a state of flux. From the relatively low sequence similarities within this group of organisms and the phylogenetic depth of the more distantly related (often mono-specific) genera, it is clear this is a loose association of bacteria. Because of these distant relationships, and the range of phenotypic characteristics, the majority of restructuring to date has been based on chemotaxonomic studies prior to the widespread use of molecular methodologies. The application of 16S rRNA gene analyses validated the restructuring of this group of organisms and in some cases extended our knowledge of interrelationships between the new genera that have arisen. Indeed, work in our own laboratory and others (Wilson and Blitchington, 1996) have revealed a huge diversity, with many organisms isolated from faecal material belonging to established genera within the bacteroides group but also representing many new lines of descent.

3.2   CLOSTRIDIUM AND THEIR RELATIVES

In comprehensive studies of Moore and Holdeman (1974) and Finegold *et al.*, (1983), *Clostridium* species were always recovered from human faecal material. In all, over 40 species have been reported to have been isolated. Indeed, clostridial species have been isolated from humans within the first 24h of life and may act as precursors for the installation of other, more fastidious, anaerobes in the colon (Bezirtzoglou and Richmond, 1990).

The genus *Clostridium* is based on a low number of characteristics, *i.e.* anaerobic, gram-positive, spore-forming rods which do not reduce sulphate. Even these simple criteria are not universally adhered to by all described species. The result is a very large group of organisms exhibiting a wide number of diverse traits. Johnson and Francis (1975) demonstrated considerable heterogeneity within the genus by using DNA-rRNA pairing methods, although it was not until the use of oligonucleotide cataloging that the very considerable phylogenetic incoherence of the genus was realised (Fox *et al.*, 1980; Ludwig *et al.*, 1988; Tannner *et al.*, 1981,1982). The technique has demonstrated deeply branching lineages within the clostridia, which included nonclostridial species and provided an insight into the phylogenetic interrelationships of this group.

Full rRNA sequences provide far greater precision for constructing phylogenetic trees and, using this technique, complexities of the clostridia were being uncovered (Lawson *et al.*, 1993). This proceeded to a major study by Collins *et al.* (1994) who undertook an extensive restructuring of the genus *Clostridium,* producing a hierarchical framework for future classifications. This study demonstrated beyond doubt the huge diversity and intermixing of organisms previously thought to have little relationships in common. An unexpected finding was that the gram-negative, non-sporeforming genera of *Fusobacterium* and *Propionigenium* were found to cluster with the gram-positive, spore-forming *C. rectum* (cluster XIX). Although without question, the Gram stain is an extremely important tool in the preliminary identification and subsequent assignment of organisms to particular taxa; the pivotal role of the technique in description of taxonomic units from an evolutionary point of view is now being questioned (Shah *et al.*, 1997).

Fig. 3 shows the phylogenetic interrelationships of species reported to have been isolated from the human gastrointestinal tract. From this habitat, the sequence similarities range from as low as 78 to over 99% and demonstrate an enormous diversity. The majority of species are found within Cluster 1, *Clostridium* sensu stricto, members of which exhibit relatively high levels of sequence similarities (> 90%). Markedly different phenotypes are contained within the group. For example, saccharolytic and proteolytic species, psychrophiles, mesophiles, and thermophiles are represented, leading to much internal structure in the topology of the phylogenetic tree.

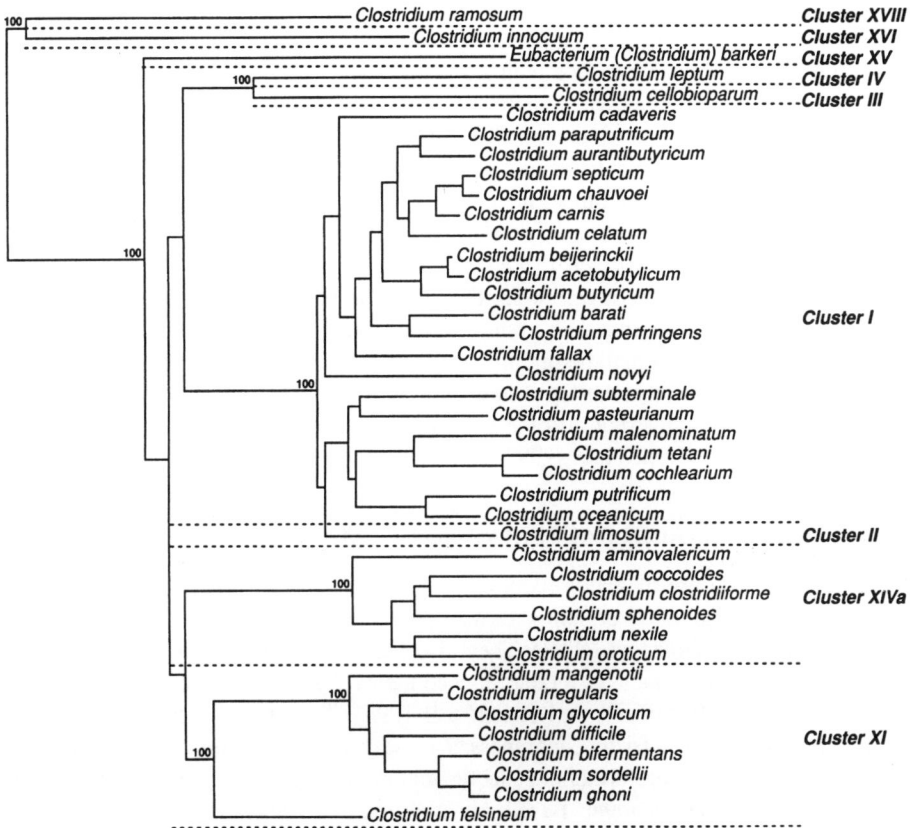

Figure 3. Unrooted phylogenetic tree for the clostridia species isolated from the human gastrointestinal tract. Bootstrap values, expressed as a percentage of 500 replications, are given at branching points.

The majority of the remaining colonic clostridia fall into two taxonomically heterogeneous groups (XI and XIVa according to the nomenclature of Collins *et al.*, 1994).

Clostridium bifermentans, C. difficile, C. ghoni, C. glycolicum, C. mangenotii, C. irregularis and *C. sordelli* make up a phylogenetically well defined (levels of similarity, > 93%, bootstrap significance, 100%) and distinct subgroup within cluster XI. *Clostridium felsineum* clearly forms a distinct line of decent, from the topology of the tree these two subclusters clearly represent new genera. Considerable internal structure was also present within Cluster XIV (*C. aminovalericum, C. clostridiiforme, C. coccoides, C. nexile, C. oroticum* and *C. sphenoides*), the overall diverse phenotypes and the wide range of evolutionary distances, indicate that this cluster is probably a suprageneric cluster. Taxonomic inconsistencies are highlighted in both these clusters with the presence of peptostreptococci, streptococci, eubacteria and ruminococci species. The remaining clostridia fall into separate subclusters each representing the nuclei of new genera (Collins *et al.*, 1994).

Clostridia therefore represent a phenotypically and phylogenetically extremely complex group of organisms forming a plethora of lines and groups of various complexities and depths. A comprehensive taxonomic rearrangement of clostridia will result in a major proliferation of new genera. Some clostridial species form phylogenetically distinct groups with sufficient phenotypic coherence that they can be readily equated with genera. However many others have only loose associations with other organisms forming individual lines of decent where the taxonomic rank of these organism is far more problematic. Collins *et al.* (1994) proposed a number of new genera and put forward a new hierarchical structure for the clostridia and close relatives. The authors did however acknowledge that their system would be modified over time but felt it more prudent not to make formal taxonomic recommendations until suitable phenotypic circumscription and differentiation of new genera was possible. This is still recognised as a problem in the construction of a naturally based classification of the clostridia (Collins *et al.*, 1994; Hippe *et al.*, 1992; Lawson *et al.*, 1993).

3.3   THE LACTIC ACID BACTERIA AND RELATIVES

Lactic acid bacteria are of major economic importance to the food industry, predominating in the natural flora of many fermented foods acting both in preservation or spoilage (Stiles and Holzapel, 1997). They also play an important role in the digestive tract of humans and animals, as a consequence their roles as probiotics is of increasing interest (Fuller, 1989; Gibson and Roberford, 1995). The term "lactic acid bacteria" has traditionally been used to encompass a loosely defined group of Gram-positive, non-sporing, carbohydrate-fermenting, lactic-acid-producing organisms, which are acid tolerant and catalase-negative. Based on traditional phenotypic criteria, including the production of lactic acid and absence of catalase and cytochromes: and on phylogenetic considerations, the LAB should encompass the genera *Carnobacterium, Enterococcus, Lactobacillus, Lactococcus, Leuconostoc, Oenococcus, Pediococcus, Streptococcus, Tetragenococcus, Vagococcus and Weissella* (Aguirre and Collins, 1993).

Because of their beneficial role in the dairy industry *Bifidobacterium, Propionibacterium* and *Brevibacterium* are often included in this group of organisms. However, these bacteria contain a high G+C content and are phylogenetically part of the *Actinomycetes* and should therefore be treated separately. *Bifidobacterium*, in particular, have received much attention because of their potential therapeutic uses (Fuller, 1989; Gibson and Roberfroid, 1995). It is therefore surprising to find that at the present time, the 16S rRNA sequence database (EMBL/GenBank) for bifidobacteria is highly fragmented and incomplete. However, many studies have exploited the use of rRNA/DNA probes in the study of the colonic microflora and are discussed in Chapter 10.

During the past decade, knowledge of the taxonomic interrelationships of the gram-positive catalase-negative cocci has markedly improved. Much of this has resulted from using a range of phenotypic methods (*e.g.* miniaturised biochemical testing, protein profiling) in concert with molecular genetic approaches, notably comparative 16S rRNA gene sequencing. The taxonomy of the LAB organisms in the context of the food industry and their role in human clinical infections as been reviewed in detail (Aguirre and Collins, 1993; Stiles and Holzapfel, 1997).

Fig. 4 shows phylogenetic relationships of the LAB commonly reported to have been isolated from the human intestinal tract. For comparison, the genera that encompass the majority of the remaining LAB group of organisms are included. 16S rRNA sequence analyses have shown several genera including *Lactobacillus* and *Streptococcus* to be phylogenetically heterogeneous (Collins *et al.*, 1991; Bentley *et al.*, 1991). In addition some taxa, *e.g.* pediococci and lactobacilli, which had previously been considered separate have been shown to be closely related (Collins *et al.*, 1991).

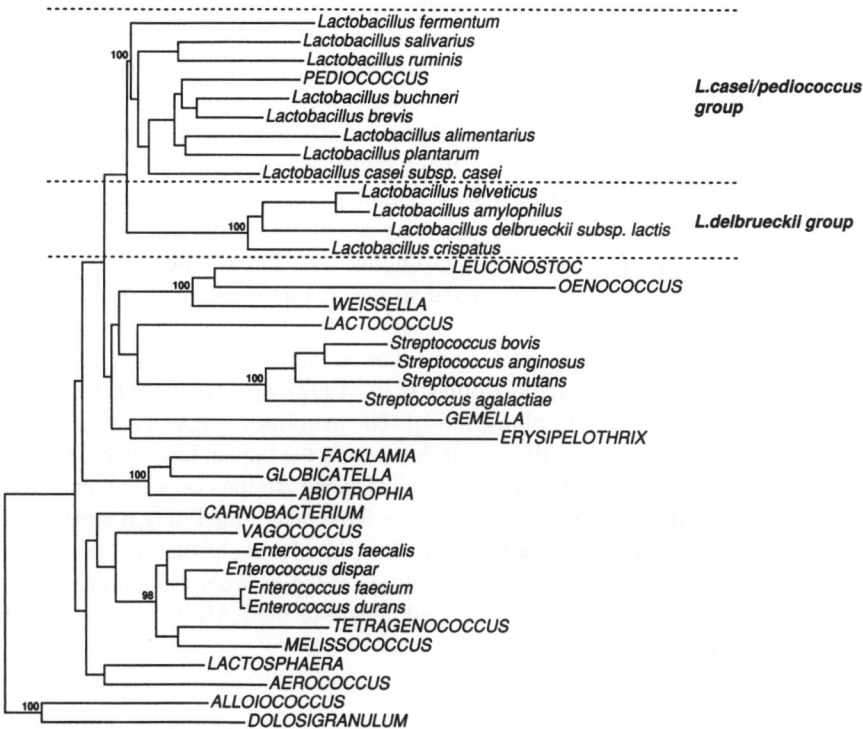

Figure 4. Unrooted phylogenetic showing the interrelationships of some of the lactic acid bacteria. The genera (shown in capitals) that have not as yet been recognised in the human colonic microbiota are by their type strains. Bootstrap values, expressed as a percentage of 500 replications, are given at branching points.

Streptococci were among the earliest bacteria to be recognised by microbiologists because of their involvement in a large number of human and animal diseases. The genus *Streptococcus* was originally described from morphological, serological, physiological and biochemical characteristics and comprised a wide range of organisms with complex nutritional requirements (Jones, 1978). Based on a molecular approach, the streptococci have undergone major revisions, being subdivided into three genetically distinct groups: *Streptococcus* sensu stricto, *Enterococcus* and *Lactococcus* (Schleifer and Kilpper-Balz, 1984, 1987).

Most, but not all, of the Lancefield group N lactic have been transferred to the genus *Lactococcus* (Schleifer, 1985; Schleifer and Kilpper-Balz, 1984). The motile group N streptococci were shown to be more closely related to the genera *Enterococcus* and *Carnobacterium* and thus included in the genus *Vagococcus* (Collins *et al.*, 1989). Bacteria of the genus *Enterococcus* have been recognised since Thiercelin (1899) but the taxonomy of this group has been vague. There are no phenotypic characteristics that separate the genus from other genera of gram-positive, catalase negative cocci. The application of genetic techniques to differentiate the enterocci has resolved many of the uncertainties resulting in the genus *Enterococcus;* being revived by Schleifer and Kilpper-Balz (1984) to include the Lancefied group D (faecal) steptococci, *S. faecalis* and *S. faecium*, as *E. faecalis* and *E. faecium*.

The principle groups derived from 16S rRNA studies of the lactobacilli can be summarised as follows: (i) *Lactobacillus debruekii* group including in the main, but not exclusively, the homofermentative lactobacillus; (ii) the *Lactobacillus casei-Pediococcus* group; (iii) the *Leuconotoc* group that includes some obligately heterofermative lactobacilli. This has subsequently been divided into three genera: *Leuconostoc, Oenococcus* and *Weissella*. Even with these revisions many of the afore mentioned groups contain a number of DNA-DNA homology clusters (Johnson *et al.*, 1980; Lauer *et al.*, 1980; Weiss *et al.*, 1991) and may constitute the nuclei of new genera.

More recently, the use of molecular taxonomic tool has resulted in the discovery and description of a plethora of genera and species comprising gram-positive catalase-negative cocci [*e.g. Aerococcus urinae* (Aguirre and Collins, 1992a), *Alloiococcus* (Aguirre and Collins, 1992b), *Dolosigranulum* (Aguirre *et al.*, 1993), *Facklamia* (Collins *et al.*, 1997), *Globicatella* (Collins *et al.*, 1992), *Lactosphaera* (Janssen *et al.*, 1995) *Melissococcus* (Cai and Collins, 1994)]. The majority of organisms that belong to the LAB group that have presently been isolated from the human large intestine, belong to the *Lactobacillus* subcluster, the remainder being streptococci and enterococci. As already stated, it is now recognised that we have determined only a small fraction of the organisms found in any one habitat. It is highly likely that as the molecular techniques are applied to the community analyses of the gastrointestinal tract, bacterial isolates will be assigned to present described genera of the LAB or form the nuclei of new closely related genera.

4   Conclusions and future perspectives

It is becoming increasingly clear that our understanding of the natural microbial world is rudimentary with sequence-based gene analysis providing a way to survey biodiversity. Indeed, the molecular revolution has had profound and remarkable affects on microbial identification, taxonomy and classification. Problems may arise in the immense amount of data being deposited within the nucleotide databases (Olsen *et al.*, 1992), the challenge will be the processing this and incorporating it into useful classification frameworks. With the advent of this molecular revolution, it is reasonable to project that cloned sequences deposited in the databases will far exceed those represented from cultured, living organisms. However, the microbial ecologist is interested in the organism itself, its physiology, and the relationship with its neighbours and ultimately its role in the community. Furthermore, generic lines will not necessarily follow the convenient lines that arose from morphological and physiological characteristics as well as biochemical pathways. Rather than replacing the classical culture-based approach the new molecular approaches must be integrated in the existing systems and reservations of the traditional microbiologists must be overcome.

Never before has the microbial ecologist possessed the tools to understand their goals of describing the biodiversity and interrelationships of members of any given community. Gut microbiologists must fully embrace these new techniques in these most exciting times where appreciation of individual organisms and their community may have far reaching results on our understanding of the gastrointestinal tract both as an organ of health and disease processes.

References

Aguirre, M. and Collins, M.D. (1992a) Phylogenetic analysis of some *Aerococcus*-like organisms from urinary tract infections: description of *Aerococcus urinae* sp. nov. *Journal General of Microbiology* **138**, 401-5.

Aguirre, M. and Collins, M.D. (1992b) Phylogenetic analysis of an unknown bacterium from human middle ear fluid: description of *Alloiococcus otitis* gen.nov., sp. nov. *International Journal of Systematic Bacteriology* **42**, 79-83.

Aguirre, M. and Collins, M.D. (1993) Lactic acid bacteria and human clinical infection. *Journal of Applied Bacteriology* **75**, 95-107.

Aguirre, M.,D., Morrison, D. and Cookson, B.D. (1993) Phenotypic and phylogenetic characterization of some *Gemella*-like organisms from human infections: description of *Dolosigranulum pigrum* gen. nov., sp. nov. *Journal of Applied Bacteriology* **75**, 608-12.

Amann, R.I., Krumholz, L. and Stahl, D.A. (1990) Fluorescent-oligonucleotide probing of whole cell for determinative, phylogenetic and environmental studies in microbiology. *Journal of Bacteriology* **172**, 762-70.

Amann, R.I., Ludwig, W. and Schleifer, K.H. (1995) Phylogenetic identification and in situ detection of individual microbial cells without cultivation. *Microbiological Reviews* **59**, 143-69.

Bentley, R.W. Leigh, J.A. and Collins, M.D. (1991) Intrageneric structure of *Streptococcus* based on comparative analysis of small-subunit rRNA sequences. *International Journal of Systematic Bacteriology* **41**, 487-94.

Bezirtzoglou, E. and Richmond, C. (1990) Apparition of *Clostridium* sp. and *Bacteroides* sp. in the intestine of the newborn delivered by caesarean section. *Comparative Immunology, Microbiology and Infectious Diseases* **13**, 217-21.

Booth, S.J., Van Tassell, R.L., Johnson, J.L. *et al.* (1979) Bacteriophages of *Bacteroides*. *Reviews in Infectious Diseases* **1**, 325-28.

Britschgi, T.B. and Giovannoni, S.J. (1991) Phylogenetic analysis of a natural marine bacterioplankton population by ribosomal-RNA gene cloning and sequencing. *Applied and Environmental Microbiology* **57**, 1707-13.

Brondz, I., Carlsson, J., Sjostrum, M. *et al.* (1989) Significance of cellular fatty acids and sugars in defining the genus *Porphyromonas*. *International Journal of Systematic Bacteriology* **39**, 314-18.

Brondz, I., Olsen, I., Haapasalo, M. *et al.* (1991) Multivariate analyses of fatty acid data from whole-cell methano-lysates of *Prevotella*, *Bacteroides* and *Porphyromonas* spp. *Journal of General Microbiology* **137**, 1445-52.

Busse, H.J., Denner, E.B.M. and Lubitz, W. (1996) Classification and identification of bacteria: current approaches to an old problem. Overview of methods used in bacterial systematics. *Journal of Biotechnology* **47**, 3-38.

Cai, J. and Collins, M.D. (1994) Evidence for a close phylogenetic relationship between *Melissococcus pluton*, the causative agent of European foulbrood disease, and the genus *Enterococcus*. *International Journal of Systematic Bacteriology* **44**, 365-67.

Collins, M.D., Aguirre, M., Facklam, R.R. *et al.* (1992) *Globicatella sanguis* gen. nov., sp. nov., a new gram-positive catalase-negative bacterium from human sources. *Journal of Applied Bacteriology* **73**, 422-37.

Collins, M.D., Ash., C., Farrow, J.A.E. *et al.* (1989) 16S ribosomal ribonucleic acid sequence analysis of lactococci and related taxa. Description of *Vagococcus fluvialis* gen. nov., sp. nov. *Journal of Applied Bacteroiology* **67**, 453-60.

Collins, M.D., Falsen, E., Lemozy, J. *et al.* (1997) Phenotypic and phylogenetic characterization of some *Globicatella*-like organisms from human sources: description of *Facklamia hominis* gen. nov., sp. nov. *International Journal of Systematic Bacteriology* **47**, 880-82.

Collins, M.D., Lawson, P.A., Willems, A. *et al.* (1994) The Phylogeny of the genus *Clostridium*: Proposal of five new genera and eleven new species combinations. *International Journal of Systematic Bacteriology* **44**, 812-26.

Collins, M.D., Rodrigues, U.M., Aguirre, M. *et al.* (1991) Phylogenetic analysis of the genus *Lactobacillus* and related lactic acid bacteria as determined by reverse transcriptase sequencingof 16S rRNA. *FEMS Microbiology Letters* **77**, 5-12.

Collins, M.D. and Shah, H.N. (1986) Reclassification of *Bacteroides praeacutus* Tissier (Holdeman and Moore) in a new genus, *Tisserella*, as *Tisserella praecuta* comb. nov. *International Journal of Systematic Bacteriology* **36**, 461-63.

Collins, M.D., Shah, H.N. and Mitsuoka (1985) Reclassification of *Bacteroides* microfusus (Kaneuchi and Mitsuoka) in a new genus *Rikenella*, as *Rikenella microfusus* comb. nov. *Systematic and Applied Microbiology* **6**, 79-81.

DeLong, E.F. (1992) Archaea in coastal marine environments. *Proceedings of the National Academy of Sciences, USA* **89**, 5685-89.

Drasar, B.S. and Barrow, P.A. (1985) Intestinal Microbiology, American Society for Microbiology, Washington, D.C.

Drasar, B.S. and Roberts, A.K. (1990) Control of the large bowel microflora. in *Human Microbial Ecology*, (eds. M.J. Hill and P.D. Marsh), CRC Press Inc. Boca Rata, Fl, pp. 87-111.

Dore, J., Pochart, P., Bernalier, A *et al.* (1995) Enumeration of H₂-utilizing methanogenic archaea, acetogenic and sulphate-reducing bacteria from human feces. *FEMS Microbiology Ecology* **17**, 279-84.

Felsenstein, J. (1982) Numerical methods for inferring evolutionary trees. *Quarterly Review of Biology* **57**, 379-404.

Finegold, S.M., Sutter, V.L. and Mathisen, G.E. (1983) Normal indigenous intestinal flora, in *Human Intestinal Microflora in Health and Disease* (ed. D.J. Hentges,) pp. 3-31. Academic Press, London.

Fitch, W.M. and Margoliash, E. (1967) Construction of phylogenetic trees: a method based on mutation distances as estimated from cytochrome c sequences is of general applicability. *Science* **155**, 279-84.

Fox, G.E., Stackebrandt, E., Hespell, R.B. *et al.* (1980) The phylogeny of the prokaryotes. *Science* **209**, 457-63.

Fuller, R. (1989) Probiotics in man and animals: a review. *Journal of Applied Bacteriology* **66**, 365-78.

Gherna, R. and Woese, C.R. (1992) A partial phylogenetic analysis of the "Flavobacter-*Bacteroides*" phylum: basis for taxonomic restructuring. *Systematic and Applied Microbiology* **15**, 513-21.

Gibson, G.R. and Roberfroid, M.B. (1995) Dietary modulation of the human colonic, microbiota: introducing the concept of prebiotics. *Journal of Nutrition* **125**, 1401-12.

Haeckel, E. (1866) Generlley morphologiy der organismen-allemeiny grundzugy der organischen formen-wissenschaft, mechanisch begrundet durch die von Charles Darwin reformirte descendenz-theorie. Georg Reimer, Berlin.

Hippe, H., Andreesen, J.R. and Gottschalk, G. (1992) The genus *Clostridium*-nonmedical, in *The Prokaryotes*, vol. 2 (eds. A.Balows, H.G. Tuper, M. Dworkin, W. Harder, and K.H. Schliefer). Springer-Verlag, New York, pp. 1800-66.

Holdeman, L.V., Cato, E.P. and Moore, W.E.C. (1972) Anaerobe manual, 4th ed., Virginia Polytechnic Institute and State University, Blacksburg.

Holdeman, L.V., Good, I.J. and Moore, W.E.C. (1976) Human fecal flora: variation in bacterial composition within individuals and a possible effect of emotional stress. *Applied and Environmental Microbiology* **31**, 359-75.

Holdeman, L.V., Kelly, R.W. and Moore, W.E.C. (1984) Genus I. *Bacteroides* Castellani and Chalmers 1919, 959[AL], in *Bergey's Manual of Systematic Bacteriology*, vol. 1. (eds. N.R. Krieg and J.G. Holt), The Williams & Wilkins Co., Baltimore, pp. 604-31.

Hugenholtz, P. and Pace, N.R. (1996) Identifying microbial diversity in the natural environmental: a molecular phylogenetic approach. *Tibtech* **14**, 190-97.

Janssen, P.H., Evers, S., Rainey, F.A. *et al.* (1995) *Lactosphaera* gen. nov., a new genus of lactic acid bacteria and transfer of *Ruminococcus pasteurii* Schink 1984 to *Lactosphaera pasturii* comb. nov. *International Journal of Systematic Bacteriology* **30**, 565-71.

Jensen, N.S. and Canale-Parola (1986) *Bacteroides pectinophilus* sp. nov. and *Bacteroides galacturonicus* sp. nov.: two new pecinolytic bacteria from the human intestinal tract. *Applied and Environmental Microbiology* **52**, 880-87.

Johnson, J.L. (1978) Taxonomy of *Bacteroides*. I. Deoxyribonucleic acid homologies among *Bacteroides fragilis* and other saccharolytic *Bacteroides*. *International Journal of Systematic Bacteriology* **28**, 245-56.

Johnson, J.L. and Francis, B.S (1975) Taxonomy of the clostridia: ribosomal ribonucleic acid homologies among the species. *Journal of General Microbiology* **88**, 229-44.

Johnson, J.L. and Harich, G. (1986) Ribosomal ribonucleic acid homology among species of the genus *Bacteroides*. *International Journal of Systematic Bacteriology* **36**, 71-79.

Johnson, J.L., Phelps, C.F., Cummins, C.S., London, J. and Gasser, F. (1980) Taxonomy of the *Lactobacillus acidophilus* group. *International Journal of Systematic Bacteriology* **30**, 53-68.

Jones, D. (1978) Composition and differentiation of the genus *Streptococcus*. in *Streptococci*, (eds. F.A. Skinnner and L.B. Quesnel) Academic Press, London, pp. 1-49.

Jukes, T.H. and Cantor, C.R. (1969) Evolution of protein molecules, in *Mammalian Protein Metabolism* (ed. H.N. Munro) Academic Press, New York, NY, pp. 21-132.

Kowalchuk, G.A., Gerards, S. and Woldendorf, J.W. (1997) Detection and characterisation of fungal infections of *Ammohila arenaria* (Marram grass) roots by denaturing gradient gel electrophoresis of specifically amplified 18S rDNA. *Applied and Enviromental Microbiology* **63**, 3858-65.

Lambe, D.J., Jr. (1974) Determination of *Bacteroides melaninogenicus* serogroups by fluorescent antibody staining. *Applied Microbiology* **28**, 561-67.

Lane, D.J., Pace, B., Olesen, G.J. *et al.* (1985) Rapid determination of the 16S ribosomal RNA sequences for phylogenetic analysis. *Proceedings of the National Academy of Sciences*, USA **82**, 6955-59.

Lauer, E., Helming, C. and Kandler, O. (1980) Heterogeneity of the species *Lactobacillus acidophilus* (Moro) Hansen and Mocquot as revealed by biochemical characteristics and DNA-DNA hybridisation. Zentralblatt fur Bakteriologie Mikrobiologie und Hygiene I Abteilung Referate. *Original C* **1**, 150-68.

Lawson, P.A., Perez, P.L., Hutson, R.A. *et al.* (1993) Towards a phylogeny of the clostridia based on 16S rRNA sequences. *FEMS Microbiological Letters* **113**, 87-92.

Liesack, W. and Stackebrandt, E. (1992) Occurrence of novel groups of the domain bacteria as revealed by analysis of genetic material isolated from an Australian terrestrial environment. *Journal of Bacteriology* **174**, 5072-78.

Ludwig, W., Wiezenegger, M., Kilpper-Balz, R. *et al.* (1988) Phylogenetic relationships of anaerobic streptococci. *International Journal of Systematic Bacteriology* **38**, 15-18.

Maniatis, T., Fritsh, E.F. and Sambrook, J. (1982) Molecular cloning: a laboratory manual. Cold Spring Habor Laboratory, Cold Spring Habor, N.Y.

Mayberry, W.R., Lambe Jr., D.W. and Ferguson, K.P. (1982) Identification of *Bacteroides* species by cellular fatty acid profiles. *International Journal of Systematic Bacteriology* **32**, 21-27.

Miyagawa, E., Azuma, R. and Suto, T. (1979) Distribution of sphingolipids in *Bacteroides* species. *Journal of General and Applied Microbiology* **25**, 41-52.

Montgomery, L., Flesher, B. and Stahl, D. (1988) Transfer of *Bacteroides succinogenes* (Hungate) to *Fibrobacter* gen. nov. as *Fibrobacter succinogenes* comb. nov. and description of *Fibrobacter intestinalis* sp. nov. *International Journal of Systematic Bacteriology* **38**, 430-35.

Moore, W.E.C. and Holdeman, L.V. (1974) Human fecal flora: the normal flora of 20 Japanese-Hawaiians. *Applied Microbiology* **27**, 961-79.

Olsen, G.J, Overbeck, R., Larsen, N. *et al.* (1992) The Ribosome Database Project. *Nucleic Acids Research* **20**, 2199-200.

Pace, N.R. (1996) New perspectives on the natural microbial world: molecular microbial ecology. *ASM News* **62**, 463-70.

Pace, N.R., Stahl, D.A., Lane, D.J. and Olsen, G.J. (1985) The analysis of natural microbial populations by ribosomal RNA sequences. *Advances in Microbial Ecology* **9**, 1-55.

Paster, B.J., Dewhirst, F.E., Olsen, I. *et al.* (1994) Phylogeny of *Bacteroides, Prevotella*, and *Porphyromonas* spp. and related bacteria. *International Journal of Systematic Bacteriology* **176**, 725-32.

Paster, B.J., Ludwig, W., Weisburg, W.G. *et al.* (1985) A phylogenetic grouping of *Bacteroides*, cytophagas and certain flavobacteria. *Systematic and Applied Microbiology* **6**, 34-42.

Reysenbach, A.L, Wichham, G.S. and Pace, N.R. (1994) Phylogenetic analysis of the hyperthermophilic pink filament community in Octopus Spring, Yellowstone-National -Park. *Applied and Environmental Microbiology* **60**, 2113-19.

Schleifer, K.H. (1995) Phylogeny of the genus *Lactobacillus* and related genera. *Systematic and Applied Microbiology* **18**, 461-67.

Schleifer, K.H. and Kilpper-Balz, R. (1984) Transfer of *Streptococcus faecalis* and *Streptococcus faecium* to the genus *Enterococcus* nom. rev. as *Enterococcus faecalis* comb. nov. and *Enterococcus faecium* comb. nov. *International Journal of Systematic Bacteriology* **34**, 31-34.

Schleifer, K.H. and Kilpper-Balz, R. (1987) Molecular and chemotaxonomic approaches to the classification of streptococci, enterococci and lactococci: a review. *Systematic and Applied Microbiology* **10**, 1-19.

Shah, H.N. and Collins, M.D. (1982) Reclassification of *Bacteroides multiacidus* (Mitsuoka, Terada, Watanbe and Uchida) in a new genus *Mitsuokella*, as *Mitsuokella multiacidus* comb. nov. Zentralbl. Bakeriol. Parasitekd. Infektionskr. Hyg. Abt. 1 Orig. Reihe C **3**, 491-94.

Shah, H.N. and Collins, M.D. (1988) Proposal for recassification of *Bacteroides asaccharolyticus, Bacteroides gingivalis, and Bacteroides endodontalis* in a new genus, *Porphyromonas*. *International Journal of Systematic Bacteriology* **38**, 128-31.

Shah, H.N. and Collins, M.D. (1989) Proposal to restrict the genus *Bacteroides* (Castillani and Chalmers) to *Bacteroides fragilis* and closely related species. *International Journal of Systematic Bacteriology* **39**, 85-87.

Shah, H.N. and Collins, M.D. (1990) *Prevotella*, a new genus to include *Bacteroides melanogenicus* and related species formerly classified in the genus *Bacteroides*. *International Journal of Systematic Bacteriology* **39**, 85-87.

Shah, H.N., Gharbia, S.E. and Collins, M.D. (1997) The Gram stain: a declining synapomorphy in an emerging evolutionary tree. *Reviews in Medical Microbiology* **8**, 1-8.

Shah, H.N. and Williams, R.A.D. (1982) Dehydrogenases patterns in the taxonomy of *Bacteroides*. *Journal of General Microbiology* **128**, 2955-65.

Stackebrandt, E. and Hippe, H. (1986) Transfer of *Bacteroides amylophilus* to a new genus *Ruminobacter* gen. nov., nom. rev. as *Ruminobacter amylophilus* comb. nov. *International Journal of Systematic Bacteriology* **8**, 204-7.

Stahl, D.A., Flesher, B., Mansfiled, H.R. *et al.* (1988) Use of phylogenetically based hybridization probes for studies of ruminal microbial ecology. *Applied and Environmental Microbiology* **54**, 1079-84.

Stiles, M.E. and Holzapfel, W.H. (1997) Lactic acid bacteria of foods and their current taxonomy. *International Journal of Food Microbiology* **36**, 1-29.

Tanner, A.C.R., Listgarten, M.A., Ebersole, J.L. *et al.* (1986) *Bacteroides forsythus* sp. nov., a slow-growing, fusiform *Bacteroides* sp. from the human oral cavity. *International Journal of Systematic Bacteriology* **36**, 213-21.

Tanner, R.S., Stackebrandt, E., Fox, *et al.* (1981) A phylogenetic analysis of *Acetobacteium woodii, Clostridium barkeri, Clostridium butyricum, Clostridum litueburense, Eubacterium limosom,* and *Eubacterium tenue. Current Microbiology* **5**, 35-38.

Tanner, R.S., Stackebrandt, E., Fox, *et al.* (1982) A phylogenetic analysis of anaerobic eubacteria capable of synthesizing acetate from carbon dioxide. *Current Microbiology* **7**, 127-32.

Thiercelin, M.E. (1899) Sur un diplocoque saprofyte de l'intestin susceptible de devenir pathogene. C.R. *Society of Biology* (Paris) **51**, 269-71.

Vandamme, P., Pot, B., Gillis, M., De Vos, P., Kersters, K. and Swings, J. (1996) Polyphasic taxonomy, a consenus approach to bacterial systematics. *Microbiological Reviews* **60**, 407-38.

van Neil, C.B. (1946) The classification and natural relationships of bacteria. *Cold Spring Habor Symposium of Quanititative Biology* **11**, 285-301.

Van Steenbergen, T.J.M., Vlaanderen, C.A. and de Graaff, J. (1982) Deoxyribonucleic acid homologies among strains of *Bacteroides melaninogenicus* and related species. *Journal of Applied Bacteriology* **53**, 269-76.

Vos, P., Hogers, R., Bleeker, M. *et al.* (1995) AFLP™: a new technique for DNA fingerprinting. *Nucleic Acids Research* **23**, 4407-14.

Ward, D.M., Weller, R. and Bateson, M.M. (1990) 16S rRNA sequences reveal numerous uncultured microorganisms in a natural community. *Nature* **345**, 63-65.

Weiss, N. (1991) The genera Pediococcus and Aerococcus. in The Prokaryokes, Vol II, 2nd edition, (eds. A. Balows, H.G. Truper, M.Dworkin, W. Harder and K.-H. Schleifer), Springer-Verlag, New York, pp. 1502-7.

Wilson, K.H. and Blitchington, R.B. (1996) Human colonic biota studied by ribosomal DNA sequence analysis. *International Journal of Systematic Bacteriology* **62**, 2273-78.

Woese, C.R. (1987). Bacterial Evolution. Microbiological Reviews **51**, 221-71.

Woese, C.R., Fox, G.E., Zablen, L. *et al.* (1975) Conservation of primary structure in 16S ribosomal rRNA. *Nature* **254**, 83-86.

Zuckerkandl, E. and Pauling, L. (1965) Molecules as documents of evolutionary history. *Journal of Theoretical Biology* **8**, 357-66.

CHAPTER 10

Application of Taxonomy and Systematics to Molecular Techniques in Intestinal Microbiology

RICHARD SHARP[1] AND CHERIE J. ZIEMER[2]
[1]*Medical Research Council Dunn Clinical Nutrition Centre, Hills Road, Cambridge, UK*
[2]*Microbiology Department, Institute of Food Research, Reading, UK*

1   Introduction

The microbial community in the large intestine is extremely complex, both in terms of numbers of organisms and in its diversity (Macfarlane and Gibson, 1994; Conway, 1995). The interactions of these gut micro-organisms with their environment have far reaching implications for the health of the host organism. While a great deal of information has been gained from cultural techniques, it has long been recognised that many micro-organisms are fastidious, resulting in the inability to cultivate the majority. Dependence on enrichment culturing techniques is limiting as microbial actions, interactions and natural representations are distorted (Woese, 1994).

Carl Woese and co-workers revolutionised the classification of life in the mid-1970's by using small subunit rRNA sequences to estimate genetic diversity (Woese *et al.*, 1985). Phylogenetic classification of bacteria has been significantly altered by the application of 16S rRNA sequencing. Furthermore, molecular characterisation has emphasised our current underestimates of microbial diversity, which may be of several orders of magnitude. Considering that there are at least 800,000 distinct insect species and a minimum of 10% of insect species harbour obligate microbial symbionts, which are most likely distinct species, Amann *et al.* (1995) have suggested the existence of about 80,000 microbial species, a far larger number than the approximately 5,000 species that have been described.

We now recognise that bacterial systematics requires a combination of phenotypic and genotypic methods. This also holds true for the study of natural, intestinal microbial communities. Whilst we gain much information from the study of enrichment cultures and colony counts on selective media, the use of molecular methods allows for increasing

G.R. Gibson and M.B. Roberfroid (eds.), Colonic Microbiota, Nutrition and Health, 167-190.
© 1999 *Kluwer Academic Publishers. Printed in the Netherlands.*

identification of micro-organisms and communities without the need for cultural isolation. In this chapter we will discuss molecular techniques which are being applied to more detailed investigations of gut micro-organisms as part of a complex community. Combining these molecular methods with cultural and biochemical techniques should help to elucidate interactions and effects of the microbial community on the host.

2   Predominant bacterial groups of the colon

Most current knowledge of the intestinal microbial community has been obtained using selective nutrient media to enumerate bacteria excreted in faeces. Table 1 presents a list of predominant bacterial groups with numerical ranges. To date, molecular methodologies have not changed this picture to any great extent. However, phylogenetic relationships and methodologies to investigate community structure are rapidly evolving. Comparisons of partial 16S rRNA sequences have already provided the most far-reaching of any phylogenetic explorations. The ability to detect and classify bacteria with or without culturing them has profoundly altered the phylogeny of colonic bacteria.

Table 1.   The numerically predominant anaerobic micro-organisms in the human colon*

Microbial Group	Range in Log Counts (g dry wt^{-1})
Bacteroides	9.2 - 13.5
Eubacteria	5.0 - 13.3
Bifidobacterium	4.9 - 13.4
Clostridia	3.3 - 13.1
Lactobacilli	3.6 - 12.5
Ruminococci	4.6 - 12.8
Peptostreptococci	3.8 - 12.6
Peptococci	5.1 12.9
Streptococci (anaerobic)	7.0 - 12.3
Methanobrevibacter	7.0 - 10.3
Desulfovibrios	5.2 - 10.9

*From Macfarlane and Gibson (1994).

The process consists of either amplification or cloning of rDNA from a bacterial population, sequencing of this rDNA and phylogenetic analysis of the sequences obtained. This approach has been applied successfully to bacteria causing several infectious diseases such as bacillary angiomatosis, human ehrlichiosis, Whipple's disease and

Tyzzer's disease (Wilson, 1994). This approach is now being applied to the human colonic microbiota (Wilson and Blitchington, 1994), however, the difficulties of extracting DNA from many organisms are well documented. Application of any generalised extraction protocol to a mixed, native microbial community such as that in faeces must surely bias, to some extent, the species composition of the DNA isolated. The previous chapter has detailed the taxonomy and systematics of predominant genera of gut bacteria. However, it is relevant that this review also summarises how phylogenetic relationships of human gut bacteria have evolved.

2.1   *BIFIDOBACTERIUM*

In human bacterial populations of the large intestine, the genus *Bifidobacterium* is considered the third most common after *Bacteroides* and *Eubacterium* (Finegold *et al.*, 1983). Members of the genus are saccharolytic and play a key role in the breakdown of complex carbohydrates that are resistant to hydrolysis by human digestive enzymes (Macfarlane and Gibson, 1994). Bifidobacteria may comprise up to 25% of the total cultivable gut flora, the most common isolates from adult humans are *Bif. adolescentis* and *Bif. longum*, while *Bif. infantis* is usually found in infants. This genus is of considerable interest due to its potential to promote health and colonisation resistance (Van der Waaij, 1989). In contrast to the genera *Bacteroides* and *Eubacterium*, *Bifidobacterium* form a monophyletic cluster, based on 16S rRNA sequences, within the high G+C Gram-positives (Larsen *et al.*, 1993). This has facilitated development of a genus-specific 16S rRNA-targeted oligonucleotide DNA probe for use in molecular ecological studies (Langenjidk *et al.*, 1995).

2.2   *LACTOBACILLUS*

Classification of the genus *Lactobacillus* highlights common discrepancies between results of traditional phenotypic tests and phylogenetic insights gained from 16S rRNA sequencing. At present, the genus is comprised of 56 species, five of which contain at least two subspecies. During the last few years, based on results of rRNA sequencing or DNA-rRNA hybridizations, nine *Lactobacillus* species were transferred to other species within the genus and at least 14 species were transferred to new or existing genera (Schleifer and Ludwig, 1995). The lactobacilli can now be divided into three main subgroups on the basis of 16S sequencing: *Lactobacillus delbreuckii* group, *Lactobacillus casei-Pediococcus* group and *Leuconostoc* group (Collins *et al.*, 1991). Predominant isolates from the large intestine fall into the first two groupings; the homofermentative *L. acidophilus* into the *L. delbreuckii* group while the heterofermentative *L. paracasei* is classified in the *L. casei-Pediococcus* group. Intestinal species such as *L. acidophilus* and *L. fermentum* are typically found at 10^9 CFU (CFU) per gram faeces. Additionally, this genus is of some importance in the human small bowel.

2.3   BACTEROIDES

Inclusion of fermentation acids in the taxonomy of the genus *Bacteroides* by Holdeman and Moore (1974) redefined this genus from inclusion of all non-sporing Gram-negative aerobes with rounded ends. The redefinition separated the genus *Fusobacterium*, whose major metabolic end-product is butyric acid. Phylogenetically, the nearest related genera to *Fusobacterium* and *Bacteroides* are *Flavobacterium* and *Cytophaga*, as defined by their 16S rRNA sequences (Weisburg *et al.*, 1985) The relationship, although not close, is specific enough that all these organisms are placed together into one of the 10 bacterial phyla of Woese *et al.* (1985); *Bacteroides* and *Fusobacterium* constitute one subdivision while *Flavobacterium* and *Cytophaga* form another. However, there are incoherent groupings within the *Fusobacterium* and *Bacteroides* sub-phyla, for example *B. succinogenes* was unrelated to any of the other 10 defined groups (Woese *et al.*, 1985) and was reclassified into the specifically named genus *Fibrobacter* (Amann *et al.*, 1992). Such inconsistencies highlight the phylogenetic problems which still exist in the classification of *Bacteroides*.

Colonic anaerobes classified as *Bacteroides* are probably the most metabolically important group of bacteria in the gut (Shah and Collins, 1989; Salyers *et al.*, 1985), representing approximately 30% of all culturable faecal isolates. Despite their predominance, currently available selective culture media contain additives, such as antibiotics, that are likely to lead to an underestimation of some species. The *B. fragilis* group of organisms (*B. fragilis, B. caccae, B. distasonis, B. eggerthii, B. merdae, B. ovatus, B. stercoris, B. thetaiotaomicron, B. uniformis* and *B. vulgatus*) contains the predominant species isolated from the human large intestine (Collins and Shah, 1987). Over 90% of faecal isolates belong to the *B. fragilis* group with the *B. vulgatus* and *B. fragilis* species considered to be the most numerous. Not surprisingly, most of the ground breaking applications of molecular techniques to detect and quantitate human gut and rumen anaerobes have involved this group.

Species-specific DNA hybridization probes were used to enumerate *Bacteroides* species in human faeces (Kuritza and Salyers, 1985; Kuritza *et al.*, 1986). Early application of the phylogenetic framework provided by 16S rRNA sequencing (Stahl *et al.*, 1988) to the analysis of microbial populations involved rumen isolates of *Fibrobacter* (*Bacteroides*) *succinogenes* (Amann *et al.*, 1992). The same framework is being applied to the human intestinal microflora. Manz *et al.* (1996) have reported a set of probes, including one specific for the *Bacteroides-Prevotella* phylogenetic group, applicable for *in situ* staining of target organisms in human faecal suspensions. More recently, a 16S-targeted hybridization probe for the *Bacteroides-Porphyromonas-Prevotella* group has been applied to quantitation of rRNA in total faecal rRNA (Doré *et al.*, 1988a) and for monitoring changes in faecal microbial populations in chemostats (Sharp *et al.*, 1998).

2.4   FUSOBACTERIUM

This genus is not as numerically significant as the *Bacteroides*, averaging about 10^8 CFU per gram of faeces, with numbers affected by diet (Finegold *et al.*, 1974). However, recent PCR and cloning experiments of the 16S rDNA gene from faecal community DNA have demonstrated that this genus could account for a greater proportion of the colonic bacterial population (Doré, personal communication). The classification of this genus has been in a state of almost constant flux. *Fusobacterium* are generally classified as Gram-negative, non-spore forming rods that produce butyric acid as their major fermentation end-product, lack respiratory quinones and contain glutamate dehydrogenase (Gharbia and Shah, 1988). Several species within this genus have now been transferred to other genera. For example, *F. symbiosum* produces spores and is now included as a *Clostridium*, whilst *F. plautii* has a Gram-positive cell wall and has been transferred out of the genus. Currently, 15 members of the genus are recognised, the most common species isolated from human faeces being *F. gonidiaformans, F. russi, F. necrophorum* and *F. necrogenes*.

2.5   *CLOSTRIDIUM, EUBACTERIUM* AND *RUMINOCOCCUS*

Up to 30 species of clostridia can be isolated from human faeces; some of the more prevalent being *C. inocuum, C. clostridioformi, C. perfringens, C. butyricum, C. aminovalericum* and *C. bifermentans. Clostridium inocuum* and *C. clostridioformi* are frequently isolated at the level of 10^{10} CFU per gram of faeces (Finegold *et al.*, 1974). Due to the nutritional diversity of this genus, which renders use of any one selective media impractical, it is possible that the metabolic significance of clostridia in the large intestine is underestimated. The taxonomy of the genus *Clostridium* was extremely fragmented until the application of complete 16S rRNA sequences was used to construct phylogenetic trees. A comprehensive phylogenetic analysis has been detailed by Collins *et al.* (1994) using a combination of new and old 16S rRNA sequences. The results confirmed the heterogeneity of the genus, many species are intermixed with other spore-forming and nonspore-forming genera. A hierarchical structure was proposed for the genus *Clostridium* and their close relatives, including five new genera and 11 new species combinations. The heterogeneity of this genera is accentuated by the difficulty in using biochemical tests to determine phylogenetic positions of cultural isolates; many common bacterial isolates from the large intestine are members of *Clostridium* or related genera such as *Eubacterium* and *Ruminococcus*.

Sequences of 16S rRNA show that all members of the genus *Ruminococcus* fall within the radiation of the genus *Clostridium*, but are also isolated from the genus. The *Ruminococcus* species are found to group in two distinct phylogenetic lineages (Rainey and Janssen, 1995). The 'true' ruminococci (*R. flavefaciens, R. callidus, R. albus* and *R. bromii*) form a group neighbouring the *Clostridium* IV cluster (Collins *et al.*, 1994), which includes *C. leptum* and *C. cellulosi*. The phylogenetic position of the second group (*R. torques, R. productus* and *R. hanesii*) falls within the XIVa cluster (Collins *et al.*, 1994), a large group including clostridia such as *C. polysaccharolyticum, C. nexile* and *C. sybiosum* as well as species from *Eubacterium* and *Lachnospira* genera. Rainey and Janssen (1995) added *R. obeum* and *R. gnavus* to this cluster. Common isolates from the large intestine include *R. albus, R. bromii, R. flavefaciens* and *R. (Pepto-streptococcus) productus*, with colony counts as high as 10^{10} per gram of faeces for some species (Finegold *et al.*, 1974).

Similarly, the phylogeny of the genus *Eubacterium* is in a state of re-organisation based on 16S rRNA sequence information. *Eubacterium* species are scattered throughout the genus *Clostridium*. One group branches between clusters IV and III (Collins *et al.*,

1994), cluster III contains cellulolytic *Clostridium* species. *Eubacterium*, as a genus, are quite metabolically important in the human large intestine, second only to *Bacteroides*. Seven of the 25 most prevalent species in the survey by Finegold *et al.* (1974) were from this genus, with *Eubacterium rectale* fourth most prevalent. The genus was proportionally greater on the Japanese diet than the Western diet (Finegold *et al.*, 1974).

3   Application of PCR technology to molecular microbial ecology of the large intestine

The polymerase chain reaction is used to amplify a target section of DNA. Amplification is usually by a factor of at least 10^5 and uses two synthetic oligonucleotide primers and DNA polymerase. The use of PCR has opened up a number of molecular based mechanisms to identify micro-organisms and evaluate *in situ* community structure.

3.1   rDNA SAMPLING OF HUMAN FAECES

In the large intestine, the ability to investigate microbial community composition by using phylogenetic analysis of 16S rDNA sequences offers another perspective that gained from traditional culture isolation methods. Molecular based methods have the additional advantage of providing sequences which are directly comparable, via phylogenetic tree diagrams, among different studies and samples. In contrast to most studies of other natural communities of bacteria, direct sampling of colonic microbiota rDNA by PCR and cloning techniques detects more or less the same organisms as culture-based studies.

Wilson *et al.* (1997) have compared established culture-based methods with direct amplification of partial sequencing of cloned 16S rRNA genes from a human faecal sample. Colonies and cloned amplicons were classified by comparing their rDNA sequences with sequences of known phylogeny. Quantitative culture recovered 58% of the microscopic count, the 48 colony types identified had 21 rDNA sequences. The authors estimated that 72% of rDNA sequences from the total population of culturable cells would match these 21 sampled sequences. When the number of PCR cycles was minimised, biodiversity was preserved and there was good agreement between culturing bacteria and sampling rDNA directly. The phylogenetic placement of the detected species was surprising. The vast majority of the species were found to fall into 4 phylogenetic clusters: *Bacteroides, Bifidobacterium, Clostridium coccides* and a cluster consisting of *Fusobacterium* plus *Clostridium leptum* and relatives (Wilson *et al.*, 1997). Two organisms detected by the rDNA analysis were not previously known to colonise the human intestine. The first, *Lachnospira pectinoschiza,* is readily culturable from the rumen but has not been detected in human faecal microflora before, while the other organism was most similar to planktomycetes. Conversely, two organisms abundant on culture plates were absent from the clones of rDNA; *Eubacterium rectale* and an unclassified cluster I *Clostridium* species.

3.2 ESTIMATION OF ORGANISM ABUNDANCE BY 16S rDNA GENE CONTENTS

Following sequence analysis of population contents, oligonucleotides specific for given genes can be synthesised or recovered as restricted fragments from the cloned genes. Hybridization of an organism-specific sequence to the bulk DNA derived from the natural sample is then a measure of the abundance of that gene in the original population. Total rDNA abundance can be estimated using hybridization with a primer complementary to all known 16S genes. The specific to total rDNA ratio is a measure of relative abundance and only approximates cell numbers, since gene copy numbers vary in different organisms. For example, Archaea have single copies, Bacteria commonly have 5 to 10 copies and Eukaryotes may have ·many hundred copies of rRNA genes. Nevertheless, and perhaps more importantly, the amount of a specific rDNA in a mixed population likely reflects the contribution of that organism to the local metabolic potential.

3.3 QUANTITATION OF BACTERIAL POPULATIONS USING PCR

A quantitative approach to PCR has recently been applied to the problem of detection and quantification of human faecal bacteria using serial dilutions of DNA extracted from human and animal faeces (Wang *et al.*, 1996). Primer sets based upon the sequence of the 16S rRNA gene were developed for 12 of the predominant colonic bacteria suggested by Finegold *et al.* (1974) including: *Bif. adolescentis, Bif. longum, E. limosum, F. prausnitzii, Peptostreptococcus productus, L. acidophilus, Escherichia coli, B. thetaiotaomicron, B. vulgatus, B. distasonis* and *C. clostridioformi*. Attempts to determine the proportion of each of these bacteria using culture-dependent techniques were restricted by the failure of many bacteria to be subcultured from initial mixed population isolation plates (31% did not grow in subculture). These results gave limited information; 55% of the isolates were *Bacteroides* species and 26% were *Eubacterium* species, with a small proportion of *Fusobacterium* and *Peptostreptococcus* species. The PCR method correlated well with these findings and provided more detailed information. The PCR titres (maximum dilution that yielded positive PCR results) for the same faecal sample were *B. vulgatus* 10^{-6}, *B. thetaiotaomicron* 10^{-5}, *F. prausnitzii* 10^{-6}, *P. productus* 10^{-6}, *C. clostridiofromi* 10^{-4} and *E. limosum* 10^{-4}, which correlate with previous reports of culture-dependant quantification of faecal microflora (Holdeman and Moore, 1974; Drasar and Roberts, 1990).

This application of the PCR method has advantages over culture techniques, in that PCR detects bacteria *in situ*, whereas culture methods detect bacteria after enrichment. Some species are unculturable or more difficult to grow than others making enumeration by culture methods growth dependent. In addition, PCR can detect unculturable or dead cells *in situ*. The PCR sensitivity of bifidobacteria was low, 10^4 cells for *Bif. adolescentis*, although this could have been due to the DNA isolation methods used (Wang *et al.*, 1996). Species of *Bifidobacterium* are Gram-positive and recalcitrant to lysis compared to Gram-negatives such as *Bacteroides*. Whether this type of PCR technique can replace culture-dependent methods depends upon the rigor, reproducibility, practical ease and economic viability of the technique compared to the established and accepted plate methods. Previously, the rate limiting step has been the isolation of DNA from faecal samples, however, a rapid boiling/Triton procedure described by Wang *et al.* (1996) should overcome that problem. As with all community PCR techniques the procedure is reliable to the same extent as DNA composition is representative of the microbial community in the original sample.

3.4   GENETIC FINGERPRINTING

Techniques for the genetic fingerprinting of microorganisms exploit the existence of polymorphisms. Restriction fragment length polymorphisms are formed by cutting with restriction endonucleases. This is followed by electrophoresis on an agarose gel, resulting in the specific banding patterns of the fragments. Formation of restriction fragment length polymorphisms can be done from genomic DNA or amplified fragments. Five techniques which exploit genomic polymorphisms are restriction fragment length polymorphism (RFLP), pulsed-field gel electrophoresis (PFGE), ribotyping, amplified ribosomal DNA restriction analysis (ARDRA) and amplified fragment length polymorphism (AFLP) (Table 2). With the exception of PFGE and AFLP, these techniques are often, and interchangeably, referred to as RFLP or ribotyping.

Table 2.   Summary of genetic fingerprinting techniques applicable to colonic microbiology

Technique	Mode of differentiation
Restriction fragment length polymorphism (RFLP)	Comparison of banding patterns obtained from digestion of chromosomal DNA with restriction endonuclease(s)
Pulsed-field gel electrophoresis* (PFGE)	Rare cutting restriction enzymes are utilised to simplify the RFLP (*i.e.* reducing the number of bands in RFLP)
Ribotyping*	Labelled rDNA probe is employed to highlight bands within the RFLP which contain rRNA gene sequences
Amplified ribosomal DNA restriction analysis (ARDRA)*	RFLP of rDNA amplified by PCR
Amplified fragment length polymorphism (AFLP)	Selective amplification of restriction fragments by high stringency PCR

* Specialist RFLP technique providing simplified banding patterns to accommodate interpretation.

Kullen *et al.* (1997) used ARDRA to follow the excretion of an ingested probiotic bifidobacterium. The banding patterns resulting from *Hae* III endonuclease digestion of 16S rDNA were specific enough to differentiate the fed from other bifidobacteria in the faeces. Ribotyping has been used to differentiate strains of *Bifidobacterium* (McCartney and Tannock, 1995), *Lactobacillus* (Johansson *et al.*, 1995; Ning *et al.*, 1997) and *Escherichia coli* (Tarkka *et al.*, 1994). While for most applications of RFLP and ribotyping, use of two or more restriction endonucleases, with unique target sites, is desired for adequate banding pattern discrimination (McCartney and Tannock, 1995). A number of researchers have had success with single endonucleases (Tarkka *et al.*, 1994; Ning *et al.*, 1997).

As an extension of its role in bacterial typing, PFGE has potential as an epidemiological tool. Matushek *et al.* (1996) were able to differentiate among eight to ten strains, of each, *E. coli, Klebsiella pneumoneae, Serratia marcescens, Staphylococcus aureus, Streptococcus pneumoniae, Enterococcus faecalis* and *Enterococcus faecium* in less than six days using one of two restriction enzymes (*Xba* I or *Sma* I). Pulsed-field gel electrophoresis has also been used to rapidly type clinical isolates of *Bacillus cereus* (Liu *et al.*, 1997). Whilst PFGE is generally thought to be more discriminatory than other RFLP methods, McCartney *et al.* (1996) found ribotyping to be more discriminatory for bifidobacteria isolated from human faeces, when *Xba* I was used as the PFGE cutting enzyme. For lactobacilli isolates, PFGE remained the more sensitive method for typing. Pulsed-field gel electrophoresis using *Sma* I or *Nru* I was more effective in differentiating *Clostridium difficile* strains isolated from human faeces than ribotyping with *Hind* III (Chachaty *et al.*, 1994). It should be noted that, for all of these methods, the choice of restriction enzymes is of paramount importance.

Amplified fragment length polymorphism combines the principle of restriction fragment length polymorphism analysis with highly specific PCR amplification. However, AFLP does not detect the length of the restriction fragments only their presence or absence (Huys *et al.*, 1996; Janssen *et al.*, 1996). This technique involves three steps: (i) restriction enzyme digest of genomic DNA followed by ligation of specific oligonucleotide adapters, (ii) selective amplification of some fragments in the total digest and (iii) acrylamide gel based analysis of the amplified products. Combinations of restriction enzymes are used for the initial digest often *Eco* RI - *Mse* I (for low G+C genomes), Apa I - Taq I (for high G+C genomes) or Hind III - Taq I (for G+C rich, 40 to 50 mol %, genomes) (Huys *et al.*, 1996; Janssen *et al.*, 1996; Kein *et al.*, 1997).

Advantages of AFLP as a taxonomic tool include relative speed of assay, its high resolution and reproducibility, discriminatory power, correlation with existing genomic sequences and prior sequence knowledge not being required. Amplified fragment length polymorphism has been used to differentiate among related bacterial strains such as *Xanthomonas, Aeromonas, Clostridium, Bacillus, Acinetobacter, Pseudomonas,* and *Vibrio* (Huys *et al.*, 1996; Janssen *et al.*, 1996) and among monomorphic bacterial strains, *Bacillus* (Kein *et al.*, 1997). Analysis by AFLP is more suited to investigations at the subgenomic level and is appropriate for subspecies differentiation. It may prove to be applicable to epidemiological investigations of pathogenic strains across geographical and generational isolation, as demonstrated by Kein *et al.* (1997) for *Bacillus anthracis.*

3.5   DENATURING GRADIENT GEL ELECTROPHORESIS OF HUMAN COLONIC BACTERIA

Denaturing gradient gel electrophoresis (DGGE) is a technique that separates DNA fragments of similar length but with different base-pairing, according to the melting property of each fragment. In combination with an initial PCR amplification step, this technique is an invaluable tool for examining microbial communities in complex ecosystems such as the gastrointestinal tract. The use of highly variable regions of DNA, such as the rDNA internal transcriber spacer regions 1 or 2 as the target for PCR, and DGGE maximises the potential for observing differences in highly related organisms (Kowalchuk *et al.*, 1997). The use of nested PCR, within a variable region of SSU rDNA, with a 5' 36-bp G+C rich sequence introduced by the PCR method, is increasingly used to study mixed microbial populations (Sheffield *et al.*, 1987). The GC clamp stabilises the melting behaviour of the DNA fragment, making it suitable for analysis by DGGE. Denaturing gradient gel electrophoresis has recently been used to study the ecology of bacterial communities from several different environments and has helped to identify important community members after PCR with general bacterial 16S primers (Muyzer *et al.*, 1993), primers specifically targeted to the 16S rRNA gene of monophyletic bacterial groups (Kowalchuk *et al.*, 1997) or primers directed at relevant structural genes (Wawer and Muyzer, 1995). Community fingerprints can subsequently be interpreted by comparison with reference patterns of known phylogenetic origin by hybridization with oligonucleotide probes targeted for internal band sites or by sequence analysis of excised bands.

Community fingerprinting using general 16S bacterial primers has been applied to bacterial samples obtained from the lumen and mucosa of the stomach small intestine, caecum and large intestine of the porcine gastrointestinal tract (Simpson *et al.*, 1997) and the rumen (Kocherginskaya *et al.*, 1997). The reproducibility of extraction, amplification and separation were high when general bacterial primers for the rDNA gene were used. Results showed that age, location (mucosa versus lumen) and gut site generated different banding patterns (Simpson *et al.*, 1997). These banding patterns were analysed and quantified using image analysis software to estimate relative abundance of bacterial assemblages from the gastrointestinal tract. The diversity in human colonic samples may often be too high for clear banding of a representative DNA sample using general bacterial rDNA primers; DGGE is probably better adapted to studies with a smaller defined mixture of bacteria, for example members of an individual genus.

One potential drawback of using a PCR-based approach is the introduction of bias due to the preferential amplification of certain templates. The use of two rounds of PCR could augment this further, thus a single PCR step is preferable. As with any electrophoresis-based technique, the DGGE analysis suffers from the limitation that although different band mobilities show non-identity, similar band mobilities do not prove identity. A DGGE banding pattern alone cannot be used to identify and phylogenetically place recovered PCR fragments. Furthermore, due to the relatively broad specificity of the given PCR strategy, especially in the absence, or very low, concentrations of some rDNA's, certain DGGE bands might represent DNA fragments not originating from the sample. In addition, DGGE band excision and sequence determination were necessary for full interpretation of the patterns.

4 Ribosomal RNA-targeted oligonucleotide DNA probes

A complete quantitative analysis of a microbial community in complex samples can be achieved by hybridizations with oligonucleotide probes complementary to the 16S rRNA of individual and phylogenetically coherent groups of organisms (Stahl *et al.*, 1988; Raskin *et al.*, 1994). This approach requires the extraction of RNA from the sample. The most common method of hybridization uses membrane supports to immobilise the recovered RNA. A labelled probe is allowed to hybridize to the immobilised 16S rRNA and the extent of hybridization is used as a measure of abundance of the 16S rRNA target population and, consequently, provides an estimate for corresponding organism abundance in the original sample. In addition, since rRNA content varies in proportion to growth rate, the extent of hybridization may also provide information on population activity. Radioisotope labelling is still preferred over non-radioactive techniques when high sensitivity (Stahl and Amann, 1991) and response linearity are important. Since these issues are generally of concern when using hybridizations for quantitative applications in microbial community characterisation, most of the methods described here relate to the use of radioisotope labels (^{32}P).

The general 16S rRNA approach was developed for the study of the cellulolytic rumen micro-organism *Fibrobacter succinogenes* (Stahl *et al.*, 1988). Species-specific 16S rRNA targeted oligonucleotide probes were developed to enumerate strains of *Lachnospira multiparus* and *F. succinogenes* in the bovine rumen before, during and after perturbations of this environment with ionophores. Changes in the 16S rRNA levels of these populations were monitored successfully using the 16S rRNA-targeted probes. A parallel, culture-based attempt to enumerate *F. succinogenes* was largely unsuccessful due to low numbers of the organisms present. The results demonstrated the ability of 16S rRNA probes to monitor shifts in population abundance and/or changes in population activity in complex microbial communities

4.1 PROBE DESIGN AND LABELLING

Currently, the 16S rRNA database contains more than 4000 sequences and this allows the design of DNA probes, usually consisting of 18-20 nucleotides that can detect a particular sequence in the 16S rRNA molecule (Maidak *et al.*, 1993). Information regarding published probes is now available on a database (Alm *et al.*, 1996). The procedure for developing a genus specific probe involves the following steps: (i) comparison of sequences by alignment, (ii) identification of a target sequence which is unique for the genus yet encompasses all member species, (iii) synthesis and labelling of complementary nucleic acid probes and (iv) experimental evaluation of the probe. Differences in the secondary and tertiary structure of the 16S molecule may result in sterical hindrance which prevents effective probe binding, so careful probe evaluation is vital. A detailed review of probe design and usage is provided by Amann *et al.* (1990).

The oligonucleotide probe database (OPD) has been established to circumvent difficulties often encountered when relying on published literature for information about rRNA probes and primers (Alm *et al.*, 1996). The OPD centralises information related to the design and use of oligonucleotide probes and primers. For each probe or primer, information in the OPD includes design and characterisation data important for probe and primer use; including standardised name (nomenclature format is included), probe sequences, nucleotide position within the target gene, optimal hybridization and wash conditions (or annealing conditions for PCR primers), intended target group specificity and original citations.

4.2   rRNA EXTRACTIONS AND HYBRIDIZATION

4.2.1 *rRNA extraction*
Mechanical disruption using a reciprocation shaker with zirconium beads is used for ribosomal RNA extraction from gut samples, a slightly modified method from that previously described by Stahl *et al.* (1988). Samples in screw cap tubes are initially subjected to direct phenol extraction with mechanical disruption by zirconium beads. All glassware and solutions should be prepared using conventional techniques to maintain an RNase free working environment. Mechanical disruption is followed by another extraction with phenol saturated with Tris-Cl Buffer, two sequential phenol: chloroform: isoamylalcohol and chloroform: isoamylalcohol extractions each. Total rRNA is precipitated at -20 °C for 3h with ammonium acetate. After two washes in ethanol, pellets can be resuspended in double-distilled water. Concentrations of nucleic acid are estimated spectrophotometrically by assuming that an OD value of 1 at 260 nm indicates a concentration of 40 µg RNA per ml. The quality of the extracted RNA is evaluated using polyacrylamide gel electrophoresis.

4.2.2 *rRNA preparation and blotting*
Nucleic acids are denatured and diluted as described by Raskin *et al.* (1994). Samples are applied in triplicate specially charged membranes using a slot- or dotblot device under a slight vacuum. Membranes are air dried then baked at 80 °C for 2 h. Baked membranes are prewetted in hybridization buffer (Sambrook *et al.,* 1989) and placed in hybridization tubes. Membranes are incubated with approximately 10 ml of hybridization buffer for 2 h at 40 °C in a rotating incubator. The first hybridization buffer is discarded and labelled probe added by inclusion in a volume of hybridization buffer equal to that added the first time. Incubation is then continued at 40 °C for 16 h to 20 h. After incubation with the probe, membranes are washed in the hybridization tubes with a SDS/SSC solution for 2 h at 40 °C. Membranes are then removed from hybridization tubes and washed twice at the experimentally determined dissociation temperatures (Td) for individual probes (Stahl and Amann, 1991). The synthetic HPLC purified oligonucleotide probes used are 5' end labelled with ^{32}P as previously described (Raskin *et al.*, 1994).

4.2.3 *Imaging of hybridization signal*
Hybridization signal on air dried membranes is quantified using an imager such as Instant Imager (Canberra Packard, Pangbourne, UK) or 400A PhosporImager, which requires Phosphor screens (Molecular Dynamics, Sunnyvale, CA, USA). The time of exposure varies depending upon the intensity of the ^{32}P signal. Analysis of the signal is performed by software like ImageQuant (Molecular Dynamics). Abundances of specific groups of organisms are presented as both percentage of total bacterial SSU rRNA

in the sample and as total amounts of RNA per unit sample mass. Standard curves are calculated from reference RNA by linear regression.

4.3   DETERMINING COMPOSITION OF BACTERIAL POPULATIONS

Ribosomal RNA analysis confirms the domination of the colonic microflora with the domain Bacteria, accounting for some 78.2% of total rRNA in an average sample (Doré *et al.*, 1998b). The domains Archaea and Eukarya show relatively minor contributions to total rRNA (0.28 and 2.02 % respectively). The Achaeal signal is likely attributable to populations of methanogens commonly found in some individuals (Gibson *et al.*, 1988) since these are the only Archaea that have been isolated from the large intestine of any animal. The suite of probes for the methanogenic Archaea described by Raskin *et al.* (1994) has been applied to human faeces by Doré *et al.* (1998b) and this showed the genera *Methanobacteriacaea* to be the predominant group, accounting for approximately 60% of the Archaeal hybridization signal. The most commonly isolated faecal methanogens, such as *Methanobacterium smithii,* are encompassed by this probe group. The Eukaryotic signal was suspected to come from residual rRNA of sloughed enterocytes, although yeasts have been isolated from human faeces.

The full extent of the diversity of the bacterial populations in the human large intestine is far from clear, therefore a suite of probes cannot be designed to completely circumscribe predominant populations which constitute the bacterial domain in the colon. Sequencing of 16S rDNA has provided more insight into the diversity. Using this information, probes can be designed for groupings of bacteria that appear to be important. The lack of phylogenetic information for metabolically significant groupings of bacteria available for probe design is exacerbated by the lack of phylogenetic coherence for the few established groups. However, probes are now available and have been tested for key groups such as *Bacteroides* and *Bifidobacterium.*

The *Bacteroides-Porphyromonas-Prevotella* group probe described by Doré *et al.* (1998a) and adapted from Manz *et al.* (1996) was demonstrated as a useful molecular marker for the phylogenetic cluster which includes many of the *Bacteroides* from the large intestine. These include *B. vulgatus, B. fragilis, B. thetaiotaomicron, B. ovatus, B. uniformis* and *Prevotella distasonis.* Molecular quantitation data (Fig. 1) is consistent with previously published population levels for *Bacteroides* spp. in human faeces (Finegold *et al.*, 1983; Holdemann and Moore, 1974) determined using culture-based techniques. However, culture-based determinations of total *Bacteroides,* were much lower than values previously reported or achieved using 16S rRNA hybridizations (unpublished data). The genus accounted for 48.9 and 36.7% of the total ribosomal abundance for chemostats inoculated with human faeces and operated at dilution rates of 0.30 and 0.03 h^{-1}, respectively. The results are consistent with molecular data described by Doré *et al.* (1998a). Using the same probe sequence; a study of infant faeces revealed that this group accounted for 30% of the relative proportion of 16S rRNA shortly after weaning. These results also confirmed the metabolic significance of the genus in the colonic microflora as a whole.

The probe targeting the genus *Bifidobacterium* developed by Langejidk *et al.* (1995) for fluorescent labelling studies in faeces has had its specificity determined against a range of colonic bacterial species (unpublished data). We have determined the precise Td of the probe for quantitative hybridization studies with community 16S rRNA (62 °C). The abundance of rRNA corresponding to this genus was surprisingly low in chemostats inoculated with human faeces (4.8 and 4.1% (w/v) for dilution rates of 0.03 and 0.3 h^{-1}, respectively) (Fig. 2). This compared to proportions of 15.8 and 12.6% determined using plate counts to quantify *Bifidobacterium* and total anaerobes. However, bifidobacteria can be overestimated by culture-dependent methods. *Bifidobacterium* in faeces are easily culturable, much more reliably than the diverse group which constitute total anaerobes. In the studies of Langenjidk *et al.* (1995) and Welling *et al.* (1997), fluorescent probe-based estimates of bifidobacteria have equalled culture counts, whereas total anaerobes have been underestimated by a factor of 10 using culture-based techniques.

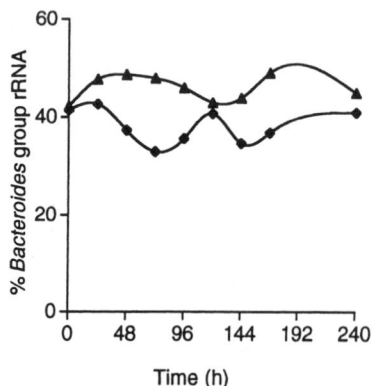

Figure 1. Abundance of *Bacteroides* group rRNA in fermenters with resistant starch as the main source of carbohydrate and operated at dilution rates of 0.03 (◆) and 0.30 h^{-1}(▲). Ribosomal RNA samples were blotted onto nylon membranes and hybridized with ^{32}P labelled probes for total bacterial rRNA (S-D-Bact-0338-a-A-18) and *Bacteroides* rRNA (S-G-Bac-1080-a-A-18), in triplicate for each sample.

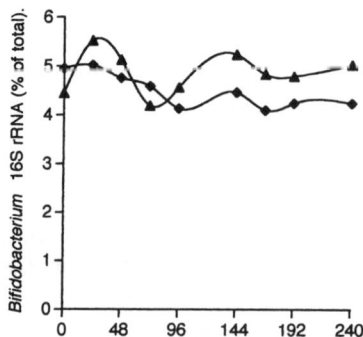

Figure 2. Abundance of *Bifidobacterium* genus rRNA in fermenters with resistant starch as the main source of carbohydrate and operated at dilution rates of 0.03 (◆) and 0.30 h^{-1} (▲). Ribosomal RNA samples were blotted onto nylon membranes and hybridized with ^{32}P labelled probes for total bacterial rRNA (S-D-Bact-0338-a-A-18) and *Bifidobacterium* genus rRNA (S-G-Bif-162-a-A-18), in triplicate for each sample.

More recently, Doré *et al.* (1998b) achieved similar results applying the same technique to human faeces and found that the hybridization signal obtained with their *Bifidobacterium* genus probe accounted for approximately 5% of the total signal, compared to approximately 10% using culture plates.

The advent of increasing numbers of sequences in databases from rDNA cloning of colonic microbiota has increased our knowledge base for the design of rRNA targeted oligonucleotide probes. A paradigm for the interpretation of sequence data obtained by shotgun cloning into novel 16S rRNA-targeted oligonucleotide probes for the assessment of the metabolic significance of new groups is provided by Doré *et al.* (1998a). Low G+C Gram-positives; which include genera such as *Streptococcus*, *Lactococcus*, *Enterococcus*, *Lactobacillus*, *Clostridium*, *Eubacterium*, *Ruminococcus* and others, are of great metabolic significance. Sequencing experiments of shotgun cloning of 16S genes amplified from faecal DNA have shown that this group usually accounts for the majority of isolates. This group can only be currently accounted for by a broad group probe which has mismatches for some representatives of all genera concerned. It could therefore only give a partial measure of low G+C Gram-positives. The rRNA represented 24.2 ± 5.5% of bacterial rRNA based on this probe. Improved phylogenetic characterisation of these groups will substantially improve the capacity for probe design and metabolic characterisation.

Attempts have previously been made to classify the bacteria of the large intestine into two groups based upon their G+C content. Two additive probes targeting the same region of, in this instance the 23S rRNA gene (Roller *et al.*, 1994), the one specific for high G+C Gram-positives and the other for all other bacteria, hybridized with 1.4% and 80.6% of bacterial rRNA respectively. The average sum of the corresponding signals was 80.2 ± 8.4% of total bacterial rRNA (Doré *et al.*, 1998a). A further probe was also used in the experiments of Doré *et al.* (1998a) for the enterics, this group constituted 2.7 ± 0.7% of total bacterial rRNA. Therefore using the currently available suite of probes, covering the genera *Bacteroides* and *Bifidobacterium*, and the enteric groups and low G+C Gram-positives, approximately 65% of the bacterial rRNA could be attributed to phylogenetic groups (Table 3).

Table 3. Ribosomal RNA abundances for the genus *Bifidobacterium*, the *Bacteroides Porphyromonas-Prevotella* cluster, the enterics group and the phylum of low G+C Gram-positives measured by quantitative dot blot hybridizations using the Bacteria domain probe as reference.

Bacterial Group	% of Total bacterial rRNA
Bifidobacteria	5.1 (± 1.1)
Bacteroides	35.5 (± 4.5)
Enterics	2.7 (± 0.7)
Low GC Gram-positives	24.2 (± 5.5)
Sum of probed groups	67.9

Data taken from Doré *et al.* (1998a).

It is likely that further sequence information will yield information needed to more completely circumscribe the phylogenetic groups of the colon with a suite of probes. Doré *et al.* (1998a) speculate that the bacterial rRNA not accounted for by the probes used in their study may be essentially from low G+C Gram-positives. These would then account for 47 to 67% of bacterial rRNA. Alternatively, the faecal flora that has to date not been cultured could also account for a part of this discrepancy between the domain probes targeting a conserved region and group probes based on more variable regions of sequences derived from cultured organisms.

4.4   MONITORING ENVIRONMENTAL CHANGE

In studies of the compositional dynamics and metabolic activity of the complete gut flora, genus-specific probes for the identification of distinct phylogenetic groups combine rapid monitoring with useful taxonomic resolution. That is, when applied in concert, a limited number of genus-specific probes can be used to describe the overall composition and metabolic potential of the total population. An approach to validating the robustness of such a technique for monitoring microbial populations, is to verify the utility of the method for interpretation of the perturbation of a microbial community, as commonly seen in ruminants. When large quantities of readily fermentable carbohydrates, such as starch, are added to the diet, ruminal pH decreases to between 5.5 and 4. This is characteristic of sub-acute acidosis, thought to be associated with the overgrowth of *Streptococcus bovis* and the inability of lactate utilising bacteria to utilise all the lactic acid produced.

In an unpublished study to validate the rRNA hybridization technique for studying rumen microbial populations; populations of methanogens, protozoa, total bacteria and composite populations of fibre-degrading bacteria (*Fibrobacter*) and *Streptococcus bovis* were determined using rRNA hybridizations to monitor ribosomal abundance for each population. Lactating dairy cows were used for this study, they were fed every 12 h for 5 d, sampled for 12 h, fasted for 12 h and then fed the acidotic diet for 24 h. The acidotic diet (20% chopped alfalfa hay and 80% grain mix) was the inverse of the basal diet (80% chopped alfalfa hay and 20% grain mix).

The relative abundance of bacterial rRNA in the rumen was negatively correlated with pH, while eukaryote (predominantly ciliate protozoa in the rumen) rRNA abundance was positively correlated with the decline in pH (Fig. 3). The amount of archaeal rRNA was not statistically correlated with pH but was affected by the overall process of acidosis, values declined between 4 and 6 h after the induction of acidosis. From 12 to 16 h, archaea returned to levels equivalent to those prior to induction of acidosis but began to fall again by 20 h (Fig. 4). This final reduction was one of the striking features of the study, that the abundance of archaeal rRNA declined from 1.43 to 0.37 µg per gram of ruminal contents between the 0 and 24 h time points and remained low for the remaining 48 h of the study. During this period, the composite methanogen populations changed such that the proportion of *Methanobacteriaceae* increased and the significance of the group encompassed by the families *Methanoplanaceae*, *Methanomicrobiaceae* and *Methanocorpusculaceae* declined (Fig. 5). The genus *Fibrobacter* was the population most positively correlated with pH changes. *Fibrobacter* rRNA was virtually undetectable between 16 and 24 h after the induction of acidosis (Fig. 6). This is a clear reflection of the relation between the metabolic activities of microbial populations and fermentation processes. Fibre degradation is negatively correlated with rumen pH, acutely so during periods of acidosis. Surprisingly, the abundance of *Streptococcus bovis* was not correlated with pH, but was associated positively with total SCFA concentration, populations increased up to 20 h and remained elevated in comparison to levels measured prior to fasting and induction of acidosis when it was less than 2% of total rRNA (Fig. 6).

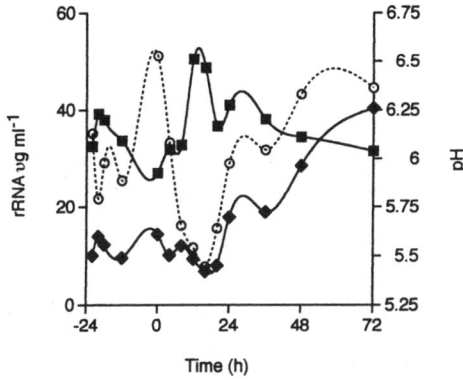

Figure 3. Abundance of bacterial (■) and eucaryotic (◆) rRNA and pH (○) in the rumen of cows (n = 4) before, during and after the induction of acidosis. Ribosomal RNA samples were blotted onto nylon membranes and hybridized with ^{32}P labelled probes for total bacterial rRNA (S-D-Bact-0338-a-A-18) and eucaryotic rRNA (S-D-Euca-0502-a-A-16), in triplicate for each sample.

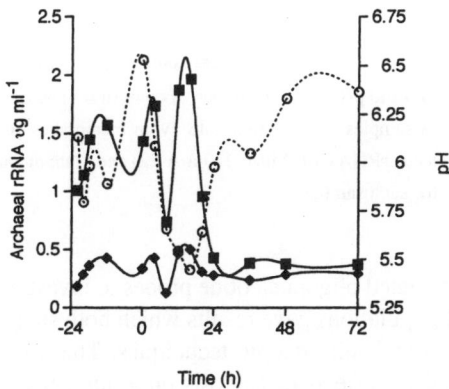

Figure 4. Abundance of archaeal (■) and *Methanobacteriaceae* rRNA (◆) and pH (○) in the rumen of cows (n = 4) before, during and after the induction of acidosis. Ribosomal RNA samples were blotted onto nylon membranes and hybridized with ^{32}P labelled probes for archaeal rRNA (S-D-Arc-0915-a-A-20) and *Methanobacteriaceae* rRNA (S-F-Mbac-310-a-A-22), in triplicate for each sample.

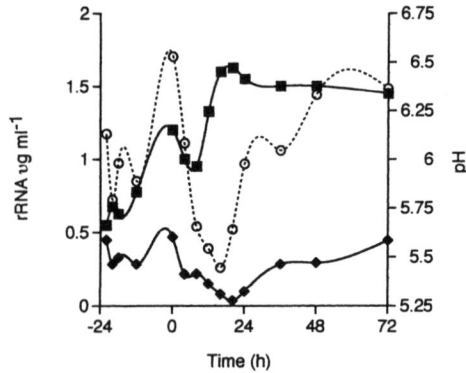

Figure 5. Abundance of *Streptococcus bovis* rRNA (■) and *Fibrobacter* rRNA (◆) and pH (○) in the rumen of cows (n = 4) before, during and after the induction of acidosis. Ribosomal RNA samples were blotted onto nylon membranes and hybridized with ^{32}P labelled probes for archaeal rRNA (S-D-Arc-0915-a-A-20) and *Methanobacteriaceae* rRNA (S-F-Mbac-310-a-A-22), in triplicate for each sample.

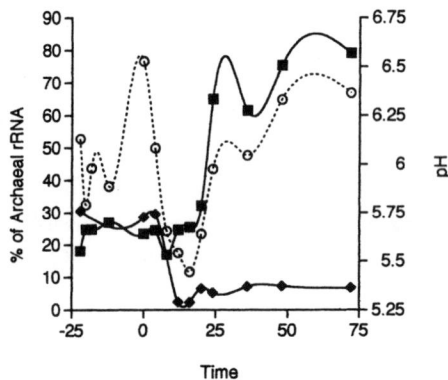

Figure 6. Abundance of *Methanobacteriaceae* rRNA (■) and *Methanogenium* rRNA (◆) (expressed as a proportion of total archaeal rRNA) and pH (○) in the rumen of cows (n = 4) before, during and after the induction of acidosis. Ribosomal RNA samples were blotted onto nylon membranes and hybridized with ^{32}P labelled probes for *Methanobacteriaceae* rRNA (S-F-Mbac-310-a-A-22) and *Methanomicrobiaceae* rRNA(S-F-Mmic-1200-a-A-21), in triplicate for each sample.

The use of rRNA-targeted oligonucleotide probes to investigate the effects of acidosis on ruminal microbial populations gave results which both supported and extended those reported using cultural and microscopic techniques. The use of probes gave a higher resolution and allowed the effects of acidosis on a number of populations to be determined simultaneously. By following changes that occurred in community structure during the course of acidosis, we confirmed the validity of 16S rRNA-targeted probes as a way to monitor microbial populations changes during both the perturbation and recovery of a complex microbial community.

4.5 QUANTITATIVE FLUORESCENT WHOLE CELL HYBRIDIZATIONS

Advances in light microscopy, particularly fluorescent *in situ* hybridization (FISH), and confocal microscopy combined with advances in molecular biology and image analysis software has revolutionised the way microbiologists can observe microbial communities. Single cells can be identified following hybridization of fixed specimens to DNA probes complementary to the rRNAs. The bacterial cell wall, with the exception of some thick cell walled micro-organisms, is made permeable to DNA probes labelled with fluorescent dyes. The correlation between cellular ribosome content and growth rate was established by Scheacter *et al.* (1958). Probe conferred fluorescence corresponds to individual cell ribosome content and, by inference, to single cell activity. However, growth rate related changes in ribosome abundance has been examined for relatively few environmentally relevant micro-organisms. Biases can be introduced by variable probe penetration and possible growth rate associated changes in cell wall permeability. This technique has been applied in a qualitative manner to a number of mixed communities of bacteria, marine planktonic bacteria (Lee and Kemp, 1994), plant rhizospheres (Macnaughton *et al.*, 1996), fresh water biofilms (Szewzyk *et al.*, 1994), endosymbionts of rumen protozoa (Lloyd *et al.*, 1996), soil bacteria (Thomas *et al.*, 1997) and human faeces ((Langendijk *et al.*, 1995). Whilst this has provided great insight into the spatial arrangement of bacterial populations within communities this technique can also be used to enumerate bacteria that are resistant to enumeration by culture-based techniques.

Quantitative fluorescence *in situ* hybridization can detect whole cell to a limit of 10^6 bacteria per gram of faeces (Langendijk *et al.*, 1995). The contribution of an individual genus to the total bacterial population can be determined if total numbers of bacteria are determined using a fluorescently labelled probe for the bacterial domain and those of a particular genus are quantified using a genus specific probe. Image analysis software is required to measure fluorescence parameters such as the level of fluorescence per cell and the hybridization percentage (Jansen *et al.*, 1993). This technique was used by Langendijk *et al.*, (1995) to enumerate bifidobacteria in human faeces. Virtually all the bacteria that were present in the samples as determined by fluorescent in situ hybridization could be cultured, however, the technique could be applied to a genus or species of bacteria that is much more recalcitrant to culturing.

There are however drawbacks to this probe technique. Firstly, a probe is only as reliable as the available sequence information. Once a good probe has been identified, the labelled probe still has to reach the target sequence of the rRNA molecule. Therefore, the bacterial cell wall has to be penetrated. This has proved difficult for Gram-positive bacteria like lactobacilli and bifidobacteria. The FISH procedure can be different for various genera of bacteria, thus in a mixed faecal culture the worse case scenario for cell wall penetration has to be assumed. Neither of these problems are insurmountable and the technique should be an invaluable tool for monitoring the composition of the intestinal flora in health and disease. It should be recognised, however, that data obtained using fluorescently labelled probes will provide the same information as any cell counting technique. Indices of ribosomal abundances will provide a clearer picture of the metabolic potential of a bacterial population.

5   Conclusions

Within the bacterial domain, available probes are restricted by the state of the phylogeny of the major genera in the colon. Sequencing of the 16S rDNA gene from cultured bacteria and shotgun cloning of molecular isolates is a laborious procedure, yet it will increase our understanding of phylogeny of the major genera. Using this information, molecular ecology studies can advance to a point where it will be possible to circumscribe a significant proportion of gut bacterial rRNA with a limited number of probes. These probes can then be used to assess environmental changes and the metabolic potential of mixed populations of bacteria.

The techniques detailed here are either based upon the detection of the genomic DNA or the analysis of rRNA itself. The detection of genes reveals the microbes present, including both dormant and inactive organisms. Because of higher numbers of ribosomes in metabolically active cells than in dormant cells, it is assumed that the analysis of rRNA reflects the active organisms. A more specific approach to monitor microbial activity and to understand the function of the community members, however, is the analysis of mRNA. Detection of mRNA can ensure that the target gene, and therefore the corresponding micro-organism is active. This approach has been used to study the expression of the (NiFe) hydrogenase gene of *Desulfovibrio* spp. in environmental samples (Wawer *et al.*, 1997) but has yet to be applied to gut microbial communities.

References

Alm, E., Oether, D.B., Larsen, N., Stahl, D.A. and Raskin, L. (1996) The oligonucleotide probe database. *Applied and Environmental Microbiology* **62**, 3557-59.

Amann, R.I., Krumholz, L. and Stahl, D.A. (1990) Fluorescent-oligonucleotide probing of whole cells for determinative, phylogenetic and environmental studies in microbiology. *Journal of Bacteriology* **172**, 767-70.

Amann, R.I., Lin, C.H., Key, R., Montgomery, L. and Stahl, D.A. (1992) Diversity among *Fibrobacter* isolates: Towards a new phylogenetic classification. *Systematic and Applied Microbiology* **15**, 23-31.

Amann, R.I., Ludwig, W. and Schleifer, K.-H. (1995) Phylogenetic identification and in situ detection of individual microbial cells without cultivation. *Microbiological Reviews* **59**, 143-69.

Chachaty, E., Saulnier, P., Martin, A., Mario, N. and Andremont, A. (1994) Comparison of ribotyping, pulsed-field gel electrophoresis and random amplified polymorphic DNA for typing *Clostridium difficile* strains. *FEMS Microbiology Letters* **122**, 61-68.

Collins, M.D. and Shah, H.N. (1987) Recent advances in the taxonomy of the genus *Bacteroides*. In *Recent Advances in Anaerobic Microbiology*. Proceedings of the fourth anaerobe discussion group symposium. pp. 248-258. Martinus NijHoff Publishers, Dordrecht.

Collins, M.D., Rodrigues, U.M., Ash, C., Aguirre, M., Farrow, J.A.E., Matinez-Murcia, A., Phillips, B.A., Williams, A.M. and Wallbanks, S. (1991) Phylogenetic analysis of the genus *Lactobacillus* and related lactic acid bacteria as determined by reverse transcriptase sequencing of 16S rRNA. *FEMS Microbiology Letters* **95**, 235-40.

Collins, M.D., Lawson, P.A., Willems, A., Cordoba, J.J., Fernandez-Garayzbal, J., Garcia, P., Cai, J., Hippe, H. and Farrow, J.A.E. (1994) The phylogeny of the genus *Clostridium*: Proposal of five new genera and eleven new species combinanations. *International Journal of Systematic Bacteriology* **44**, 812-26.

Conway, P.L. (1995) Microbial ecology of the human large intestine. In *Human Colonic Bacteria: Role in Nutrition, Physiology, and Pathology*. Gibson, G.R. and Macfarlane, G.T. (Eds.). pp. 1-24. CRC Press, Boca Raton, FL.

Doré, J., Sghir, A., Hannequart-Gramet, G., Corthier, G. and Pochart, P. (1998a) Design and evaluation of a 16S rRNA-targeted oligonucleotide probe for specific detection and quantitation of human faecal *Bacteroides* populations. *Systemic and Applied Microbiology* (in press).

Doré, J., Gramet, G., Goderel, I. and Pochart, P. (1998b) Culture independent characterisation of the human fecal flora using rRNA-targeted hybridization probes. *Genetics Selection and Evolution* (in press).

Drasar, B.S. and Roberts, A.K. (1990) Control of the large bowel microflora. In *Human Microbial Ecology*. Hill, M.J. and Marsh, P.D. (Eds.). pp. 95-100. CRC Press Inc., Boca Rata, FL.

Finegold, S.M., Atterby, H.R. and Sutter, V.I. (1974) Effect of diet on human fecal flora: Comparison of Japanese and American diets. *American Journal of Clinical Nutrition* **27**, 1456-69.

Finegold, S.M., Sutter, V.L. and Mathison, G.E. (1983) Normal indigenous intestinal flora. In *Human Intestinal Microflora in Health and Disease*. Hentges, D.J. (ed.). pp. 3-31. Academic Press, New York.

Gharbia, S.E. and Shah, H.N. (1988) Growth responses to glucose and protein hydrolysates by *Fusobacterium* species. *Current Microbiology* **17**, 229-34.

Gibson, G.R., Cummings, J.H. and Macfarlane, G.T. (1988) Use of a three-stage continuous culture system to study effects of mucin on dissimilatory sulfate reduction and methanogenesis by mixed ruminal popula-tions of human gut bacteria. *Applied and Environmental Microbiology* **54**, 2750-55.

Holdemann, L.V. and Moore, W.E.C. (1974) Genus I. *Bacteroides* (Castellani and Chalmers, 1919). In *Bergeys Manual of Determinative Bacteriology*. 8th ed. Buchanan, R.E. and Gibbons, N.E. (Eds.). p. 385. Williams and Wilkins, Baltimore.

Huys, G., Coopman, R., Janssen, P. and Kersters, K. (1996) High-resolution genotypic analysis of the genus *Aeromonas* by AFLP fingerprinting. *International Journal of Systematic Bacteriology* **46**.

Jansen, G.J., Wilkinson, M.H.F., Deddens, B. and Van der Waaij, D. (1993) Characterisation of human fecal flora by means of an improved fluoromorphometrical method. *Epidemiology and Infection* **111**, 265-72.

Janssen, P., Coopman, R., Huys, G., Swings, J., Bleeker, M., Vos, P., Zabeau, M. and Kersters, K. (1996) Evaluation of the DNA fingerprinting method AFLP™ as a new tool in bacterial taxonomy. *Microbiology* **142**, 1881-93.

Johansson, M.-L., Molin, G., Pettersson, B., Uhlén, M. and Ahrné, S. (1995) Characterization and species recognition of *Lactobacillus plantarum* strains by restriction fragment length polymorphism (RFLP) of the 16S rRNA gene. *Journal of Applied Bacteriology* **79**, 536-41.

Keim, P., Kalif, A., Schupp, J., Hill, K., Travis, S.E., Richmond, K., Adair, D.M., Hugh-Jones, M., Kuske, C.R. and Jackson, P. (1997) Molecular evolution and diversity in *Bacillus anthracis* as detected by amplified fragment length polymorphism markers. *Journal of Bacteriology* **179**, 818-24.

Kocherginskaya, S.A., Simpson J.M. and White, B.A. (1997) Microbial community structure of the rumen as assessed by denaturing gradient gel elctrophoresis of polymerase chain-reaction amplified genes. p. 37. *Proceedings of the Evolution of the Rumen Microbial Ecosystem. Rowett Research Institute, Aberdeen, UK*.

Kowalchuk, G.A., Gerards, S. and Woldendorf, J.W. (1997) Detection and characterisation of fungal infections of *Ammohila arenaria* (Marram grass) roots by denaturing gradient gel elctrophoresis of specifically amplified 18S rDNA. *Applied and Environmental Microbiology* **63**, 3858-65.

Kullen, M.J., Amann, M.M., O'Shaughnessy, M.J., O' Sullivan, D.J., Busta, F.F. and Brady, L.J. (1997) Differentiation of ingested and endogenous bifidobacteria by DNA fingerprinting demonstrates the survival of an unmodified strain in the gastrointestinal tract of humans. *Journal of Nutrition* **127**, 89-94.

Kuritza, A.P. and Salyers, A.A. (1985) Use of species specific DNA hybridization probes for enumerating *Bacteroides vulgatus* in human faeces. *Applied and Environmental Microbiology* **50**, 958-64.

Kuritza, A.P., Shaughnessy, P. and Salyers, A.A. (1986) Enumeration of polysaccharide degrading *Bacteroides* species in human feaces using species-specific DNA probes. *Applied and Environmental Microbiology* **51**, 385-90.

Larsen, N., Olsen, G.J., Maidak, B.L., McCaughey, M.J., Overbeek, R., Macke, T.J., Marsh, T.L. and Woese, C.S. (1993) The ribosomal RNA database project. *Nucleic Acids Research* **21**, 3021-23.

Langenjidk, P.S., Schut, F., Jansen, G.J., Raangs, G.C., Kamphius, G.R., Wilkinson, M.H.F. and Welling, G.W. (1995) Quantitative fluorescence in situ hybridization of *Bifidobacterium* sp. with genus-specific 16S rRNA-targeted probes and its application to fecal samples. *Applied and Environmental Microbiology* **61**, 3069-75

Lee, S.H. and Kemp, P.F. (1994) Single-cell RNA content of natural marine planktonic bacteria measured by hybridization with multiple 16S ribosomal-RNA-targeted oligonucleotide probes. *Limnology and Oceanography* **39**, 869-79.

Liu, P.Y.-F., Ke, S.-C. and Chen, S.-L. (1997) Use of pulsed-field gel electrophoresis to investigate a pseudo-outbreak of *Bacillus cereus* in a pediatric unit. *Journal of Clinical Microbiology* **35**, 1533-35.

Lloyd, D., William, A.G., Amann, R., Hayes, A.J., Durrant, L. and Ralphs, J.R. (1996) Intracellular prokaryotes in rumen cilioate protozoa: Detection by confocal scanning microscopy after in situ hybridization with fluorescent 16S rRNA probes. *European Journal of Protistology* **32**, 523-31.

Macfarlane, G.T. and Gibson, G.R. (1994) Metabolic activities of the normal colonic microflora. In *Human Health: The Contribution of Microorganisms*. Gibson, S.A.W. (ed.). pp. 17-52. Springer -Verlag, London.

Macnaugthon, S.J., Booth, T., Embley, T.M. and O'Donnell, A.G. (1996) Physical stabilisation and confocal microscopy of bacteria on roots using 16S rRNA-targeted fluorescent-labelled oligonucleotide probes. *Journal of Microbiological Methods* **26**, 279-85.

Maidak, B.L., Larsen, N., McCaughey, M.J., Overbeek, R., Olsen, G.J., Fogel, K., Blandy, J. and Woese, C.R. (1994) The Ribosomal Database Project. *Nucleic Acids Research* **22**, 3485-87.

Manz, W., Amann, R.I., Ludwig, W., Vancanneyt, M. and Schleifer, K.H. (1996) Application of a suite of 16S rRNA probes designed to investigate bacteria of the phylum cytophaga-flavobacter-bacteroides in the natural environment. *Microbiology* **142**, 1097-1106.

Muyzer, G., De Waal, E.C. and Uitterlinden, A.G. (1993) Profiling complex microbial populations by denaturing gradient gel electrophoresis analysis of polymerse chain reaction-amplified genes coding for 16S rRNA. *Applied and Environmental Microbiology* **59**, 695-700.

Ralney, F.A. and Janssen, P.H. (1995) Phylogenetic analysis by 16S rDNA sequence comparison reveals two unrelated groups of species within the genus *Ruminococcus*. *FEMS Microbiology Letters* **129**, 69-74.

Raskin, L., Stromley, J.M., Rittman, B.E. and Stahl, D.A. (1994) Group-specific 16S rRNA hybridization probes to describe natural communities of methanogens. *Applied and Environmental Microbiology* **60**, 1232-40.

Roller, C., Wagner, M., Amann, R., Ludwig, W. and Schleifer, K.H. (1994) In situ probing of Gram-positive bacteria with high DNA G+C content using 23S rRNA targeted oligonucleotides. *Microbiology* **140**, 2849-58.

Salyers, A.A., Kuritza, A.P. and McCarthy, R.E. (1985) Influence of dietary fibre on the intestinal environment. *Proceedings of the Society for Experimental Biology in Medicine* **180**, 415-21.

Sambrook, J., Fritsch, E.F. and Manniatis, T. (1989) *Molecular Cloning: a Laboratory Manual*. Cold Spring Harbor Laboratory Press, Cold Spring Harbor, New York.

Scheacter, M., Maaloe, O. and Kjeldgard, N.O. (1958) Dependency on medium temperature of cell size and chemical composition during balanced growth of *Salmonella typhimurium*. *Journal of General Microbiology* **19**, 592-606.

Schleifer, K.H. and Ludwig, W. (1995) Phylogenetic relationships of lactic acid bacteria. In *The Genera of Lactic acid Bacteria*. The lactic acid bacteria. Vol. 2. Wood, B., J., B. and Holzapel, W., H. (Eds.). p. 7-18. Blackie Academic and Professional Publishers, Glasgow, Scotland.

Shah, H.N. and Collins, M.D. (1989) Proposal to restrict the genus *Bacteroides* (Castellani and Chalmers) to *Bacteriodes fragilis* and closely related species. *International Journal of Systematic Bacteriology* **39**, 85-87

Sharp, R. and Macfarlane, G.T. (1998) Microbial ecology of resistant starch breakdown in chemostats inoculated with human faeces. *Microbiology* (submitted).

Sheffield, V.C., Cox, D.R., Lerman, S.L. and Myers, R.M. (1987) Attachment of a 40 base pair G+C rich sequence (GC clamp) to genomic DNA fragments by the polymerase chain reaction results in improved detection of single base pair changes. *Proceedings of the National Academy of Science* USA **86**, 232-36.

Simpson, J.M., Kocherinskaya, S.A., Stroot, P.G., White, B.A. and Mackie, R.I. (1997) Use of DGGE to study bacterial communities in the gasto-intestinal tract. *Bioscience and Microflora* **16**, 22.

Stahl, D.A., Flesher, B., Mansfield, H.R. and Montgomery, L. (1988) Use of phylogenetically based hybridization probes for studies of ruminal microbial ecology. *Applied and Environmental Microbiology* **54**, 1079-84

Stahl, D.A. and Amann, R.I. (1991) Development and application of nucleic acids probes. In *Nucleic Acid Techniques in Bacterial Systematics*. Stackebrandt, E. and Goodfellow, M. (Eds.). pp. 205-48. John Wiley and Sons Ltd., New York.

Szewzyk, U., Manz, W., Amann, R., Schleifer, K.H. and Stenstrom, T.A. (1994) Growth and in situ detection of a pathogenic *Escherichia coli* in biofilms of a heterotrophic water-bacterium by use of 16S-rRNA directed and 23S-rRNA directed fluorescent oligonucleotide probes. *FEMS Microbiology Ecology* **13**, 169-75.

Thomas, J.C., Desrosiers, M., St Pierre, Y., Lirette, P., Bisaillon, J.G., Beaudet, R. and Villemur, R. (1997) Quantitative flow cytometric detection of specific microorganisms in soil samples using rRNA targeted fluorescent probes and ethidium bromide. *Cytometry* **27**, 224-32.

Van der Waaij, D. (1989) The ecology of the human intestine and its consequences for overgrowth by pathogens such as *Clostridium difficile*. *Annual Reviews in Microbiology* **43**, 67-87.

Wang, R.F., Cao, W.W. and Corniglia, C.E. (1996) PCR detection and quantitation of predominant anaerobic bacteria in human and animal fecal samples. *Applied and Environmental Microbiology* **62**, 1242-47.

Wawer, C. and Muyzer, G. (1995) Genetic diversity of *Desulfovibrio* spp. in environmental samples analused by denaturing gradient gel electrophoresis of (NiFe) hydrogenase gene fragments. *Applied and Environmental Microbiology* **61**, 2203-10.

Wawer, C., Jetten, M.S.M. and Muyzer, G. (1997) Genetic diversity and expression of the [NiFe] hydrogenase large-subunit gene of *Desulfovibrio* spp. in environmental samples. *Applied and Environmental Microbiology* **63**, 4360-69.

Weisburg, W.G., Oyaizu, Y., Oyaizu, H. and Woese, C.R. (1985) Natural relationship between *Bacteroides* and flavobacteria. *Journal of Bacteriology* **164**, 230-36.

Welling, G.W., Elfferich, P., Raangs, G.C., Wilderboer-Veloo, C.M., Jansen, G.J. and Degener, J.E. (1997) 16S rRNA Targeted oligonuceotide probes for monitoring of intestinal tract bacteria. *Scandinavian Journal of Gastroenterology* **63**, 17-19.

Wilson, K.H. (1994) Detection of culture-resistant bacterial pathogens by amplification and sequencing of rDNA. *Clinical Infectious Diseases* **18**, 958-62.

Wilson, K.H. and Blitchington, R.B. (1996) Human colonic biota studied by ribosomal DNA-sequence analysis. *Applied and Environmental Microbiology* **62**, 2273-78.

Wilson, K.H., Ikeda, J.S. and Blitchington, R.B. (1997) Phylogenetic placement of community members of human colonic biota. *Clinical Infectious Diseases* **25**, S114-16.

Woese, C.R. (1994) Microbiology in transition. *Proceedings of the National Acadamy of Sciences*, USA **91**, 1601-03.

Woese, C.R., Stackebrandt, E., Macke, T.J. and Fox, G.E. (1985) A phylogenetic definition of the major eubacterial taxa. *Systematic and Applied Microbiology* **6**, 133-42.

CHAPTER 11

The Molecular Biology of Bifidobacteria

SIMON F. PARK

Microbiology Department, Institute of Food Research, Reading, UK

1   Summary

Bacteria belonging to the genus *Bifidobacterium* are amongst the most abundant species of the human and animal intestinal microflora and consequently, can play a fundamental role in the ecology of the gastrointestinal tract in both health and disease. However, attempts to improve the characteristics of bifidobacteria for practical applications, and gain a detailed knowledge of the physiology of these organisms, have been severely limited by the lack of molecular tools. Some progress made over the past decade, concerning characterisation of the bifidobacterial genes, the development of cloning vectors, and genetic transformation systems is discussed here. These developments now provide the basis for genetic approaches, that will eventually lead to a better understanding of these important organisms and the potential development of strains with improved characteristics.

2   Introduction

Bifidobacteria were first discovered in 1899 by Tissier at the Pasteur Institute, Paris, France. They are classified as Gram-positive, anaerobic, catalase negative, fermentative rods, often with a Y- or V- shaped morphology. The GC content of bifidobacteria is unusually high (usually over 55%) and consequently, they belong to a subclass of Gram-positive bacteria containing very GC-rich genomes (Scardovi, 1986). Other members of this high GC group include *Mycobacterium*, *Corynebacterium* and *Streptomyces*. From 16S rRNA analysis it is known that bifidobacteria are closely related to the actinomycete group and accordingly, the genus *Bifidobacterium* is included in the family *Actinomycetaceae* (Stanier *et al.*, 1987), rather than as part of the lactic acid group (*e.g. Lactobacillus* and *Lactococcus*)

G.R. Gibson and M.B. Roberfroid (eds.), Colonic Microbiota, Nutrition and Health, 191-200.
© 1999 *Kluwer Academic Publishers. Printed in the Netherlands.*

Bifidobacteria are numerically important among the human and animal microbiota. In the human large intestine, for example, bifidobacteria can be present at concentrations of 10^{10} per gram of gut content (Mitsuoka, 1992) and may represent up to 25% of the culturable population of the gut microflora of the adult, and 95% of that in new-borns (Kawaze *et al.*, 1981) where their numbers are especially high in breast-fed infants. Among the major genera of human colonic bacteria, bifidobacteria are not considered to be pathogenic. Moreover, studies have concentrated upon the advantageous physiological functions of these bacteria in the human colon and it is now widely believed that bifidobacteria can have beneficial properties for their host. For example, established populations of *Bifidobacterium* are thought to form a barrier prohibiting the invasion of pathogens, either by the production of acids as metabolic end products, or via the excretion of other antibacterial compounds (Wang and Gibson, 1993). Other so-called probiotic effects include reducing the risk of cancer (Reddy and Rivenson, 1993) and modulating the immune response (Yasui *et al.*,1989).

Whilst some of the health promoting aspects of bifidobacteria remain speculative, their economic importance is unquestionable and there is increasing interest in the incorporation of these species into fermented milk products as probiotics. However, attempts to improve the characteristics of bifidobacteria for practical applications, and to gain a detailed knowledge of the physiology of these organisms have been severely limited by a lack of efficient molecular tools. Only in the last few years are molecular techniques, which have been applied successfully to the study of many other Gram-postive bacteria, finding application with bifidobacteria. The progress made over the past decade, concerning genetic developments in the genus *Bifidobacterium* is discussed in this chapter.

3   Genome size and structure

Pulse-field gel electrophoresis (PFGE) provides a means of separating DNA fragments of up to 700 kb in size and, thereby, provides a means of studying the organisation of bacterial genomes. The GC content of bifidobacteria is between 55 and 64 mol % (Scardovi, 1986) and, accordingly, restriction enzymes with recognition sequences containing only A and T nucleotides would be expected to cleave the genome at relatively few locations. Using PFGE, following the digestion of bifidobacterial DNA with such restriction endonucleases Bourget *et al.*, (1993) have been able to produce physical maps of the genome of *Bifidobacterium breve* and have estimated the size of the genome to be approximately 2.1 Mb. This is 46 % the size of the *Escherichia coli* genome, but is comparable to those of other bacteria producing lactic acid and their close relatives, the size of which varies between 1.7-2.4 Mb (Le Bougeois *et al.*, 1991).

When the genomes of various strains of *B. breve* have been compared by PFGE significant heterogeneity is observed, when rare cleaving restriction enzymes are used to produce the electrophoretic profiles (Bourget *et al.*, 1993). This raises the possibility that differences in restriction site distribution in the strains is due to significant chromosomal rearrangements such as translocations or inversions. PFGE has also been used to demonstrate significant genomic variation in strains of various commercially important species of *Bifidobacterium* (Roy *et al.*, 1996). The greatest degree of genetic heterogeneity appears to occur with industrial strains of *B. longum,* since only one out of fifteen strains tested gave a profile similar to the type strain for this species. In contrast, commercially-available strains of *B. animalis* were all found to be identical to the reference strain for this species.

4   Structure and analysis of cloned bifidobacterial genes

The acquisition of complete microbial genome sequences is markedly changing the manner in which bacteria are studied. By comparing genes from different bacteria, we can gain insight into which are likely to be essential for particular functions. Unfortunately, to date, only a very limited number of genes have been characterised from *Bifidobacterium* species.

The comparison of rRNA sequences is a powerful tool for deducing phylogenetic and evolutionary relationships among bacteria. In eubacterial DNA, the rRNA genetic loci (*rrn*) include, 16S, 23S and 5S rRNA genes, which are separated by internally transcribed spacer regions. A number of these genes have now been isolated from bifidobacteria and have proved useful for identification and phylogenetic purposes (Frothingham *et al.*, 1993; Leblond-Bourget *et al.*, 1996). In addition, it has also been shown that for *B. breve*, at least, that there are three *rrn* loci per genome (Bourget *et al.,* 1993). Other genes from *Bifidobacterium* species have also been exploited as targets for DNA based methodologies for speciation and identification. Like the 16S rRNA gene, the *recA* gene, which encodes a multifunctional enzyme that plays a role in recombination by promoting DNA strand transfer, is considered to be universally present in bacteria and shows a high degree of sequence conservation. A 300 bp DNA fragment of the *recA* gene from six species of *Bifidobacterium* has been isolated using PCR with primers directed to two universally conserved regions of the gene (Kullen *et al.*, 1997). Subsequently, the *recA* gene was shown to be useful in the intrageneric phylogenetic analysis of bifidobacteria. Moreover, this may form the starting point for studies concerning the characterisation of this important gene in bifidobacteria. In a separate study concerned with the development of DNA based speciation methods, randomly cloned fragments from the chromosomal DNA of a number of bifidobacteria were used as hybridisation probes (Mangin *et al.*, 1995). Nucleotide sequence analysis of one of the cloned fragments has revealed an open reading frame which shows significant homology with the lambda integrase family of site specific recombinases. Given the similarity, it is likely that this protein catalyses some form of one site specific recombination in bifidobacteria.

Lactate dehydrogenase (LDH) is a key enzyme in the fermentation of milk by lactic acid bacteria used in the dairy industry. The gene encoding this enzyme has been cloned from *B. longum* aM101-2 into *E. coli* following its identification with a hybridisation probe derived from the N-terminal amino sequence of the purified LDH (Minowa *et al.*, 1989) and is one of a very limited number of loci from bifidobacteria for which the complete nucleotide sequence is known. As such, it provides important information on the general structure and organisation of genes in bifidobacteria. An evolutionary study of the LDH sequences for most bacteria clearly divides the prokaryotic form from the eukaryotic enzyme (Griffin *et al.*, 1992). Interestingly, the *B. longum* LDH groups anomalously with the eukaryotic enzymes. In addition, the nucleic acid sequence has provided information which allowed identification of signals required for both transcriptional and translational initiation in bifidobacteria. In this context, the transcription start points for the gene have been identified and the sequence GTAGCAA-(14bp)-TTATAGA assigned as the promoter. Unlike the promoter sequence of the gene encoding ß-D- glucosidase, which has been identified as TTGGAA-(15bp)-TAATCT (Nunoura *et al.*, 1997), this sequence is unlike the consensus for *E. coli* promoters. However, since LDH activity was detected in the absence of a correctly orientated *E. coli* promoter, either this sequence or sequences upsteam of this region must function as promoters *in E. coli*. In addition, the sequence AGAGAGGT, which is highly homologous to the *E. coli* Ribosomal Binding Site (RBS), was designated as the RBS for *ldh*.

Bifidobacteria are known to produce a number of enzymes, such as ß-D-glucosidase, ß -D-fucosidase (Nunoura *et al.*, 1996) and ß -fructofuranosidase (Imamura *et al.*, 1994) which are involved in the degradation of oligosaccharides. These enzymes are likely to play important roles in the nutrition of these organisms within the gut and may also be of considerable industrial value. The genes encoding two such enzymes have recently been cloned and characterised, namely, the ß -galactosidase from *B. breve* (Patent Application JP5146296) and the ß-D-glucosidase gene from *B. breve* clb (Nunoura *et al.*, 1996). Little is known about the mechanisms which enable regulation of gene expression in bifidobacteria, but analyses of the gene sequence for ß-D-glucosidase has provided some insight. The expression of ß-D- glucosidase is regulated in response to availability of its substrate. Accordingly, when cells of *B. breve* are acclimated to cellobiose the activity of this enzyme is increased (Nunoura *et al.*, 1997). An operator-like sequence which has significant homology with the *lac* operator of *E. coli* has been identified between the transcriptional start point and the RBS, and since the *E. coli* sequence contributes towards regulation of expression of the *lac* operon, it is highly likely that this bifidobacterial sequence is involved in regulation of expression of ß-D-glucosidase in response to cellobiose.

5 Extrachromosomal elements and plasmid vectors

Plasmids have been observed in representatives of virtually all bacterial genera examined and *Bifidobacterium* species are no exception. In an early study, involving 24 different species of bifidobacteria, 20% of isolates were found to contain detectable plasmids (Sgorbati *et al.*, 1982). Moreover, the presence of more than one plasmid was found to be common amongst isolates of *B. longum* and *B. asteroides*. More recently, the presence of endogenous plasmids has been demonstrated in strains of *B. breve, B. longum, B. asteroides, B. indicum* and *B. globosum* (Sgorbati *et al.*, 1986ab; Iwata and Morishita 1989; Bourget *et al.*, 1993; Park *et al.,* 1997). Despite the identification of numerous bifidobacterial plasmids, their functions remain cryptic. However, the complete nucleotide sequence of the 1847 b plasmid pMB1 from *B. longum* B2577 has recently been determined (Rossi *et al.*, 1996) and the information has provided an important insight into the nature of plasmids from the genus *Bifidobacterium*. Analysis of the plasmid sequence revealed a high degree of homology with endogenous plasmids found in *Corynebacterium glutamicum* and *Mycobacterium fortuitum*, which in addition to bifidobacteria, are included with Gram-positive bacteria belonging to the high GC subdivision. The plasmid was also found to encode two open reading frames which essential for plasmid replication. Furthermore, the significant homology between one of these, designated *orf1,* and colE-type Rep proteins, suggest that ORF1 may be the pMB1 Rep protein which is the trans-acting product responsible for initiation of replication (Rossi *et al.*, 1996).

A means for introducing and expressing enzymes of technological importance is essential if any applied improvement of bifidobacterial strains is to be brought about using molecular technologies. Progress towards this goal, however, has been hampered by a lack of many of basic tools for genetic manipulation - such as plasmids vectors. The small size of pMB1 and the fact that its nucleic acid sequence has been determined, has made it a popular starting point for the construction of recombinant plasmid able to replicate in bifidobacteria. Accordingly, a number of shuttle vectors have been constructed that contain the origin of this plasmid and various antibiotic markers expressed (Table 1).

Table 1. *Bifidobacterium* plasmid vectors

Plasmid	Size (kb)	Resistance markers	Comments	Reference
pNC7	4.9	Cm	Derived from *B. longum* cryptic plasmid pMB1	Rossi *et al.*, (1996)
pBLES100	9.1	Sp	Derived from *B. longum* cryptic plasmid pTB6	Matsumura *et al.*, (1997)
pDG7	7.3	Cm	Derived from plasmid pMB1	Matteuzzi *et al.*, (1990)
pDGE7	NA	Cm, Er	As pDG7 but contains an erythromycin resistance gene from *Staphylococcus aureus*	Rossi *et al.*, (1996)
pRM2	6.0	Sp	Derived from pMB1	Missich *et al.*, (1994)
pEBM3	9.6	Km, Cm	Contains a replicon from *Corynebacterium* species	Argnani *et al.,* (1996)
pECM2	10.3	Km, Cm	Contains a replicon from *Corynebacterium* species	Argnani *et al.*, (1996)

Abbreviations: Cm, chloramphenicol; Sp, Spectinomycin; Er, erythromycin; Km, kanamycin

An alternative strategy for identifying plasmids that are suitable for use as cloning vectors in bifidobacteria, has been to examine whether pre-existing plasmids, that have been built for use in other closely related species, are able to transform. Using this approach, Argnani *et al.,* (1996) have been able to transform *B. animalis* with pEBM3 and pECM2 (Table 1), plasmid vectors which originate from *Corynebacterium* species. In contrast, vectors such as pGK12 and pLP825, which carry replicons from lactic acid bacteria, are unable to transform bifidobacteria. Given the significant homology between the native *Bifidobacterium* plasmid pMB1 and those from *Corynebacterium*, and the lack of genetic similarity with those from lactic acid bacteria, the result is perhaps not surprising. Indeed, it may reflect the fact that the replication functions of some plasmids of the GC-rich branch of Gram-positive bacteria are recognised, specifically by bacteria of this subgroup, as are replicative functions of plasmids from the AT-rich branch recognised by their corresponding group (Argnani *et al.,* 1996).

For other bacteria bacteriophages, in addition to plasmids, have proved to be extremely useful tools for genetic manipulation. When fourteen strains of *B. longum* were tested for the presence of bacteriophages, four were found to contain phage-like particles (Sgorbati *et al.,* 1983). However, although bacteriophages with specificity for bifidobacteria exist, their study is extremely limited and consequently, bacteriophages are unlikely to be adapted for use as genetic vectors in the near future.

6   Genetic transformation systems

The availability of protocols for the introduction of DNA into the bacterial cell is a prerequisite for application of sound molecular technologies. For many Gram-positive bacteria, the cell wall forms a major barrier to incoming plasmid DNA and that of bifidobacteria, which is very thick and complex (Fischer, 1987), does not prove any exception. Consequently, the aim of many techniques for introducing plasmid DNA into Gram-positive bacteria is to circumvent this barrier. In this respect, a number of techniques for generating protoplasts, in which the cell wall has been removed following treatment with lytic enzymes, have been developed for bifidobacteria (Brigidi *et al.,* 1986). However, to date, none of these methods have actually been used for transformation.

More recently, protoplasting techniques have largely been superseded as a method for the introduction of plasmid DNA by electroporation, which has proven widely applicable for the genetic transformation of bacterial cells (Luchansky *et al.,* 1988) and is increasingly being utilised to facilitate transformation of bifidobacteria.

The first report concerning genetic transformation of a *Bifidobacterium* species appeared in 1994 (Missich *et al.,* 1994) and described the introduction of plasmid pRM2 into *B. longum* B2577. Using an applied voltage of 10.0 kV/cm, the authors were able to obtain transformation frequencies of 3.8×10^2 transformants per µg of plasmid DNA. Since this initial report, a number of studies have sought to identify conditions for increasing transformation efficiency and extend this technique into other species. Generally, the effect of a higher applied voltage, increases permeability to plasmid DNA but also results in significant cell death. Thus, to attain optimal transformation efficiencies, a voltage which produces a compromise between these two disparate processes must be found. In this context, the optimum voltage for transformation of bifidobacteria is generally considered to be 12.5 kV/cm (Matsumara *et al.*, 1997; Rossi *et al.*, 1997) and under this condition 50% of the cells lose viability during discharge of the electric pulse. In addition, the efficiency of transformation can be improved by incorporating carbohydrates into the growth medium used to culture the electro-competent cells. For instance, the inclusion of Actilight P™ (a specified mixture of fructooligosaccharide molecules) or raffinose results in a 100-fold increase in trans-formation efficiency (Rossi *et al.*, 1997).

Even when electroporation is use to permeabilise Gram-positive cells to plasmid DNA, the cell wall can still form a significant barrier for uptake of exogenous DNA molecules. Consequently, the efficiency of transformation in these bacteria can be dramatically improved, by treatment of cells with penicillin G or muralytic enzymes which reduce the integrity of the cell wall (Park and Stewart, 1990; Powell *et al.*, 1988). In this context, pre-incubation of bifidobacterial cells at 4 °C for several hours prior to electroporation dramatically increases the transformation efficiency (Argnani *et al.*, 1996) and it is possible that this is a consequence of limited autolysis of the cell wall. Furthermore, an alternative procedure, which involves freezing the cells at -135 °C may also facilitate electrotransformation by freeze/thaw induced autolysis (Missich *et al.*, 1994).

The presence of restriction and modification systems is widespread amongst bacteria. In the classical system, cellular DNA is protected from restriction by a DNA- methyl-transferase, which modifies adenosyl or cytosyl residues within sequences recognised by the restriction enzymes. The main biological role for restriction/modification systems is to protect cells against phage infection. They can, however, provide a serious barrier to the introduction of plasmid DNA into bacteria by transformation systems. Two type-II restriction endonulceases, *Blo*I and *Blo*II, have recently been detected in *B. longum* and these recognise the sites RGATCY and CTGCAG, respectively (Hartke *et al.*, 1996). The identification of this restriction system will allow its role in effecting the frequency of transformation, following the introduction of plasmid DNA by electroporation. It may also facilitate the construction of strains resistant to bacteriophage infection as potential starter culture constituents.

7   Conclusion

Members of the genus *Bifidobacterium* play a unique role in the ecology of the gas-trointestinal tract and, as such, they may offer unique opportunities for regulating gut health. Attempts to improve the characteristics of bifidobacteria for practical applications, and to gain a detailed knowledge of the physiology of these organisms have been very restricted by the lack of molecular tools. However, the number of investiga-tors applying molecular methodologies to *Bifidobacterium* species is increasing and given the relative ease with which the limited number of genes characterised to date have been cloned and expressed in *E. coli*, it seems likely that over the next few years many more genes will be cloned and sequenced. Substantial progress has also been made in the development of plasmid cloning vectors and genetic transformation systems, and realistic opportunities for applying molecular methodologies to these bacteria now exist.

Bifidobacteria play such an important role in the ecology of the gut that there continued study is assured. Undoubtedly, investigations of their molecular biology will continue and will contribute substantially, during the next decade, to our understanding and ability to manipulate these bacteria for important applied purposes.

7   References

Argnani, A., Leer, R.J., van Luijk, N. and Pouwels, P. (1996) A convenient and reproducible method to genetically transform bacteria of the genus *Bifidobacterium*. *Microbiology* **142**, 109-14.

Bourget, N., Simonet, J.-M. and Decaris, B. (1993) Analysis of the genome of five *Bifidobacterium breve* strains: plasmid content, pulsed-field gel electrophoresis genome size estimation and *rrn* loci number. *FEMS Microbiology Letters* **110**, 11-20.

Brigidi, P., Matteuzzi, D. and Crociani, F. (1986) Protoplast formation and degeneration in *Bifidobacterium*. *Microbiologica* **9**, 243-48.

Fischer, W. (1987) Analysis of the lipoteichoic acid-like macroamphiphile from *Bifidobacterium bifidum* subspecies *pennslyvanium* by one- and two-dimensional 1H- and 13C-NMR spectroscopy. *European Journal of Biochemistry* **165**, 647-52.

Frothingham, R., Duncan, A.J. and Wilson, K.H. (1993) Ribosomal DNA sequences of bifidobacteria: implications for sequenced based identification of human colonic flora. *Microbial Ecology in Health and Disease* **6**, 23-27.

Griffin, H.G., Swindell, S.R. and Gasson, M.J. (1992) Cloning and sequence analysis of the gene encoding L-lactate dehydrogenase from *Lactococcus lactis*: evolutionary relationships between 21 different LDH enzymes. *Gene* **122**, 193-97.

Hartke, A., Benachour, A., Boutibonnes, P. and Auffray, Y. (1996) Characterisation of a complex restriction/modification detected in a *Bifidobacterium longum* strain. *Applied Microbiology and Biotechnology* **45**, 132-36.

Imamura, L., Hisamitsu, K. and Kobashi, K. (1994) Purification and characterization of ß-fructofuranosidase from *Bifidobacterium infantis*. *Biological and Pharmaceutical Bulletin* **17**, 596-602.

Iwata, M. and Morishita, T. (1989) The presence of plasmids in *Bifidobacterium breve*. *Letters in Applied Microbiology* **9**, 165-68.

Kawase, K., Suzuki, T., Kiyosawa, I Okoonogi, S., Kawashima, T. and Kuboyama, M. (1981) Effects of composition of infant's formulas on the intestinal microflora of infants. *Bifidobacteria Microflora* **2**, 25-31

Kullen, M.J., Brady, L.J. and O'Sullivan, D.J. (1997) Evaluation of using a short region of the *recA* gene for rapid and sensitive speciation of dominant bifidobacteria in the human large intestine. *FEMS Microbiology Letters* **154**, 377-83.

Leblond-Bourget, N., Herve, P., Mangin, I. and Decaris, B. (1996) 16S rRNA and 16S to 23S internal transcribed spacer sequence analysis reveal inter- and intraspecific *Bifidobacterium* phylogeny. *International Journal of Systematic Bacteriology* **46**, 102-11.

Le Bourgeois, P., Mata, M. and Ritzenthaler, P. (1991) Pulsed-field gel electrophoresis as a tool for studying the phylogeny and genetic history of lactococcal strains. In *Genetics and Molecular Biology of*

Streptococci, Lactococci and Enterococci, (eds. G.M. Dunny, P.P. Cleary, and L.L. McKay), American Society for Microbiology, Washington DC. pp. 140-45.

Luchansky, J.B., Muriana, P.M. and Klaenhammer, T.R. (1988) Application of electroporation for transfer of plasmid DNA to *Lactobacillus, Lactococcus, Leuconostoc, Listeria, Pediococcus, Bacillus, Staphylococcus, Enterococcus and Propionobacterium. Molecular Microbiology* **2**, 637-46.

Mangin, I., Bourget, N., Simonet, J.-M. and Decaris, B. (1995) Selection of DNA probes which detect strain restriction polymorphisms in four *Bifidobacterium* species. *Research in Microbiology* **146**, 59-71.

Matsumura, H., Takeuchi, A. and Kano, Y. (1997) Construction of *Escherichia coli-Bifidobacterium longum* shuttle vector transforming *B. longum* 105-A and 108-A. *Bioscience, Biotechnology and Biochemistry* **61**, 1211-12.

Matteuzzi, D., Brigidi, P., Rossi, M. and Di, D. (1990) Characterization and molecular cloning of *Bifidobacterium longum* plasmid pMB1. *Letters in Applied Microbiology* **11**, 220-23.

Minowa, T., Iwata, S., Sakai, M. Masaki, H. and Ohata, T. (1989) Sequence and characteristics of the *Bifidobacterium longum* gene encoding L-lactate dehydrogenase and the primary structure of the enzyme: a new feature of the allosteric site. *Gene* **85**, 161-68.

Missich, R., Sgorbati, B. and Leblanc, D.J. (1994) Transformation of *Bifidobacterium longum* with pRM2, a constructed *Escherichia coli-B. longum* shuttle vector. *Plasmid* **32**, 208-11.

Mitsuka, T. (1992) The human gastrointestinal tract, *in The Lactic Acid Bacteria in Health and Disease* (ed. B.J.B. Wood) Elsevier Applied Science, London, pp. 69-114.

Nunoura, N., Ohdan, K., Tanaka, K., Tamaki, H., Yano, T., Inui, M., Yukawa, H., Yamamoto, K. and Kumagi, H. (1996) Cloning and nucleotide sequence of the ß-D-glucosidase gene from *Bifidobacterium breve* clb, and expression of ß-D-glucosidase activity in *Escherichia coli. Bioscience, Biotechnology and Biochemistry* **60**, 2011-18.

Nunoura, N., Ohdan, K., Yamamoto, K. and Kumagi, H. (1997) Expression of the ß-D-glucosidase I gene in *Bifidobacterium breve* 203 during acclimation to cellobiose. *Journal of Fermentation and Bioengineering* **83**, 309-14.

Park, M.S, Lee K.H. and Li, G.E. (1997) Isolation and characterization of two plasmids from *Bifidobacterium longum. Letters in Applied Microbiology* **25**, 5-7.

Park, S.F. and Stewart, G.S.A.B. (1990) High efficiency transformation of *Listeria monocytogenes* by electroporation of penicillin treated cells. *Gene* **94**, 129-32.

Powell, I.B., Achen, M.G., Hillier, A.J. and Davidson, B.E. (1988) Simple and rapid method for genetic transformation of lactic streptococci by electroporation. *Applied and Environmental Microbiology* **54**, 655-60.

Reddy, B.S. and Rivenson, A. (1993) Inhibitory effect of *Bifidobacterium longum* on colon, mammary and liver carcinogenesis induced by 2-amino-3-methylimidazol [4,5-f]quinilone, a food mutagen. *Cancer Research* **53**, 3914-18.

Rossi, M., Brigidi, P. and Matteuzzi, D. (1997) An efficient transformation system for *Bifidobacterium* spp. *Letters in Applied Microbiology* **24**, 33-36.

Rossi, M., Brigidi, P., Gonzalez Vara y Rodriguez, A. and Matteuzzi, D. (1996) Characterization of the plasmid pMB1 from *Bifidobacterium longum* and its use for shuttle vector construction. *Research in Microbiology* **147**, 133-43.

Roy, D., Ward, P. and Champagne, G. (1996) Differentiation of bifidobacteria by use of pulse-field gel electrophoresis and polymerase chain reaction. *International Journal of Food Microbiology* **29**, 11-29.

Scardovi, V. (1986) Genus *Bifidobacterium*, in *Bergey's Manual of Systematic Bacteriology*, vol 2, (eds. P.H.A. Sneath, N.S. Mair, M.E. Sharpe and J.G. Holt), Williams and Wilkins, Baltimore, pp. 1418-34.

Sgorbati, B., Scardovi, V. and Leblanc, D. (1982) Plasmids in the genus *Bifidobacterium. Journal of General Microbiology* **128**, 2121-31.

Sgorbati, B., Scardovi, V. and Leblanc, D. (1986a) Related structures in the plasmid profiles *Bifidobacterium longum, Microbiologica* **9**, 415-22.

Sgorbati, B., Scardovi, V. and Leblanc, D. (1986b) Related structures in the plasmid profiles of *Bifidobacterium asteroides, B. indicum* and *B. globosum. Microbiologica* **9**, 443-56.

Sgorbati, B. Smiley, M.B. and Sozzi, T. (1983) Plasmids and phages in *Bifidobacterium longum. Microbiologica* **6**, 169-73.

Stanier, R.Y., Ingraham, J.L., Wheelis, M.L. and Painter, P.R. (1987) The classification and phylogeny of bacteria, in *General Microbiology*, 5th Edition, Macmillan, London, pp. 311-29.

Wang, X. and Gibson, G.R. (1993) Effects of the *in vitro* fermentation of oligofructose and inulin by bacteria growing in the large intestine. *Journal of Applied Bacteriology*, **75**, 373-80.

Yasui, H., Mike, A. and Ohawki, M. (1989) Immunogenicity of *Bifidobacterium breve* and change in the antibody production in Peyer's patches. *Journal of Dairy Science*, **72**, 30-35

CHAPTER 12

Intestinal Microflora and the Mucosal Mechanisms
of Protection

EDUARDO J. SCHIFFRIN AND DOMINIQUE BRASSART
*Nestle Research Centre. P.O. Box 44. Vers-chez-les-Blanc. 1000 Lausanne
26. SWITZERLAND*

1   Introduction

Mucosal epithelial surfaces such as the respiratory system, the gastrointestinal and urogenital tracts are sites where allochthonous microorganisms can encounter, and in some instances, infect the host. Some mucosal surfaces, such as the heavily colonised distal intestine, are physiologically colonised by a commensal microflora that entertains symbiotic interactions with its host. Others, such as the distal respiratory airways, remain relatively sterile. The commensal intestinal microflora can be considered an organ with a hugh diversity of (bacterial) cells that, in turn, perform different important functions for the body (Brassart and Schiffrin, 1997).

This chapter will highlight some of the interactions between the microflora and defence mechanisms of the host. Protection afforded by the symbiotic microflora falls into two major types, a) direct antagonism against pathogens called the "barrier effect" of the microflora, or colonisation resistance, and b) modulation of the immune response. This last aspect implies very rich interactions between prokaryotes and the eukaryotic immune-competent cells.

Since antigen specific, cellular or humoral responses require some time to fully mature, the host has a very effective first line of defence (innate immunity) to eliminate potential pathogens when they are encountered for the first time. At the systemic level, this comprises neutrophils, macrophages and natural killer cells, complement factors and acute phase reactants. Mucus secretion, glycolipids, cytoprotective products and antibiotic-like substances (Bonan, 1995), such as defensins, are probably the most conspicuous constituents of the innate mucosal immune system.

Innate and adaptive immune responses act in concert, most of the time, to effect destruction of pathogens. They can probably discriminate between mounting an energetic

G.R. Gibson and M.B. Roberfroid (eds.), Colonic Microbiota, Nutrition and Health, 201-211.
© 1999 *Kluwer Academic Publishers. Printed in the Netherlands.*

response against pathogens or remain tolerant to the commensal microflora in colonised organs. In contrast, any microbial presence at sterile surfaces will engender an active defensive response. The epithelial cell reaction against pathogens - a type of innate immunity- might be an example of this dichotomy. In fact, epithelial cells participate in the early events of mucosal infections. Their cytokine response to microorganisms seems to depend on the site of infection and specific pathogen. While non-invasive bacteria can induce pro-inflammatory cytokines at sterile or weakly colonised mucosal organs (lung, bladder, stomach), invasive bacteria are important at highly colonised organs (Rasmussen *et al.*, 1997).

The specific mucosal immune system functions through two main pathways, a) secretory antibody production for immune exclusion and, b) tolerance to innocuous components of the intestine (oral tolerance). The gut associated lymphoid tissue (GALT) is characterised by a clear anatomical distinction between inductive and effector sites of the immune response. Peyer's patches are the major inductive structures constituted by aggregates of lymphocytes and accesory cells covered by the follicular associated epithelium (FAE) which contains M cells specialised for antigenic sampling (Neutra *et al.*, 1996). Upon stimulation, antigen-specific lymphoblasts leave the patch, circulate through the lymph and peripheral blood to finally home back into mucosal compartments. This selective homing is brought about by decreasing the expression of L-selectin, a molecule that favours interaction with, and homing into, high endothelial venules of peripheral lymphoid tissues and increasing expression of $\alpha 4$ $\beta 7$ integrin which interacts with the mucosal addressin cell molecule (MadCAM-1) expressed on mucosal blood vessels (Berlin *et al.*, 1993).

The hallmark of a mucosal immune response is the detection of antigen-specific secretory immunoglobulins (Ig) in the external secretions (Underdown and Schiff, 1986). IgA and IgM are synthesised by terminally differentiated B-lymphocytes (plasma cells) in the lamina propria and transported trans-epithelially by the polymeric immunoglobulin receptor (pIgR) or secretory component (sc). IgA is the main secretory antibody isotype in humans. IgA plasma cells are not present in the lamina propria of the newborn, and salivary IgA is very low until 6 months of age. Some bacterial components, such as *Escherichia coli* and spp., are highly stimulatory for their production and it has recently been shown that probotic lactic acid bacteria (LAB) modulate their synthesis in the human without any detrimental effect (Link-Amster *et al.*, 1994).

2   The gut microflora: a postnatal acquired organ

2.1   THE ONTOGENY

During gestation, the uterus provides a sterile environment for the developing foetus, but upon birth, a diverse microbial inoculum is derived from the mother's genital-tract and faeces, skin microflora and the environment. This results in an initial colonisation of the intestine and meconium (Haenel, 1970) during the few first hours of life. This is then followed by a more organised microbial development by major components of the adult microflora. All the individual factors that dictate this organisation are not known, although the global intestinal microenvironment as well as the competition between different bacterial components, is of obvious importance. It therefore follows, that colonisation of germ free (GF) mice by human flora is probably more representative of the normal mouse flora than the human microflora (Hirayama *et al.*, 1995). The persistence of some bacterial species in this model is dependent on the initial inoculum.

A very important factor in neonatal colonisation of the intestine is the type of nutrition. Enterobacteria, streptococci and clostridia can all be detected in faeces during the first day of life. Around the third day, bacteroides, bifidobacteria and clostridia are found in about half of babies, irrespective of the type of nutrition. Thereafter, bifidobacteria become the predominant microbial component of breast-fed infants while in formula-fed infants levels of enterobacteria are higher or similar to those of bifidobacteria (Haenel, 1970, Mitsuoka 1992) One month after birth, bifidobacteria are the dominant microorganisms in both groups, with no major species differences being detected. Generally, the microflora of the breast-fed infant is less complex with only 1% of coliforms, enterococci and lactobacilli, whilst that of formula-fed infant has higher levels of bacteroides, eubacteria, clostridia, enterobacteria and lactobacilli. These differences in composition parallel those seen through metabolic activities.

Putrefactive bacteria, and their related metabolic activity, are below the normal levels of detection and there are low pH and high redox potential values in the breast-fed population. However, in formula-fed babies the putrefactive flora (*Bacteroides, Clostridium, Proteus*, etc.) are found at higher levels and therefore their metabolic activity becomes more evident through a higher pH, lower redox-potential and the presence of ammonium and amines in faeces.

These observations support the contention that a predominant bifidobacterial flora can modify the physiology of the digestive tract and may even have a beneficial role in human health. Indeed, bifidobacteria added to the diet have been shown to alter the colonic microflora in children (Langhendreis *et al.*, 1995).

2.2   DEVELOPMENT OF IMMUNOLOGICAL TOLERANCE TO EXOGENOUS MICROBIAL ANTIGENS

Microbes that constitute indigenous microflora of the gastrointestinal ecosystem are those which colonise the host early in life and attain high levels after colonisation (Berg and Savage, 1975). It is difficult to identify host factors that participate in the organogenesis of this postnatally acquired organ but, as stated previously, the intestinal environment, with all its complexity, plays a major role. It is this changing environment that guides the composition and adaptation of the microflora (Luckey, 1970).

In addition to the type of nutrition, determinant factors include intimate asssociation with the bacteria at the mucosal surface, as well as an immune response against the bacteria themselves. The former mainly depends on regional glycosylation patterns. The mucosal immune response, although less well-anatomically defined, also exhibits regional differences, depending on the load of luminal bacteria (Kjell *et al.*, 1995). The immune response seems an important selective force of microflora composition (Berg and Savage, 1975). It has been shown, in the gnotobiotic mouse model, that mice monocolonised with non-indigenous bacteria exhibit a higher immune response against bacterial antigens than those associated with an indigenous flora (Berg and Savage, 1972). It has been proposed that antigenic similarity between indigenous microorganisms and host tissues could account for this "immunological indifference" (Foo and Lee, 1974), an example of immune-tolerance, that may help the intestinal colonisation.

Antigenic challenge of the neonate immature system and/or the action of "tolerogenic" cytokines (TGF-ß or IL-10) may also be responsible for such tolerance. It has been suggested that a strong immune response against a particular bacterium will prevent colonisation. In fact, it has recently been reported that the human indigenous flora is only partially covered by IgA specific antibodies and even less so by IgG and IgM (Van der Waaij et al., 1996). An important proportion of the microflora bacteria is not covered by antibodies at all. Secretory IgA coating or total unresponsiveness to the colonic microflora, either humoral or T cell unresponsiveness, could play a homeostatic role in symbiotic interactions between the host and intestinal microflora and prevent an inflammatory type of immune response. Indeed, in inflammatory bowel disease (IBD, e.g. Crohn's disease, ulcerative colitis) the host-microflora immune interactions are altered (Duchmann et al., 1995, MacDonald, 1995). Although the aetiology of IBD remains unknown, a shift from an IgA towards IgG pro-inflammatory response against resident bacterial antigens has been reported and could be responsible for inflammatory tissue damage (MacPherson et al., 1996).

2.3   MICROFLORA COMPOSITION AND AGEING

Lower numbers of bifidobacteria and greater numbers of clostridia, lactobacilli and enterobacteria have been reported in the elderly compared to a younger (e.g. < 55 years) population (Mitsuoka, 1992). In postmenopausal women, an increase in fungi, clostridia, aerobic lactobacilli, *Escherichia coli* and other Enterobacteriaceae was observed. Although age seems to be a major factor, differences in hormonal status could also account for the observed changes (Bertazzoni Minelli, 1993). Alterations in the microflora with ageing could be the result of variability in the intestinal micro-environment including immune reactivity against components of the microflora. In turn, bacterial changes in the intestine could also result in different modulatory activities on the GALT and explain modifications observed in immune function in later life.

3   Gut microflora/mucosal immune system cross talk: a homeostatic factor of the gut immune system

The intraepithelial lymphocyte population (IEL) is an important component of the mucosal immune system. Most IELs are T lymphocytes of the CD8+ (suppressor/cytotoxic) phenotype. There are γδ and αß T cell receptors (TCR) bearing T cells in this compartment. γδ TCR+ IEL population seems to be independent of bacterial challenge but αß TCR+ IEL increase on bacterial colonisation of GF animals. The intraepithelial lymphocyte population express α^E ß7 integrin that is a ligand for

E-cadherin expressed on the basolateral membrane of the enterocytes (Cepek *et al.,* 1994) and is responsible for lymphocyte homing into the intraepithelial compartment. The IELs show high cytolytic activity and as such may eliminate infected, transformed or seriously damaged epitheliall cells. The size and function of the IEL compartment depends on the presence of the intestinal microflora (Abreu-Martin and Targan, 1996; Lefrancois and Goodman, 1989). Thus, germ-free (GF) animals have a smaller IEL population with a low level of activation. Upon mono-colonisation with probiotic strains, a clear expansion of the IEL population has been reported (Link *et al.,* 1995).

Germinal centres (GC) of the GALT are the main lymphopoietic compartment of IgA B cells. Their development depends on antigenic challenge especially by bacterial antigens. B cells move into the follicles and the GC depending upon their affinity for the antigen. Thereafter, B cells undergo somatic mutations of their antigen receptors, which permit greater affinity. Furthermore, follicular dendritic cells retain immune complexes, which are oriented in such a fashion that the antigen is exposed and B cells with the correct mutation are selected. Low affinity B cells are deleted by apoptosis. Immunoglobulin isotype switching then leads to production of a predominantly IgA isotype at the mucosal surface. T cell subsets expressing CD 40 ligand and producing IL-4 and IL-10, co-localise with the GC centrocytes undergoing isotype switching (Liu and Arpin, 1997) and are active participants in the process.

Monoassociation of germ-free animals with commensal bacteria initially induces the generation of Peyer's patch GC that then declines with time. Development of the B cell compartment is associated with a specific immune response against colonising bacteria. Although GC disappear after a few weeks, the IgA specific response persists longer. This suggests that observed unresponsiveness against autochthonous microflora appears after a transient immune response (Shroff *et al.,* 1995). As yet, the effect of probiotic bacteria on GC development in the GALT has not been determined. However, it is conceivable that such bacteria could sustain activation of the GC and thereby promote an IgA specific response, not only against specific bacterial antigens, but also against by-stander antigens sampled by the M cells.

4   Pathological conditions related to dysregulation of the intestinal microflora

In general, pathogens adapt to different habitats of the gastrointestinal tract (GIT) to find their ecological niches. Defence against intestinal pathogens involves both common and regional specific mechanisms of the GIT. Although no major variations have been described for secretory immunity in diverse regions, non-immune mechanisms of defence show striking geographical differences. For example, gastric acidity renders the stomach almost sterile, while bile salt secretion, antibiotic peptides produced by Paneth cells and pancreatic secretion, *i.e.* defensins, protect the small bowel. In addition to these anti-microbial mechanisms, a rapid small bowel transit time prevents the development of a stable microflora. By contrast, the more stagnant colonic content allows for establishment of a steady state and complex ecosystem in the lumen. This exerts a barrier effect through acidification and ecological competition.

Depending on the organ they infect, and their specific location within the organ, pathogens have to overcome different barriers to gain entry into the host. The micro-flora barrier effect plays a major role in the colon but is probably weak in the distal small bowel, and almost non-existent in the stomach and proximal small bowel. Therefore, *a priori* it is tempting to speculate that indigenous microorganisms are a strong antagonistic force against colonic pathogens (*e.g.* clostridia), less important for small bowel pathogens (*i.e.* ETEC, EPEC, *Salmonella*) and probably negligible for stomach pathogens (*e.g. Helicobacter pylori*).

4.1   SITES LACKING A BARRIER EFFECT: THE STOMACH

During normal physiological conditions, a low pH (ca 1-2) induced by activity of the proton pump renders the stomach almost sterile. In the normal situation, this gastric acidity is independent of bacterial regulation and offers a powerful non-specific defence. However, some pathological or sub-clinical conditions, (*e.g.* administration of antacids) may develop when the gastric acidity barrier is impaired. Furthermore, a reduced gastric microflora renders the gastric mucosa particularly susceptible to colonisation by bacteria such as *H. pylori*. This organism is touted as a co-factor in the development of gastric and duodenal ulcers and also of some gastric cancers (Blaser and Parsonnet, 1994). Rupture of the gastric barrier has been demonstrated in acute infections with *H. pylori*. The bacterium induces a transient hypochlorhydria which generally resolves in few months but which may nevertheless facilitate colonisation of the gastric mucosa by the pathogen (Mc Gowan *et al.*, 1996).

A disturbance in the gastric barrier also leads to endogenous bacterial overgrowth in the small intestine, the consequences of which are malabsorption syndromes and a higher risk of infection by Gram-negative bacteria. Indeed, bacterial translocation has been positively correlated with exposure to non-lethal challenge with LPS and protein-energy malnutrition (Deitch, 1994). Bacterial overgrowth associated with rupture of gastric acidity has been reported in the elderly (Pedrosa *et al.*, 1995) and in patients undergoing antacid therapy with Omeprazole (Fried *et al.*, 1994). Consequently, oral and faecal bacteria can be recovered from the duodenum of such patients. The effects of this bacterial overgrowth are not fully understood and as such, are the subject of some controversy. In certain cases, no effects on fat or carbohydrate absorption are detected in elderly people with bacterial overgrowth (Saltzman *et al.*, 1994). However, other studies have correlated bacterial overgrowth with decreased cobalamine absorption, vitamin B12 deficiency and lipid malabsorption (Simon and Gorbach, 1984).

4.2   SITUATIONS WHERE THE PATHOGEN HAS TO OVERCOME A WEAK
BARRIER EFFECT: THE SMALL BOWEL

The proximal small bowel (SB) bacterial ecosystem is fairly insubstantiable because of peristalsis, residual gastric acidity, luminal secretions, (Simon and Gorbach, 1995) and renewal of the epithelium (Bry *et al.*, 1996). As a consequence, any barrier effect of the upper SB is reduced. For this reason, many pathogens can target this organ. Pathogens have evolved two strategies to infect this part of the GIT.

The first is based on adhesion of the pathogen and/or of its toxins to enterocytes. The lack of a true ecological barrier effect probably favours this process and gives the pathogen a competitive advantage. This is the case for enteropathogenic *E. coli* (EPEC) which invades the host after a firm attachment to the normal epithelial cells (Donnenberg and Kaper, 1992). Enterotoxigenic *E. coli* (ETEC) utilizes the Colonisation Factor Antigen (CFA) to attach to the epithelium (Neeser *et al.,* 1989) and thereafter, secretes toxins, which stimulate intestinal secretion without cellular damage, resulting in a watery diarrhoea. Shiga toxin-producing enterohaemorragic *E. coli* (EHEC) also utilize the same mechanism in which the toxin is able at least *in vitro* to translocate across epithelial cells in culture (Acheson *et al.,* 1996). In general, the five main types of diarrhoeagenic *E. coli* (Levine, 1987) (enteroinvasive *E. coli,* enterotoxigenic *E. coli,* enteropathogenic *E. coli,* enteroaggregative *E.coli,* and the Shiga toxin-producing enterohaemorragic *E. coli*) require an adhesion step to initiate the infectious process. This is also the case for infections produced by other intestinal pathogens such as *Yersinia* spp., *Shigella* spp., *Salmonella* spp. (Hromockyj and Falkow, 1995) or even rotavirus (Svensson *et al.,* 1991).

The second strategy, used by certain pathogens, is to infect the host through antigen-sampling M cells. This is the case for *Salmonella* spp., *Shigella flexneri,* *Yersinia* spp., *Listeria monocytogenes* and certain viruses like reovirus and poliovirus (Jones and Falkow, 1996). However, it remains possible that specific cellular targeting by pathogens is not so clear cut. If the primary target for *Salmonella* spp. or other pathogens is the M cell, the bulk of the interaction, prokatyotic-eukaryotic, also involves the general columnar epithelial surface (McCormick *et al.,* 1993). Whatever the mechanism, infection is greatly facilitated by the fact that no barrier effect exists in the small bowel. When infection occurs, antibiotics are usually the treatment of choice. However, creation of a transient microflora through administration of specific LAB, may provide another form of therapy. This is related to the probiotic concept (Fuller, 1989). In order to antagonise infectious agents in the SB, probiotics have to target an environment devoid of a stable microflora. Their persistence will depend on continuous administration and/or adhesion to the gut wall. They then protect the SB by providing a barrier effect or by modulating host defence mechanisms. To attain a barrier effect, the probiotic must block interaction between the pathogen and its cellular target. Certainly, this has been demonstrated in vitro using adherent lactobacilli and bifidobacteria that induced a better mucosal barrier effect than non-adherent strains (Bernet *et al.,* 1993; Bernet *et al.,* 1994). This could also be the case in vivo since it has been demonstrated that some probiotic lactobacilli attach to enterocytes and colonocytes (Johansson *et al.,* 1993). Antibacterial substances produced by probiotics may also contribute to their protective effect. (Bernet-Camard *et al.,* in press).

In conclusion, probiotics may perform their barrier effect close to the M cell in some instances, or over the normal epithelium in others. It is not known whether a direct bacterial-viral antagonism exists or if the protection that has been observed with probiotic LAB in paediatric rotavirus infection, is associated with an immune-adjuvant effect of the probiotic strain (Isolauri *et al.,* 1991; Saavedra *et al.,* 1994). A particular case is represented by immune-deficient patients. They often suffer from opportunistic infections that find their niche both in the small bowel and the colon (cryptosporidia, cytomegalovirus, cytomegalovirus). AIDS enteropathy is an example of diarrhoea associated with SB bacterial overgrowth and ultimately provokes a wasting syndrome (Kotler et al., 1995).

4.3   SITUATIONS WHERE THE PATHOGEN HAS TO OVERCOME A STRONG BARRIER EFFECT: THE COLON

In normal physiological conditions, the colonic microflora efficiently exerts its barrier effect. However, during antibiotic treatment, an important disruption in the steady-state colonic microflora can often occur. Indeed, antibiotic associated diarrhoea (AAD) is a frequent complication in hospitalised patients. As a consequence, bacterial overgrowth of opportunistic intestinal pathogens takes place. *Clostridium difficile* is the most important organism in this situation. Up to one third of AAD patients, and virtually all cases of pseudomembranous colitis, are associated with *C. difficile* overgrowth (Lyerly and Wilkins, 1995) and toxin production. It has been reported that *C. difficile* is also associated with recurrent colitis. In fact 10-25% of patients with *C. difficile* colitis relapse after broad-spectrum antibiotic treatment, which further disrupts the colonic ecosystem. In addition, it has been recently suggested that enterotoxigenic *Bacteroides fragilis*, one of the most frequent anaerobic species isolated from human clinical specimens, can overgrow as a consequence of antibiotic treatment and induce diarrhoea (Well et al., 1996). Due to these difficulties, the use of probiotics may provide an alternative therapy to antibiotics for the treatment of AAD. To this end, *Saccharomyces boulardii* and *Lactobacillus rhamnosus* GG have shown positive effects in preventing or treating AAD (Elmer et al., 1996). It is clear that for the treatment of relapses or indeed, their prevention, it may be more appealing to utilise the probiotic approach than to continue repeated cycles of antibiotics.

References

Abreu-Martin, M.T. and Targan S.R. (1996). Regulation of immune responses of the intestinal mucosa. *Critical Reviews in Immunology* 16, 277-309.

Acheson, D.W.K., Moore, R., De Breuker, S., Lincicome, I., Jacewicz, M., Skutelski, E. and Keusch, T. (1996). Translocation of Shiga toxin across polarized intestinal cells in tissue. *Infection and Immunology* 64, 3294-00.

Berg, R.D. and Savage, D.C. (1972) Immunological responses and microorganisms indigenous to the gastrointestinal tract. *American Journal of Clinical Nutrition* 25, 1364-71.

Berg, R.D. and Savage, D.C. (1975) Immune responses of specific pathogen-free and gnotobiotic mice to antigens of indigenous and nonindigenous microorganisms. *Infection and Immunology* 11, 320-29.

Berlin, C., Berg, E.L., Briskin, M.J., Andrew, D.P., Kilshaw, P.J., Holzmann, B., Weissman, I.L., Hamann, A. and Butcher, E.C. (1993) (4 (7 integrin mediates lymphocyte binding to the mucosal vascular addressin MAdCAM-1. *Cell* 74, 185-95

Bernet, M.F., Brassart, D., Neeser, J.R. and Servin, A.L. (1993) Adhesion of human bifidobacterial strains to cultured human intestinal epithelial cells and inhibition of enteropathogen-cell interactions. *Applied and Environmental Microbiolology* 59, 4121-28.

Bernet, M.F., Brassart, D., Neeser, J.R. and Servin, A.L. (1994) *Lactobacillus acidophilus* LA 1 binds to intestinal cell lines and inhibits cell attachment and cell invasion by enterovirulent bacteria. *Gut* 35, 483-89.

Bernet-Camard, M.F., Lievin, V., Brassart,D., Neseer J.R., Servin, A.L. and Hudault, S. (1998). The *Lactobacillus acidophilus* Strain LA1 secretes a broad spectrum non-bacteriocin antibacterial substance(s) active *in vitro* and *in vivo*. *Applied and Environmental Microbiolology* In press.

Bertazzoni Minelli, E., Benini, A., Beghini, A.M., Cerutti, R. and Nardos, G. (1993) Bacterial faecal flora in healthy women of different ages. *Microbial Ecology in Health and Disease* 6, 43-51.

Blaser, M.J. and Parsonnet, J. (1994) Parasitism by the "slow" bacterium *Helicobacter pylori* leads to altered gastric homeostasis and neoplasia. *Journal of Clinical Investigation* 94, 4-8.

Bonan H.G. (1995) Peptide antibiotics and their role in innate immunity. *Annual Reviews in Immunology* 13, 61-92.

Bry, L., Falk, P.G., Midtvet, T. and Gordon, J.I. (1996) A model of host-microbial interactions in an open memmalian ecosystem. *Science* 273, 1380-83.

Cepek, K.L., Shaw, S.K., Parker, C.M., Russell, G.J., Morrow, J.S., Rimm, D.L. and Brenner, M.B. (1994) Adhesion between epithelial cells and lymphocytes mediated by E-cadherin and the $\alpha_{e} \beta_{7}$ integrin. *Nature* 372, 190-93.

Deitch E.A. (1994) Bacterial translocation: the influence of dietary variable. *Gut* Supplement 1, S23-27.

Donnenberg, M.S. and Kaper, J.B. (1992) Enteropathogenic *Escherichia coli*. *Infection and Immunology* 60, 3953-61.

Duchmann, R., Kaiser, I., Hermann, E., Mayet, W., Ewe, K. and Meyer Zum Bu schenfelde, K.H. (1995) Tolerance exists towards resident intestinal flora but is broken in active inflammatory bowel disease (IBD). *Clinical Experimental Immunology* 102, 448-55.

Elmer, G.W., Surawicz, C.M. and Mc Farland, L.V. (1996) Biotherapeutics agents: a neglected modality for the treatment and prevention of selected intestinal and vaginal infections *JAMA* 275, 870-76.

Foo, M.C. and Lee, A. (1974) Antigenic cross-reaction between mouse intestine and a member of the autochthonous microflora. *Infection and Immunology* 9, 1066-69.

Fried, M.,. Siegriest, H, Frei, R., Froehlich, F., Duroux, P., Thorens, J., Blum, A., Bille, J., Gonvers, J.J. and Gyr, K. (1994) Duodenal bacterial overgrowth during treatment in outpatients with omeprazole. *Gut* 35, 23-26.

Fuller, R. (1989) Probiotics in man and animals. *Journal Applied Bacteriology* 66, 365-78.

Haenel, H. (1970) Human normal and abnormal gastrointestinal flora. *American Journal of Clinical Nutrition* 23, 1433-39.

Hirayama, K., Miyaji, K., Kawamura, S., Itoh, K., Takahashi, E. and Mitsuoka, T. (1995) Development of intestinal flora of human-flora-associated (HFA) mice in the intestine of their offspring. *Experimental Animals* 44, 219-22.

Hromockyj, A.E. and Falkow, S. (1995) Interactions of bacteria with the gut epithelium. In *Infections of the gastrointestinal tract*. pp. 603-15. Edited by M.J. Blaser, PD. Smith, JI: Ravdin, HB., Greenberg, RL. Guerrant. Raven Press Ltd New-York.

Isolauri, E., Juntunen, M., Rautanen, T., Sillanaukee, P. and Koivula, T. (1991) A human *Lactobacillus* strain (*Lactobacillus casei* spp. strain GG) promotes recovery from acute diarrhoea in children. *Pediatrics* 88, 90-97.

Johansson, M.L., Molin, G. and Jeppsson, B. (1993) Administration of different *Lactobacillus* strains in fermented oat meal soup: In vivo colonization of human intestinal mucosa and effect on the indigenous flora. *Applied and Environmental Microbiology* 59, 15-20.

Jones, B.D. and Falkow, S. (1996) Salmonellosis: host immune responses and bacterial virulence determinants. *Annual Review of Immunology* 14, 533-41.

Kjell, K., Baklein, K., Bakken,A., Kral, J.G., Fausa, O. and Brandtzaeg, P. (1995) Instestinal B-cell isotype response in relation to local bacterial load: Evidence for immunolgobulin A subclass adaptation. *Gastroenterology* **109**, 819-25.

Kotler, D.P., Giang, T.T., Thiim, M., Nataro, J.P., Sordillo, E.M. and Orenstein, J.M. (1995) Chronic bacterial enteropathy in patients with AIDS. *Journal Infections Diseases* **171**, 552-58.

Langhendreis, J.P., Detry, J., van Hees, J., Lamboray, J.M., Darimont, J., Mozin M.J., Secretin, M.C. and Senterre, J. (1995) Effect of fermented infant formula containing viable bifidobacteria on the fecal composition and pH of healthy full-term infants. *Journal Pediatric Gastroenterology and Nutrition* **21**, 177-81.

Lefrancois, L. and Goodman, T. (1989) In vivo modulation of cytolytic activity and Thy-1 expression in TCR-gamma delta + intraepithelial lymphocytes. *Science* **243**, 1716-18.

Levine, M.M. (1987) *Escherichia coli* that cause diarrhea: enterotoxigenic, enteropathogenic, enteroinvasive, enterohemorragic and enteroadherent. *Journal Infections Diseases* **155**, 377-89.

Link, H., Rochat, F., Saudan, K.Y. and Schiffrin, E.J. (1995) Immunomodulation of the gnotobiotic mouse through colonization with lactic acid bacteria. *Advances in Mucosal Immunology*, Edited by J. Mestecky *et al.* Plenum press, New York. pp. 465-67.

Link-Amster, H., Rochat, F., Saudan, K.Y., Mignot,O. and Aeschlimann, J.M. (1994) Modulation of a specific humoral immune response and changes in intestinal flora mediated through fermented milk intake. *FEMS Immunology and Medical Microbiology* **10**, 55-64.

Liu, Y.J. and Arpin, C. (1997) Germinal centre development. *Immunological Reviews* **156**, 111-26.

Luckey, T.D. (1970) Introduction to the ecology of the intestinal flora. *American Journal of Clinical Nutrition* **23**, 1430-32.

Lyerly, D.M. and Wilkins, T.D. (1995) *Clostridium difficile*. In *Infections of the gastrointestinal tract*. pp. 867-92. Edited by M.J. Blaser, P.D. Smith, J.I: Ravdin, H.B. Greenberg, R.L. Guerrant. Raven press Ltd New-York.

MacDonald, T.T. (1995) Breakdown of tolerance to the intestinal bacterial flora in inflammatory bowel disease (IBD). *Clinical Experimental Immunology* **102**, 445-47

MacPherson, A., Khoo, U.Y. Forgacs, I., Philpott-Howard, J. and Bjarnson, I. (1996) Mucosal antibodies in inflammatory bowel disease are directed against intestinal bacteria. *Gut* **38**, 365-75.

McCormick, B.A., Colgan, S.P., Delp-Archer, C., Miller, S.I. and Madara, J.L. (1993) *Salmonella typhimurium* attachment to human intestinal epithelial monolayers: Transcellular signaling to subepithelial neutrophils. *Journal Cell Biology* **123**, 895-907.

McGowan, C.C., Cover, T.L. and Blaser, M.J. (1996) *Helicobacter pylori* and gastric acid: biological and therapeutic implications. *Gastroenterology* **110**, 926-38.

Mitsuoka, T. (1992) The human gastrointestinal tract, pp. 69-114. In *The Lactic Acid Bacteria*, Edited by B.J.B. Wood, Elsevier Applied Science, London and New York.

Neeser, J.R., Chambaz, A., Golliard, M., Link-Amster, H., Fryder, V. and Kolodziejczyk, E. (1989) Adhesion of Colonisation Factor Antigen II-positive enterotoxigenic *Escherichia coli* strains to human enterocyte-like differentiated HT-29 cells: a basis for host-pathogen interactions in the gut. *Infection and Immunology* **57**, 3727-34.

Neutra, M.R., Frey, A. and Kraehenbuhl, J.P. (1996) Epithelial M cells: gateways for mucosal infection and immunisation. *Cell* **86**, 345-48.

Pedrosa, M.C., Golner, B.B., Goldin, B.R., Barakat, S., Dallal, G.E. and Russell, R.M. (1995) Survival of yoghurt-containing organisms and *Lactobacillus gasseri* (ADH) and their effect on bacterial enzyme activity an the gastrointestinal tract of healthy and hypochlorhydric elderly subjects. *American Journal of Clinical Nutrition* **61**, 353-59.

Rasmussen, S.J., Eckmann, L., Quayle, A.L., Shen, L., Zhang, Y.X., Anderson, D.J., Fierer, J., Stephens, R.S.

and Kagnoff, M.F. (1997) Secretion of proinflammatory cytokines by epithelial cells in response to Chlamydia infections suggests a central role for epithelial cells in the chlamydial pathogenesis. *Journal of Clinical Investigation* **99**, 77-87

Saavedra, J.M., Bauman, N.A., Oung, I., Perman, J.A. and Yolken, R.H. (1994) Feeding of *Bifidobacterium* and *Streptococcus thermophilus* to infants in hospital for prevention of diarrhoea and shedding of rotavirus. *The Lancet* **344**, 1046-49.

Saltzman, J.R., Kowdley, K.V., Pedrosa, M.C., Sepe, T., Golner, B., Perrone, G. and Russell, R.M. (1994) Bacterial overgrowth without clinical malabsorption in elderly hypochlorhydric subjects. *Gastroenterology* **106**, 615-23.

Shroff, K.E., Meslin, K. and Cebra, J.J. (1995) Commensal enteric bacteria engender a self-limiting humoral mucosal immune response while permanently colonizing the host. *Infection and Immunology* **63**, 3904-13.

Simon, G.L. and Gorbach, S.L. (1984) Intestinal flora in health and disease. *Gastroenterology* **86**, 174-93.

Simon, G.L. and Gorbach, S.L. (1995) Normal alimentary tract microflora. In *Infections of the Gastrointestinal Tract*. pp. 53-69. Edited by Blaser, M.J., Smith, PD, Ravdin, J.I. and Greenberg, R.L. Guerrant. Ravenpress Ltd New-York.

Svensson, L., Finlay, BB, Bass, D., Von, B.C. and Greenberg, H.B. (1991) Symetric infection of rotavirus on polarized human intestinal epithelial (Caco-2) cells. *Journal of Virology* **65**, 4190-97.

Underdown, B.J. and Schiff, J.M. (1986) Immunlglobulin A: Strategic defence initiative at he mucosal surfaces. *Annual Reviews in Immunology* **4**, 389-417.

Van der Waaij, L.A., Lindburg, P.C., Mesander, G. and van der Waaij, D. (1996) *In vivo* Iga coating of anaerobic bacteria in human faeces. *Gut* **38**, 348-54.

Well, C.L., Van de Westerlo, E.M.A., Jechorek, R.P., Feltis, B.A., Wilkins, T.D. and Erlandsen, S.L. (1996) *Bacteroides fragilis* enterotoxin modulates epithelial permeability and bacterial internalization by HT-29 enterocytes. *Gastroenterology* **110**, 1429-37.

CHAPTER 13

Actions of Non-Digestible Carbohydrates on Blood Lipids in Humans and Animals

NATHALIE DELZENNE[1] AND CHRISTINE M. WILLIAMS[2]
[1]*Université Catholique de Louvain, Department of Pharmaceutical Sciences, Brussels, BELGIUM*
[2]*Hugh Sinclair Unit of Human Nutrition, Department of Food Science and Technology, University of Reading, Reading, UK*

1   Chemical structure and biological features of non-digestible carbohydrates

Several types of non-digestible carbohydrates (NDC) have been studied for their lipid lowering properties in both humans and animals. They can be classified according to their chemical structures and fermentability by the resident bacterial microflora (Englyst *et al.,* 1992; Cummings, 1997). Some have been studied more extensively in order to assess their influence on lipid metabolism and circulating blood lipids. This is especially the case for the fructans, resistant starch (RS), cellulose and derivatives, and soluble fibres like pectin, guar gum and ß glucan.

1.1   FRUCTANS

Fructan is a general term used for any carbohydrate in which one or more fructosylfructose links constitutes the nature of osidic bonds. *Inulin*-type fructans are present in significant amounts in miscellaneous edible fruits and vegetables. Because of the ß-configuration of the anomeric C^2 in their fructose monomers, inulin-type fructans are resistant to hydrolysis by human digestive enzymes (α-glucosidase; maltase-isomaltase; sucrase), which are specific for α-osidic linkages. Inulin-type fructans have been classified as 'non-digestible' oligosaccharides, which are mainly fermented by bifidobacteria (Delzenne and Roberfroid, 1994; Cummings and Roberfroid, 1997). Some short-chain fructans may be obtained by the enzymatic hydrolysis of inulin (as in the case of oligofructose), or synthesised from sucrose (Neosugar) (Roberfroid and Delzenne, 1998).

G.R. Gibson and M.B. Roberfroid (eds.), Colonic Microbiota, Nutrition and Health, 213-231.
© 1999 *Kluwer Academic Publishers. Printed in the Netherlands.*

1.2   STARCHES

Homopolymers of glucose, existing as amylopectin (70-80% of total starch) and amylose differ from each other by their digestibility and fermentability. The reasons for the incomplete digestion of starch may be intrinsic factors (physical form, formation of complexes) or extrinsic factors (chewing, transit time). Starch granules which show X-ray diffraction patterns B and C (*e.g.* in potato, banana and vegetables) are more resistant to pancreatic amylase than those of pattern A (in cereals) (Cummings, 1997).

1.3   CELLULOSE AND DERIVATIVES

Cellulose is a polymer of glucopyrannose in which glucopyrannosyl moieties are linked by a ß 1-2 link. Micro-crystalline cellulose is neither hydrolyzed by intestinal and pancreatic enzymes, nor fermented in some species like rats. This carbohydrate is often used as a control in the studies in which the effect of fermentable carbohydrate on lipid homeostasis is assessed (Fernandez *et al.*, 1997).

1.4   SOLUBLE FIBRES

Hetero-oligomers are often classified as soluble dietary fibres. Guar gum consists of the water soluble fraction of guar flour called guaran, which consists of linear chains of (1-4) -ß-D mannopyranosyl units with α-D galactopyrannosyl units attached by 1-6 linkages. The ratio of D-galactose to D- mannose is 1: 2. Pectin occurs naturally as the partial methyl ester of alpha (1-4) linked polygalacturonate sequences interrupted with (1-2) -L- Rhamnose residues.

2   Fermentation of non-digestible carbohydrates

The fermentation of NDC produces short chain fatty acids (mainly acetate, propionate and butyrate) and lactate which allow the host to salvage part of the energy of NDC (40-50% that of a digestible carbohydrate) and which may play a role in regulating cellular metabolism (for a review see Cummings, 1997). The pattern of short chain carboxylic acids (SCCA's) produced by fermentation differs for various NDC's and may be responsible for the varying effects of NDC's on plasma lipids. Pectin is a particularly good source of acetate whilst arabinogalactans and guar are the poorest source of acetate, but produce significant amounts of propionate. Guar gum is broken down in the large bowel, yielding fermentation products rich in propionic acid (up to 50% of the total SCCA's) (Favier *et al.,* 1997). Supplementing the diet with inulin-type fructans decreases the caecal pH, increases the size of the caecal pool of SCCA, with the relative proportions in the order acetate > butyrate > propionate (Campbell *et al.,* 1997).

3   Blood lipids as risk factors for coronary heart disease

Early studies of effects of NDC's on blood lipids measured only fasting total cholesterol levels. As epidemiological evidence emerged showing that only the LDL fraction of cholesterol was associated with increased risk of coronary heart disease (CHD), and that the HDL cholesterol fraction appeared to be protective, animal and human studies began to concentrate on measurements of total, LDL and HDL cholesterol responses to dietary NDC's. At the time there was little interest in the measurement of triglyceride concentrations, since epidemiological studies and mechanistic studies in animals provided little evidence for involvement of elevated triglyceride (TG) concentrations in the pathogenesis of CHD. Most studies appeared to show that raised TG levels were closely associated with low HDL levels and that it was the latter abnormality, rather than the TG levels themselves, which were responsible for the weak association found between fasting TG and risk of CHD. More recent findings from large scale epidemiological surveys provide evidence that elevated fasting TG levels are associated with greater risk of CHD and that their effect is independent of any association with low levels of HDL cholesterol (Hokanson and Austin, 1996). A significant number of studies have also shown that elevated postprandial triglyceride concentrations are observed in subjects with CHD (Groot *et al.*, 1991; Karpe *et al.*, 1994) and that the 6-8h postprandial TG value is a strong predictor of risk of CHD (Patsch, 1992). A number of mechanisms have been put forward to explain the association between raised postprandial triglyceride-rich lipoproteins and risk of CHD (Williams, 1997), including direct atherogenic effects of chylomicron (CM) and VLDL remnants (Zilversmit, 1979) and, more recently, the indirect atherogenic consequences of neutral lipid exchange between CM's and LDL and HDL leading to the formation of small dense LDL and HDL (Patsch, 1992). Recent evaluations of effects of NDC's on blood lipids in humans and animals have included measurements of effects on postprandial lipoproteins, both with respect to inclusion of NDC in lipid-containing test meals (Gatti *et al.*, 1984; Redard *et al.*, 1990; Cara *et al.*, 1992) and effects of NDC in the background diet (Wolever *et al.*, 1997). However, such studies remain relatively few in number (Lairon, 1996) and much of the information available with respect to effects of NDC's on TG levels are confined to measurements of fasting blood levels.

4   Modulation of cholesterol homeostasis by non-digestible carbohydrates

4.1   FRUCTANS AND RESISTANT STARCH

The effect of *fructans* on cholesterolemia is not well established in animal models; only long term (16 weeks) administration of oligofructose (OFS) has been shown to decrease total cholesterol levels in the serum of rats (Fiordaliso *et al.*, 1995). Oligofructose supplementation has also been shown to protect rats against the increase in free cholesterol level induced by a high-fat diet, without preventing the accumulation of cholesterol in liver tissue (Kok *et al.*, 1996a,b). However inulin or OFS administration does not modify faecal bile acid excretion in rats receiving a hypercholesterolemic diet containing 0.5%(w/v) cholic acid and 1%(w/v) cholesterol (N. Delzenne unpublished data). The effect of fructans on blood cholesterol levels in human volunteers are less extensively studied than is the case for other NDC. A small number of studies have been conducted in subjects with hyperlipidaemia as well as in normolipemic individuals. Subjects with non insulin dependent diabetes (NIDDM) who were administered 8g OFS (Neosugar), a synthetic oligosaccharide, for 14 days showed an 8% reduction in total, and a 10% reduction in LDL cholesterol, compared with a control group given sucrose in the same food vehicles (Yamashati *et al.,* 1984). Similar reductions in blood lipids were reported to have been observed in a group of Japanese subjects with hyperlipidemia (Hidaka *et al.,* 1986) but no data were shown to support this conclusion. More recently, Davidson *et al.* (1998) in a randomised cross over trial in subjects with modest hyperlipidemia, showed significantly lower total and LDL concentrations during inulin treatment compared with placebo phases, but the authors reported no effects on HDL cholesterol. Luo *et al.* (1996), studying normal young subjects, investigated effects of OFS (20 g /d) fed as 100g cookies a day, in a randomised cross over design with treatment periods of four weeks. No changes in serum cholesterol or apolipoproteins were observed in either the treatment or placebo periods. In contrast, Canzi *et al.* (1995) observed significantly lower cholesterol concentrations in young male volunteers who consumed 9g inulin per day added to a rice breakfast cereal for a period of four weeks. Total cholesterol and LDL cholesterol levels were reduced by 5% and 7% respectively with inulin compared with placebo. Pedersen *et al.* (1997) reported no effect on blood lipids of a daily intake of 14 g inulin added to a low fat spread for a period of four weeks in 66 young healthy women. Although HDL cholesterol and the LDL: HDL cholesterol ratio were lower at the end of both the control and test (inulin) periods, there were no significant differences in blood lipids between placebo and inulin. In a recently conducted study, 58 middle aged subjects with moderately raised blood lipid concentrations, consumed 10 g per day of inulin or placebo, in a powdered form which could be added to beverages, soups, cereal (Williams, 1998). No significant changes in concentrations of total, LDL or HDL cholesterol or in apolipoproteins B and A, were found in either of the groups over the 8 week intervention.

The administration of (*RS*) decreases the plasma free and total cholesterol concentrations in hypercholesterolemic, but not in normocholesterolemic rats (Vanhoof and DeSchrijver, 1997). The hypocholesterolemic effect of RS is not dose-dependent in hamsters, and is due to a reduction in HDL and non-HDL particles (Ranhotra *et al.,* 1997). In humans, two studies have shown modest reductions in total and LDL cholesterol in subjects fed high amylose corn starch (Reiser *et al.,* 1989; Behall *et al.,* 1989). In contrast, one other study showed higher total cholesterol after 4 weeks on a high amylose corn starch diet (Behall and Howe, 1995), whereas another showed no effect of such a diet on total LDL or HDL cholesterol levels (Heijenen *et al.,* 1996).

4.2   SOLUBLE FIBRES

There have been numerous studies which have investigated effects of soluble fibres (pectin, guar gum, psyllium and ß glucan) on blood lipid levels in animal models, the majority of which have shown cholesterol lowering effects of these additions in animals on both high and low cholesterol diets (Kritchevsky and Strory, 1993). *Guar gum*, when added in a diet containing normal (0.04%) or high levels (0.25%) of cholesterol, decreases serum and hepatic cholesterol in rats (Overton *et al.,* 1994; Fernandez, 1995). Both low viscosity and high viscosity guar gum decreases HDL-2 cholesterol when given in rats, whereas only high density guar gum depresses HDL 1 and LDL-cholesterol (Favier *et al.,* 1997). In a number of reviews of the many human studies (> 50) which have investigated effects of guar gum, the overall conclusions were that guar gum reduced total and LDL cholesterol levels by -10 to -17 % (Gatenby, 1990; Truswell, 1995; Lairon, 1996). In the majority of studies, guar gum was fed at dose levels of 10-15 g / day, but in studies which fed larger doses (> 20 g/day) reductions were no greater than those observed when guar gum was fed at the more modest intakes. Over half of the studies were conducted in subjects with diabetes and in the remainder subjects were either normolipidaemic or hypercholesterolaemic.

Pectin, given at the dose of 7.5% in the diet of rats receiving a diet enriched in cholesterol, decreased total lipids and cholesterol concentrations in the liver and in the serum (Arjamndi *et al.,* 1992). In this study, as in the case of many animal studies, cellulose was added as a control substance. The most extensively studied pectins in humans are those from citrus, apples and sugar beet. In human subjects, doses of pectin of approximately 15g/day result in reductions in total and LDL cholesterol between 8-12%, with greater responses in subjects with raised cholesterol levels at baseline (Lairon, 1996). In another review, ten out of nineteen studies were shown to result in statistically significant lower total cholesterol levels in subjects receiving supplements of pectin (Truswell, 1995).

The effects of *oat bran (β glucan)* on blood lipids and lipid metabolism have been repeatedly shown in several animal models and these data are carefully reviewed by Kritchevsky and Story (1993). In human subjects, the relationship between oat bran consumption and blood lipid risk factors has been extensively investigated since 1963. In January 1996, the USA Food and Drug Administration registered oat bran as a cholesterol-lowering food (FDA, 1996). The published data describing potential hypocholesterolaemic properties of oats is extensive, although there has been much debate suggesting that large additions of oat bran in the human diet can cause changes in other constituents (*e.g.* reductions in total and saturated fats), which may be responsible for the observed lipid changes (Swain *et al.,* 1990). At least twelve studies have demonstrated that dietary supplementation with oat products produces cholesterol-lowering in individuals with mild to moderate hypercholesterolaemia (Table 1). A meta-analysis of thirteen studies concluded that oat bran-supplemented diets could reduce TC levels modestly (-7%) and that this reduction was not caused by other dietary changes (Ripsin *et al.,* 1992). The data available for the effect of oat bran on lipid parameters in normocholesterolaemic individuals (TC less than 5.2 mmol/l) is weaker but two studies demonstrate that oat bran has the capacity to reduce cholesterol when added to the diet (Table 2). Overall, human studies show that oat bran modestly reduces total serum cholesterol and LDL-cholesterol levels as a solitary dietary addition, and also in combination with a low fat diet, without decreasing HDL cholesterol. Doses of more than 3g β glucan per day are required and the effect is more pronounced in individuals with hypercholesterolemia.

A recent meta analysis of eight published and five unpublished human studies of effects of *psyllium* enriched cereal products on plasma lipids concluded that these products resulted in only modest reductions in total and LDL cholesterol in subjects with mild to moderate hypercholesterolaemia (Olson *et al.*, 1998).

Table 1.   Human studies investigating the effects of oat bran on blood lipid parameters in hypercholesterolaemic individuals

Author	Dose / day (Soluble fibre)	Time	% change in Total-C	% change in LDL-C
Kirby *et al.*1981	100g (14.8g)		-13%	-12%
Anderson *et al.*1984	100g (17g)		-19%	-23%
Anderson *et al.*(1990)	25g (3.5g)	2 weeks	-5.4% p < 0.05	-8.5% p < 0.025
Kahn *et al.*(1990)	80g		-8% p < 0.02	-10% p < 0.02
Kestin *et al.*(1990)	95g (5.8g)	4 weeks	-4.9% p < 0.01	-6.6% p < 0.01
Anderson *et al.*(1991)	106g (7.6g)	3 weeks	-12.8%	-12.1%
Spiller *et al.*(1991)	77g (5g)	3 weeks	-3.7%	-6.6%
Van Horn *et al.*(1991)	57g	8 weeks	-6.2% p < 0.05	-9.2% p < 0.05
Whyte *et al.*(1992)	123g (10.3g)	4 weeks	-3.1% p < 0.01	-5.7% p < 0.01
Hegsted *et al.*(1993)	100g	3 weeks	-10% p < 0.001	
Braaten *et al.*(1994)	7.2g (5.8g ß-glucan)	4 weeks	-9 % p < 0.0001	-10% p < 0.001
Leadbetter *et al.*(1991)	0, 30, 60, 90g (3.7-4.2% ß-glucan)	4 weeks	NS	NS
Kashtan *et al.*(1992)	6.8g total dietary fibre/ 1000kCal	2 weeks	-4.1% p < 0.022	-10% p < 0.024
Törrönen *et al.*(1992)	11.2g ß-glucan	8 weeks	NS	NS
Poulter *et al.*(1993)	50g (2.6g)	4 weeks	-2.23% p < 0.04	-4.6 % P < 0.05

Table 2.   Human studies investigating the link between oat bran consumption and blood lipid parameters in normocholesterolaemic individuals

Author	Dose / day (Soluble fibre)	Time	%change in Total Cholesterol	% change in LDL-cholesterol
Gold and Davidson (1988)	34g (10g Total dietary fibre)	4 weeks	-5.3% p < 0.05	-8.7% p < 0.05
Marlett *et al.*(1994)	100g		-9% p < 0.01	
Swain *et al.* (1990)	87g	6 weeks	NS	NS
Beer *et al.* (1995)	9g ß glucan	14 days	NS	NS

5   Mechanisms of action of non-digestible carbohydrates on cholesterol metabolism

Several hypothesis have been proposed to explain the influence of NDC on cholesterol metabolism (Fig. 1).

5.1   MODIFICATIONS OF FAECAL EXCRETION OF CHOLESTEROL METABOLITES (NEUTRAL STEROLS AND/OR BILE ACIDS)

The possibility that NDC's might interfere with cholesterol and bile acid homeostasis by increasing faecal sterol excretion has received considerable attention and has been reviewed in a number of recent publications (Kritchevsy and Story, 1993; Cassidy and Calvert, 1993; Lairon, 1996). Numerous studies have been performed in animals which are generally consistent with the conclusion that soluble fibres increase faecal bile acid excretion but less marked and consistent effects are observed as far as cholesterol *per se* is concerned. A recent study has shown that, in rats, the cholesterol-lowering effect of guar gum appears to be mediated by an accelerated faecal excretion of steroid and a rise in the intestinal pool and biliary production of bile acids (Moundras *et al.*, 1997). Although there were adaptive modifications in hepatic enzyme activities (inducement of HMG-CoA reductase and 7-α-hydroxylase), these were insufficient to compensate for faecal steroid losses with resulting reductions in blood cholesterol levels (Moundras *et al.*, 1997). Fernandez (*et al.*, 1997) reported that guar gum reduced cholesterol absorption, but only in rats receiving a diet enriched in cholesterol. Ten%(w/v) guar gum in the diet for 4 weeks decreased serum and hepatic cholesterol in rats, via a mechanism involving intestinal bile acid binding and loss of faecal bile acids. In the liver of guar gum fed rats, 7-α-hydroxylase activity (the rate limiting enzyme of bile acids synthesis) was increased, whereas endogenous cholesterol synthesis (HMG-CoA reductase) was not modified (Overton *et al.*, 1994). Rats receiving pectin for 3 weeks had a significantly higher faecal excretion of neutral sterols (but not acidic sterols) than those fed cellulose - supplemented diet (Arjamndi *et al.*, 1992).

A higher faecal excretion of coprostanol, a form of cholesterol synthesised by intestinal bacteria, has been proposed as one of the putative mechanisms for the hypocholesterolemic effect of RS. Some studies report that total faecal bile acids excretion is not modified during RS treatment of rats (Vanhoof and DeSchrijver, 1997), whereas, in other studies, it is suggested that qualitative (increased excretion of chenodeoxycholic acid and derivatives) and quantitative modifications of faecal bile acids occur when rats are fed a diet enriched with retrograded potato RS (Chezem *et al.*, 1997). However, in rats Younes *et al.* (1995) found that RS achieved greater cholesterol-lowering than cholestyramine (a bile acid-sequestering compound), yet RS only slightly increased faecal bile acids excretion. This led the authors to postulate that large bowel fermentation, rather than a bile-acid binding effect, was the key event in the cholesterol-lowering effects of RS.

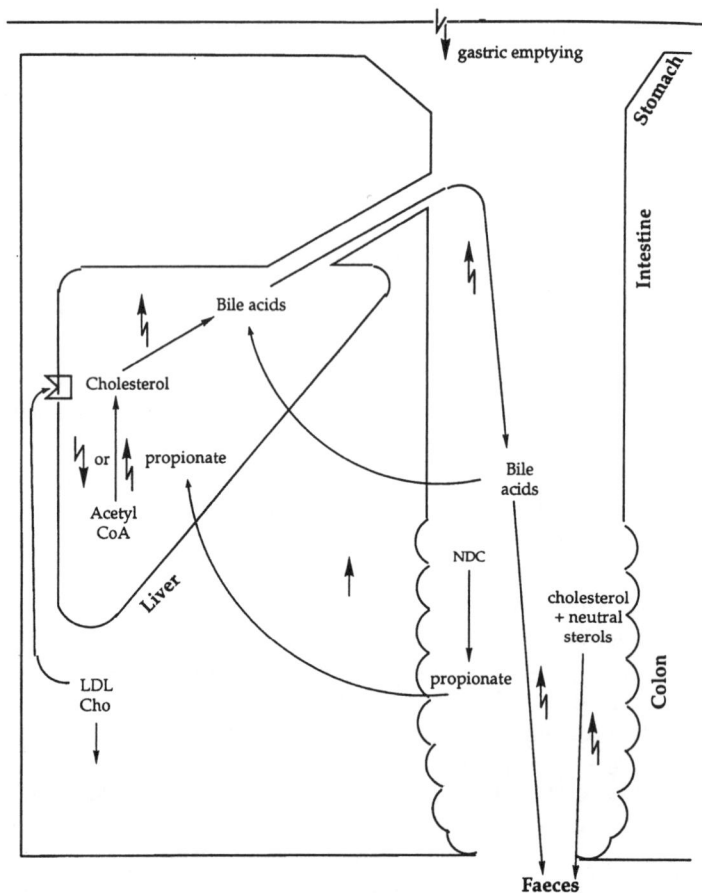

Figure 1. Putative mechanisms explaining the hypocholesterolemic effect of non-digestible carbohydrates (*e.g.* guar gum, pectin) including effects on gastric emptying; reabsorption bile acids; and fermentative production of short chain carboxylic acids.

In humans, evidence for reduced bile acid reabsorption and increased bile acid excretion with the feeding of soluble fibres is strong, with consistent evidence of elevation in faecal bile acid excretion in a significant number of the forty studies conducted to date (Lairon, 1996). The conclusions from these studies are strongly supported by the observations of Zhang *et al.* (1992) in ileostomy subjects, showing increased ileal excretion of bile acids (but not cholesterol) with pectin, oat bran, rye bran and sugar beet fibre. Although the reliability of sterol balance studies has been a limiting factor in drawing definitive conclusions in this area (Andersson and Bosaeus, 1993), more reliable data have been obtained using stable isotopic enrichment methods, which showed that feeding oat bran resulted in increased faecal bile acid excretion, increased primary bile salt synthesis and reduced total and LDL cholesterol levels (Marlett *et al.,* 1994). Lairon (1996) has suggested that individual variations in cholesterol response to NDC, and discrepancies between studies, may in part be explained by the impact which increased bile acid excretion has on individual bile acid synthesis. In one study using the soluble fibre psyllium it was shown that, whilst all subjects showed increased bile acid excretion, only in those in whom bile acid synthesis was also correspondingly increased, was there a significant reduction in LDL cholesterol (Everson *et al.*, 1992).

5.2 THE INVOLVEMENT OF FERMENTATION PRODUCTS IN THE CHOLES-TEROL-LOWERING EFFECT OF NON-DIGESTIBLE CARBOHYDRATES

It has been suggested that fermentation of soluble fibres in the large bowel, suppresses hepatic cholesterol synthesis via inhibitory effects of the SCCA which are absorbed into the portal vein. By comparing the effect of NDC and other dietary fibres differing in their bacterial fermentability, Stark and Madar (1993) have suggested that a high rate of production of SCCA by fermentation does not correlate with the inhibition of cholesterol synthesis in hepatocytes, thus suggesting that SCCA's are not the cholesterol-lowering factor of highly fermentable carbohydrate. Moreover, the role of SCCA's in the effects of NDC on cholesterol homeostasis is difficult to establish, because, either isolated or in mixture, these acids have antagonistic effects on cholesterol metabolism. Acetate is a metabolic precursor of cholesterol and has been claimed to be the cause of the hyper-cholesterolemia observed in healthy patients receiving lactulose (Jenkins *et al.*, 1991), whereas propionate, when given in the diet of rats, may decrease cholesterol synthesis, by inhibiting HMG-CoA reductase (Illman *et al.*, 1988). Whereas some authors suggest that feeding propionate redistributes cholesterol from plasma to liver (Illman *et al.*, 1988), it seems that dietary propionate has less hypocholesterolemic effect than that derived from NDC fermentation. Some studies report that calcium propionate (2.5% w/v in the diet) is unable to decrease plasma cholesterol in rats (Levrat *et al.*, 1994), whereas supplementation of the diet with inulin or cyclodextrin, leading to increased propionate in portal serum to the same extent as calcium propionate, does significantly reduce blood cholesterol. Evidence that large bowel fermentation is not a primary site for the mode of action of soluble NDC on blood cholesterol levels is supported by the fact that in germ free and heteroxenic animals, cholesterol-lowering effects of guar gum were comparable (Alvarez-Leite *et al.*, 1994), and in subjects with ileostomy, oat bran supplementation reduced LDL cholesterol to the same extent as that in subjects with an intact colon (Zhang *et al.*, 1992). These data would appear to demonstrate that cholesterol-lowering effects of the soluble fibres guar gum and ß glucan are not secondary to the production of short chain fermentation products in the colon.

5.3 LIPID DIGESTION AND ABSORPTION

The possibility that NDC's influence circulating lipids by attenuation of gastric emptying and/ or by more direct effects on lipid digestion and absorption has been considered in some detail by Lairon (1996). Some in vitro studies suggest that under highly controlled conditions, soluble fibres can inhibit pancreatic lipase, although there is limited information from in vivo studies to support this. In vitro, it is suggested that increased viscosity, leading to reduced emulsification and impaired action of both gastric and pancreatic lipase, may contribute to reduced lipolytic activity and attenuated lipid absorption in the presence of high viscosity soluble fibres. Additionally, increased intra luminal viscosity can alter small intestinal motility, reduce intraluminal mixing and increase the thickness of the unstirred water-layer, which may delay the final stage of lipid assimilation. Animal studies support the concept that chronic feeding of soluble fibres causes delayed rather than inhibited lipid uptake, since, when fat test meals were fed to animals adapted to diets high in soluble fibre, early recovery of fatty acids and cholesterol in lymph was lower, but 24h recovery was similar to that of control fed animals (Vahouny et al., 1988). More recently, in a study which employed a range of cellulose-derivatives of varying viscosity, Carr et al. (1988) suggested an independent role of intestinal contents viscosity in lowering plasma cholesterol concentration and dietary cholesterol absorption efficiency.

6   Hypotriglyceridemic effect of non-digestible carbohydrates

Several NDC are able to decrease serum TG in animals although effects in humans are less clear cut. The biochemical mechanisms underlying the hypotriglyceridaemic effects of NDC have not yet been elucidated although several hypotheses have been proposed.

6.1   FRUCTANS AND RESISTANT STARCH

The effects of *inulin-type fructans* on triglyceridemia has been studied both in human subjects and in animals (Roberfroid and Delzenne, 1998). In rats, a decrease in serum triglyceridemia (both in fed and fasted state) has consistently been reported in several studies (Kok et al., 1996b; Delzenne and Kok, 1998). Feeding rats a diet supplemented with OFS (10%w/v) significantly lowered serum triglyceride and phospholipid concentrations (Delzenne et al., 1993), but did not modify free fatty acid concentration in the serum (Fiordaliso et al., 1995). Very few studies have investigated triglyceride-lowering properties of fructans in human volunteers. In the study of Pedersen et al, (1997) and that of Luo et al. (1996), no changes in serum triglycerides, were observed with either inulin or OFS, although there was a strong trend for free fatty acid (FFA) concentrations to be reduced in the latter study. In contrast, Canzi et al. (1995), observed significantly lower blood triglycerides, with fasting concentrations reduced by 27% during inulin treatment. The recent study of Williams (1998) also found that serum triglycerides were 19% lower at 8 weeks than baseline in the inulin treated group and values were significantly lower than in the placebo group, returning to baseline values four weeks after treatment.

There is reasonably consistent evidence to suggest that *dietary RS* reduces plasma TG's in both animals and humans. The replacement of wheat starch (digestibility: 62%; content in RS: 22%) by mung-bean starch (digestibility: 40%; content in RS: 77%) reduces plasma TG and adipocyte volume in both normal and diabetic rats (Lerer-Metzger et al., 1996). This treatment decreased plasma glucose and free fatty acids levels in normal, but not diabetic, rats. Various amounts of raw or retrograded RS (amylomaize starch, potato starch, or modified high amylose starch) lowered serum TG and epididimal fat pads in rats (de Deckere *et al.,* 1995). In human subjects, evidence for TG reducing effects of RS is stronger than that for cholesterol reduction. In three separate studies, feeding diets high in amylose corn starch resulted in reductions in fasting TG's of 17%, 19% and 28% respectively (Behall *et al.,* 1989; Reiser *et al.,* 1989; Behall and Howe, 1995).

6.2   SOLUBLE FIBRE

Both high and low viscosity *guar gum* depress plasma TG in rats. Guar gum is also able to counteract the adverse influence of a high sucrose diet on systolic blood pressure and serum triglycerides in rats, whereas cellulose has no effect (Gondal *et al.,* 1996). Studies in guinea pigs have shown that guar gum, but not pectin, is able to decrease serum total TG (Fernandez *et al.,* 1997). The secretion rate of VLDL was not modified, whereas the secretion rate of Apo B was decreased by the treatment. The composition of the VLDL secreted was different (presence of larger nascent VLDL) (Fernandez *et al.,* 1997).

Evidence for effects of soluble fibres on fasting plasma TG's in humans is equivocal and this may be because many of the studies have been designed, primarily, to study effects on total and LDL cholesterol so that fasting triglycerides have either not been measured or have been measured under circumstances which do not allow clear interpretation of the findings. In one study, Noakes *et al.* (1996) have shown oat bran to be protective against increased triglyceridaemia induced by a high carbohydrate diet in hypertriglyceridimic individuals. Other studies such as that of Wolever *et al.* (1997) showed no effect on fasting triglycerides, of adding a mix of soluble fibres over a four month period, in subjects with either the Apo E3 or Apo E4 phenotype. Indeed, in this study postprandial triglycerides were actually increased on the soluble fibre diet in the E3 group. Data concerning effects on postprandial triglycerides, of soluble fibres added to a single test meal or when supplemented in the background diet, are conflicting. The supplementation of a single 70g fat meal with oat bran, (maintaining the same carbohydrate: fat ratio as a control low-fibre test meal), induced an increase in postprandial triglycerides, insulin and free phospholipids and cholesterol in men. The subsequent supplementation of the background diet with 40g/day of oat bran exacerbated the postprandial changes seen with the oat supplemented test meal (Dubois *et al.,* 1995). An increase in postprandial TG has also been reported in females receiving an oat supplemented meal (Redard *et al.,* 1990), although in this study no effect of fibre supplementation was observed in male subjects. These three studies appear to be in direct conflict with the findings of Cara *et al.* (1992) who also studied men, but observed a fall in postprandial TG when oat bran, rice bran or wheat bran were added to single meals. Differences between acute meal effects, compared with chronic diet effects of soluble fibres on postprandial triglycerides, may be explained by adaptive changes in intestinal morphology and enzyme secretion which occur in response to inhibitory effects of soluble fibre on digestive and absorptive processes (Wolever *et al.,* 1997).

7   Mechanism of action of non-digestible carbohydrates on triglycerides

The hypotheses to explain effects of NDC on the modulation of triglyceride metabolism include: indirect inhibitory effects on hepatic triglyceride synthesis or stimulation of systemic clearance, or direct effects on triglyceride absorption (Fig. 2). The very large reductions in serum triglycerides (30-40%) observed in animals fed dietary fructans mean that attenuation of digestive and absorptive processes are unlikely to account for the great majority of this response. At the present time, most evidence points towards inhibition of hepatic lipogenesis as the most likely site of inhibition by fructans, albeit through indirect mechanisms involving modulation of glucose and insulin homeostasis, stimulation of gut hormone secretion or direct metabolic effects of SCCA on enzymes of hepatic lipogenesis.

7.1   INHIBITION OF HEPATIC LIPOGENESIS - THE ROLE OF GLUCOSE/INSULIN AND SHORT CHAIN CARBOXYLIC ACIDS

The fact that inulin inhibits the increase in triglycerides observed in rats fed a carbohydrate-rich diet, suggests inhibition of hepatic synthesis rather than a higher catabolism of triglyceride-rich lipoproteins as the more likely site of action. Hepatocytes isolated from OFS-fed rats have a slightly lower capacity to esterify $[^{14}C]$-palmitate into triglycerides but a 40% decreased capacity to synthesize triglycerides from $[^{14}C]$-acetate (Fiordaliso et al., 1995; Kok et al., 1996a). These data support the hypothesis that decreased de novo lipogenesis in the liver, through a coordinate reduction of the activity of all lipogenic enzymes, is a key event in the reduction of VLDL-TG secretion in fructan-fed rats. Moreover, recent data suggest that OFS decreases the activity of key lipogenic enzymes (namely fatty acid synthase FAS) by reducing the corresponding mRNA (Delzenne and Kok, 1998a,b). The fact that inhibition of de novo lipogenesis is the most likely basis for the hypotriglyceridemic effect of fructans in rat liver might explain the difficulty in demonstrating consistent triglyceride reductions in healthy humans, eating diets much lower in carbohydrates than is the case in rodents.
Recent studies in humans using stable isopes have shown that hepatic de novo lipogenesis is extremely low in subjects eating high-fat, low-carbohydrate diets (Aarsland et al., 1996).

As in the case for fructans, RS induces a coordinate inhibition of all hepatic lipogenic enzymes (Morand et al., 1994; Goda et al., 1994), and it is likely that fructans and RS produce effects on triglycerides through similar mechanisms. There is some evidence to suggest that the putative triglyceride lowering properties of guar gum may also be associated with inhibition of hepatic lipogenesis, since Favier et al. (1997) have shown that the activity of FAS was decreased in the liver of guar gum-fed rats. However, these data are controversial since Overton et al. (1994) reported that guar gum feeding also decreased serum TG concentrations, but total hepatic lipid concentrations and hepatic and adipose tissue lipogenesis rates were unaffected. In a further study in guinea pigs, the secretion rate of VLDL was not modified, whereas that of Apo B was decreased by the treatment resulting in the presence of larger nascent VLDL (Fernandez et al., 1997).

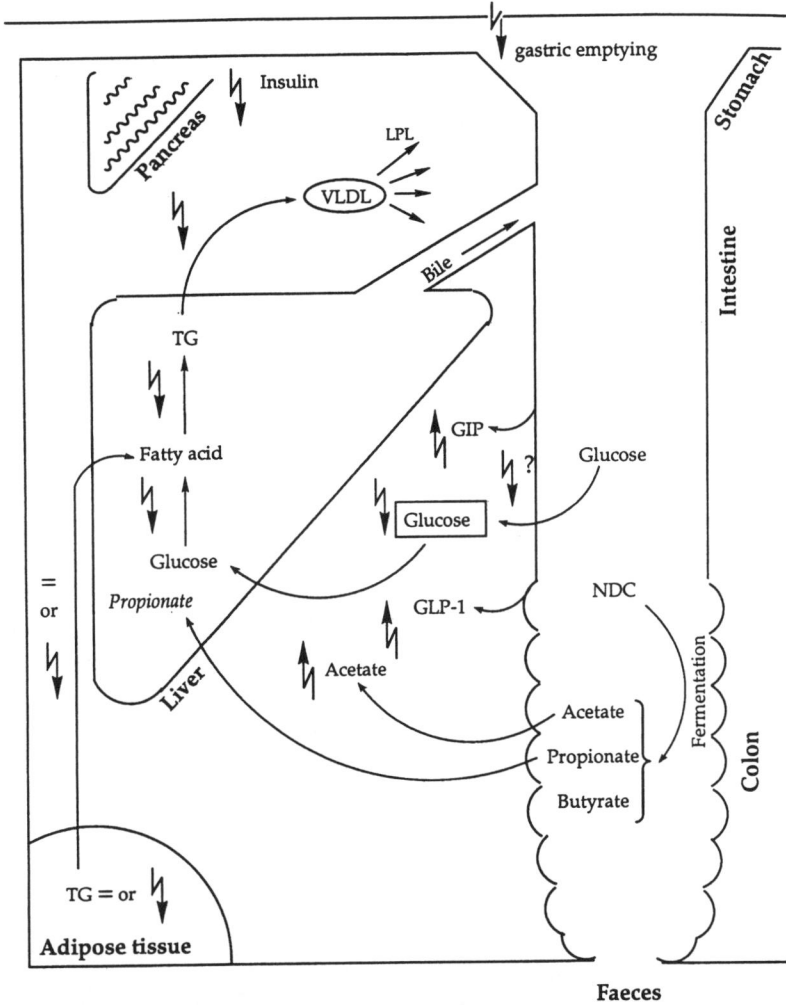

Figure 2. Putative mechanisms explaining the hypotriglyceridemic effect of non-digestible carbohydrates including effects on gastric emptying; glucose absorption; fermentative production of short chain carboxylic acids and balance of acetate: propionate: butyrate; direct and indirect effects on gut hormones (GIP and GLP-1) and consequent actions on liver, adipose tissue and pancreatic insulin output.

It has been suggested that modifications in circulating glucose and/or insulin levels are responsible for the inulin-induced inhibition of lipogenesis, since these appear to be the key physiological regulators of hepatic fatty acid synthesis. Indeed, the induction of lipogenic enzymes by glucose, occurring via an increase in gene transcription, is potentiated by insulin (Girard *et al.*, 1997). The association between reductions in glycemia/insulinemia and inhibition of hepatic triglyceride synthesis has also been demonstrated for RS which, in rats, decreases serum triglyceride concentration, reduces fatty acid synthase activity by 20%, and concomitantly lowers post-prandial insulinemia (Takase *et al.,* 1994; Morand *et al.,* 1994). This response is also clearly described in fructans fed rats: OFS, given at the dose of 10%(w/v) in the diet of rats for 30 days, reduced post prandial glycemia and insulinemia by 17 and 26%, respectively (Kok *et al.,* 1996a; Kok *et al.,* 1998a,b).

Inulin has also been shown to stimulate the production, in the large bowel, of SCCA leading to a more than two-fold increase in the portal concentration of both acetate and propionate in OFS-fed rats (Delzenne and Roberfroid, 1994). However, the involvement of SCCA is, as for the cholesterol-lowering effect, difficult to prove, as these fatty acids have antagonistic effects: propionate has been reported to inhibit fatty acid synthesis (Nishina and Freeland, 1990; Lin *et al.,* 1995), whereas acetate is a lipogenic substrate.

7.2   ACTIVATION OF LIPOPROTEIN LIPASE THROUGH THE PRODUCTION OF INCRETINS

It has been shown that several NDC are able to increase the production of incretins, namely GIP and GLP-1 in rats (Kok *et al.,* 1996b). This could be relevant, since GIP is able to lower plasma TG in response to a fat load in rats (Ebert *et al.,* 1991). This occurs through the inhibition of lipoprotein lipase activity in rat adipose tissue (Knapper *et al.,* 1995). This would suggest that incretins could be involved in a higher catabolism of TG-rich lipoproteins, a metabolic event which may participate in the hypotriglyceridemic effect of NDC given in a high-fat diet in animals (Kok *et al.,* 1998a,b). The incretins have also been reported to have effects on hepatic lipogenesis and TG-VLDL secretion although data are conflicting and these actions remain to be clarified.

8   Conclusions

Feeding NDC leads to modest reductions in blood lipids in human subjects and in animal models. The major effects of soluble fibres such as guar gum, ß glucan and psyllium appear to be on the total and LDL cholesterol fractions, whereas the fructans and RS appear to affect the triglyceride-rich fractions. These different effects may reflect the varying putative modes of action of the two types of NDC, with the former acting largely through inhibition of bile acid absorption and metabolism and the latter through effects secondary to fermentation in the large bowel. In the case of both soluble fibres and fructans and RS, variable blood lipid responses are observed in human volunteer studies and this may in part be due to the influence of other dietary components. In particular, triglyceride-lowering effects of fructans and RS may best be observed in subjects on high carbohydrate diets, whilst cholesterol-lowering effects are best observed in the presence of high dietary cholesterol. Further studies of the mechanisms underlying the actions of NDC's in humans are required.

References

Aarsland, A., Chinkes, D. and Wolfe, R.R. (1996) Contributions of de novo synthesis of fatty acids to total VLDL-triglyceride secretion during prolonged hyperglycemia/hyperinsulinemia in normal man. *Journal of Clinical Investigation* **98**, 2008-17.

Alvarez -Leite, J.I., Andrieux, C., Ferezou, J., *et al.* (1994) Evidence for the absence of participation of the microbial flora in the hypocholesteroleamic effect of guar gum in gnotobiotic rats. *Comparative Biochemistry and Physiology* **109A**, 503-10.

Anderson, J.W., Story, L., Sieling, B., *et al.* (1984) Hypocholesterolemic efects of oat-bran or bean intake for hypercholesterolemic men. *American Journal of Clinical Nutrition* **40**, 1146-55.

Andersson, H. and Bosaeus, I. (1993) Sterol balance studies in man. A critical review. *European Journal of Clinical Nutrition* **47**, 153-59.

Anderson, J.W., Spencer, D.B., Hamilton, C.C., *et al.* (1990) Oat bran cereal lowers serum total and LDL cholesterol in hypercholesterolaemic men. *American Journal of Clinical Nutrition* **52**, 495-99.

Anderson, J.W., Gilinsky, N., Deakins, D.A., *et al.* (1991) Lipid responses of hypocholesterolaemic men to oat-bran and wheat-bran intake. *American Journal of Clinical Nutrition* **54**, 678-83.

Arjamndi, B.H., Ahn, J., Nathani, S., *et al.* (1992) Dietary soluble fiber and cholesterol affect serum cholesterol concentration hepatic portal venous short-chain fatty acid concentrations and fecal sterol excretion in rats. *Journal of Nutrition* **122** (2), 246-53.

Beer, M.U., Arrigoni, E. and Amado, R. (1995) Effects of oat gum on blood cholesterol levels in healthy young men. *European Journal of Clinical Nutrition* **49**, 517-22.

Behall, K.M., Schofield, D.J., Yuhaniak, I., *et al.* (1989) Diets containing high amylose vs amylopectin starch: effects on metabolic variables in human subjects. *American Journal of Clinical Nutrition* **49**, 337-44.

Behall, K.M. and Howe, J.C. (1995) Effect of long term consumption of amylose vs amylopectin starch on metabolic variables in human subjects. *American Journal of Clinical Nutrition* **61**, 334-40.

Braaten, J.T., Wood, P.J., Scott, F.W., *et al.* (1994) Oat bran ß glucan reduces blood cholesterol concentration in hypercholesterolaemic subjects. *European Journal of Clinical Nutrition* **48**, 465-74.

Campbell, J.M., Fahey, G.C. and Wolf, B.W. (1997) Selected indigestible oligosaccharides affect large bowel mass, cecal and fecal short-chain fatty acids, pH and microflora in rats. *Journal of Nutrition* **127**, 130-36.

Canzi, E., Brighenti, F., Casiraghi, M.C., *et al.* (1995) Prolonged consumption of inulin in ready-to-eat breakfast cereals: effects on intestinal ecosystem, bowel habits and lipid metabolism. *Cost 92, Workshop on, dietary fiber and fermentation in the colon* Helsinki 15-17.

Cara, L., Dubois, C., Borel, P., *et al.* (1992) Effects of oat bran, rice bran, wheat fiber and wheat germ on postprandial lipemia in healthy adults. *American Journal of Clinical Nutrition* **55**, 81-88.

Carr, T.P., Gallaher, D.D., Yang, C.H., *et al.* (1996) Increased intestinal contents viscosity reduces cholesterol absorption efficiency in hamsters fed hydroxypropyl methylcellulose. *Journal of Nutrition* **126** (5), 1463-69.

Cassidy, M.M. and Calvert, R.J. (1993) Effects of dietary fibre on intestinal absorption of lipids. In *Spiller GA* ed. *CRC Handbook of Dietary Fiber in Human Nutrition*. 2nd ed. Boca Ranton, Fl: *CRC Press Inc* 153-62.

Chezem, J., Furumoto, E. and Story, J. (1997) Effects of resistant potato starch on cholesterol metabolism and bile acid metabolism in the rat. *Nutrition Research* **17** (11-12), 1671-82.

Cummings, J.H. (1997) The large intestine in nutrition and disease. *Danone Chair Monograph, ed.* Institut Danone, Brussels, Belgium.

Cummings, J.H. and Roberfroid, M.B. (1997) A new look at dietary carbohydrate: chemistry, physiology and health. *European Journal of Clinical Nutrition* **51**, 417-23.

Davidson, M.H., Synecki, C., Maki, K.C., *et al.* (1998) Effects of dietary inulin in serum lipids in men and women with hypercholesterolemia. *Nutrition Research* **3**, 503-17.

de Deckere E.A., Kloots W.J. and van-Amelsvoort J.M. (1995) Both raw and retrograded starch decrease serum TG concentration and fat accretion in the rat. *British Journal of Nutrition* **73** (2), 287-98.

Delzenne N. and Kok N. (1998) Effect of non-digestible fermentable carbohydrates on hepatic fatty acid metabolism. *Biochemical Society Transactions* (in press)

Delzenne, N., Kok, N., Fiordaliso, M., *et al.* (1993) Dietary fructooligosaccharides modify lipid metabolism in rats. *American Journal of Clinical Nutrition* **57**, 820S.

Delzenne, N. and Roberfroid, M.B. (1994) Physiological effects of non-digestible oligosaccharides. *Lebensm Wiss u Technology* **27**, 1-6.

Dubois, C., Armand, M., Senft, M., *et al.* (1995) Chronic oat bran intake alters postprandial lipemia and lipoproteins in healthy adults. *American Journal of Clinical Nutrition* **61**, 325-33.

Ebert, R., Nauck, M. and Creutzfeld, W. (1991) Effect of exogenous and endogenous GIP on plasma triglyceride responses in rats. *Hormone Metabolism Research* **23**, 517-21.

Everson, G.T., Daggy, B.P., McKinley, C. *et al.* (1992) Effects of psyllium hydrophylic muciloid on LDL-cholesterol and bile acid synthesis in hypercholesterolemic men. *Journal of Lipid Research* **33**, 1183-92.

Englyst, H.N., Kingman, S.M. and Cummings, J.H. (1992) Classification and measurement of nutritionally important starch fractions. *European Journal of Clinical Nutrition* **46**, S33-S50.

Favier, M.L., Bost, P.E., Guittard, C., *et al.* (1997) Reciprocal influence of fermentations and bile acid excretion on cholesterol-lowering effect of fermentable carbohydrate. *Journal of Nutritional Biochemistry* **8** (3), 127-32.

Food and Drug Administration (1996) Food labelling: Health claims; oats and coronary heart disease. *Federal Register* **61**, 296-313.

Fernandez, M.L. (1995) Distinct mechanisms of plasma LDL lowering by dietary fiber in the guinea pig: specific effects of pectin, guar gum and psyllium. *Journal of Lipid Research* **36** (11), 2394-04.

Fernandez, M.L. Vergara Jimenez, M., Conde, K., *et al.* (1997) Regulation of apolipoprotein B-containing lipoproteins by dietary soluble fiber in guinea pigs. *American Journal of Clinical Nutrition* **65** (3), 814-22.

Fiordaliso, M.F., Kok, N., Desager, J.P., *et al.* (1995) Dietary oligofructose lowers triglycerides, phospholipids and cholesterol in serum and very low density lipoproteins of rats. *Lipids* **30**, 163-67.

Gatenby, S.J. (1990) Guar gum and hyperlipidaemia - a review of the lietrature. In *Dietary Fibre Perspectives, Reviews and Bibliography.* Vol 2 eds. Leeds, A.R. and Burley, V.J., 102-15, London: John Libbey.

Gatti, E., Catenazzo, G., Camisasca, E., *et al.* (1984) Effects of guar-enriched pasta in the treatment of diabetes and hyperlipidemia. *Annals of Nutrition and Metabolism* **28**, 1-10.

Girard, J., Ferré, P. and Foufelle F. (1997) Mechanisms by which carbohydrates regulate expression of genes for glycolytic and lipogenic enzymes. *Annual Review Nutrition* **17**, 325-52.

Goda, T., Urakawa, T., Watanabe, M. *et al.* (1994) Effect of high amylose starch on carbohydrate digestive capability and lipogenesis in epididymal adipose tissue and liver of rats. *Journal of Nutritional Biochemistry* **5**, 256-60.

Gold, K.V. and Davidson, D.M. (1988) Oat bran as a cholesterol-reducing dietary adjunct in a young, healthy population. *Western Journal of Medicine* **148**, 299-302.

Gondal J.A., MacArthy, P., Myers, A.K., *et al.* (1996) Effect of dietary sucrose and fibers on blood pressure in hypertensive rats. *Clinical Nephrology* **45** (3), 163-68.

Groot, P.H.E., Van Stiphout, W.A.H.J., Krauss, X.H., *et al.* (1991) Postprandial lipoprotein metabolism in normolipidaemic men with and without coronary artery disease. *Arteriosclerosis and Thrombosis* **11**, 653-62.

Hegsted, M., Windhauser, M.M., Morris, K. *et al.* (1993) Stabilised rice bran and oat bran lower cholesterol in humans. *Nutrition Research* **13**, 387-98.

Heijenen, M.-L., van Amelsvoort, J.M.M., Deurenberg, P. *et al.* (1996) Neither raw starch nor retrograded resistant starch lowers fasting serum cholesterol concentrations in healthy normolipidemic subjects. *American Journal of Clinical Nutrition* **64**, 312-18.

Hidaka, H., Eida, T., Takizawa, T., *et al.* (1986) Effects of fructooligosaccharides on intestinal flora and human health. *Bifidobacteria Microflora* **5**, 37-50.

Hokanson, J.E. and Austin, M.A. (1996) Plasma triglyceride level is a risk factor for cardiovascular disease independent of high-density lipoprotein cholesterol level: a meta-analysis of population -based prospective studies. *Journal of Cardiovascular Risk* **3**, 213-19.

Illman, R.J., Topping, D.L., McIntosch, G.H., *et al.* (1988) Hypercholestrolemic effects of dietary propionate: studies in whole animals and perfused rat liver. *Annals of Nutrition and Metabolism* **32**, 95-107.

Jenkins, D.J.A., Wolever, T.M.S. and Jenkins, A. (1991) Specific types of colonic fermentation may raise low-density-lipoprotein-cholesterol concentrations. *American Journal of Clinical Nutrition* **54**, 141-47.

Kahn, R.F., Davidson, K.W., Garner, J. *et al.* (1990) Oat bran supplementation for elevated serum cholesterol. *Family Practice Research Journal* **10**, 37-46.

Karpe, F., Steiner, G., Uffelman, K., *et al.* (1994) Postprandial lipoproteins and progression of coronary atherosclerosis. *Atherosclerosis* **106**, 83-97.

Kashtan, H., Stern, H.S., Jenkins, D.J.A., *et al.* (1992) Wheat-bran and oat-bran supplements: effects on blood lipids and lipoproteins. *American Journal of Clinical Nutrition* **55**, 976-80.

Kestin, M., Moss, R., Clifton, P.M. *et al.* (1990) Comparative effects of three cereal brans on plasma lipids, blood pressure, and glucose metabolism in mildly hypercholesterolaemic men. *American Journal of Clinical Nutrition* **52**, 661-66.

Kirby, R.W., Anderson, J.W., Sieling, B., *et al.* (1981) Oat-bran intake selectively lower serum low-density lipoprotein cholesterol concentrations of hypercholesterolemic men. *American Journal of Clinical Nutrition* **34**, 824-29.

Knapper, J.M., Puddicombe, S.M. and Morgan, L.M. (1995) Investigations into the actions of GIP and GLP-1 on lipoprotein lipase activity in explants of rat adipose tissue. *Journal of Nutrition* **125**, 183-88.

Kok, N., Delzenne, N., Morgan, L., *et al.* (1998a) Insulin, glucagon-like-peptide 1, glucose-dependent insulinotropic polypeptide and insulin-like growth factor 1 as putative mediators of the hypolipidemic effect of oligofructose in rats? *Journal of Nutrition* (in press).

Kok, N., Taper, H. and Delzenne, N. (1998b) Oligofructose modulates lipid metabolism alterations induced by a fat-rich diet in rats. *Journal Applied Toxicology* **18** (1), 47-53.

Kok, N., Roberfroid, M., Robert, A., *et al.* (1996a) Involvement of lipogenesis in the lower VLDL secretion induced by oligofructose in rats. *British Journal of Nutrition* **76**, 881-90.

Kok, N., Roberfroid, M. and Delzenne, N. (1996b) Dietary oligofructose modifies the impact of fructose on hepatic triglyceride metabolism. *Metabolism* **45**, 1547-50.

Kritchevsy, D. and Story, J.A. (1993) Influence of dietary fiber on cholesterol metabolism in experimental animals. In *Handbook of Dietary Fiber in Human Nutrition*. ed. GA Spiller pp. 163-78. Boca Raton: CRC Press.

Lairon, D. (1996) Dietray fibres: Effects on lipid metabolism and mechanisms of action. *European Journal of Clinical Nutrition* **50**, 125-33.

Leadbetter, J., Ball, M.J. and Mann, J.I. (1991) Effects of increasing quantities of oat bran in hypercholesterolaemic people. *American Journal of Clinical Nutrition* **54**, 841-45.

Lerer Metzger, M., Rizkalla, S.W., Luo J., *et al.* (1996) Effects of long term low-glycemic index starchy food on plasma glucose and lipid concentrations and adipose tissue cellularity in normal and diabetic rats. *British Journal of Nutrition* **75** (5), 723-32.

Levrat, M.A., Favier, M.L., Moundras C., *et al.* (1994) Role of dietary propionic acid and bile acid excretion in the hypocholesterolemic effects of oligosaccharides in rats. *Journal of Nutrition* **124** (4), 531-38.

Lin, Y., Vonk, R.J., Sloof, M.J., *et al.* (1995) Differences in propionate-induced inhibition of cholesterol and triglyceride synthesis between human and rat hepatocytes in primary culture. *British Journal of Nutrition* **74**, 197-207.

Luo, J., Rizkalla, S.W., Alamovitch, C., *et al.* (1996) Chronic consumption of short-chain fructooligosaccharides by healthy subjects decreased basal hepatic glucose production but had no effect on insulin-stimulated glucose metabolism. *American Journal of Clinical Nutrition* **63**, 639-45.

Marlett, J.A., Hosig, K.B., Vollendorf, N.W., *et al.* (1994) Mechanism of serum cholesterol reduction by oat bran. *Hepatology* **20**, 1450-57.

Morand, C., Levrat, M.A., Besson, C., *et al.* (1994) Effect of a diet rich in resistant starch on hepatic lipid metabolism in the rat. *Journal of Nutritional Biochemistry* **5**, 138-44.

Moundras, C., Behr, S.R., Remesy, C., *et al.* (1997) Fecal losses of sterols and bile acids induced by feeding rats guar gum are due to greater pool size and liver bile acid secretion. *Journal of Nutrition* **127** (6), 1068-76.

Nishina, P. and Freeland, R. (1990) Effects of propionate on lipid biosynthesis in isolated hepatocytes. *Hepatology* **16**, 1350-56.

Noakes, M., Clifton, P.M., Nestel, P.J., *et al.* (1996) Effect of high-amylose starch and oat bran on metabolic variables and bowel function in subjects with hypertriglyceridemia. *American Journal of Clinical Nutrition* **64**, 944-51.

Olson, B., Anderson, S.M. and Becker, M.P. (1998) Psyllium-enriched cereals lower blood total cholesterol and LDL cholesterol but not HDL cholesterol in hypocholesterolemic adults: results of a meta analysis. *Journal of Nutrition* **127**, 1973-80.

Overton, P.D., Furlonger, N., Beety, J.M., *et al.* (1994) The effects of dietary sugar-beet fibre and guar–gum on lipid metabolism in Wistar rats. *British Journal of Nutrition* **72** (3), 385-95.

Patsch, W. (1992) Relation of triglyceride metabolism and coronary artery disease. *Arteriosclerosis and Thrombosis* **12**, 1336-45.

Pedersen, A., Sandström, B. and VanAmelsvoort, J.M.M. (1997) The effect of ingestion of inulin on blood lipids and gastrointestinal symptoms in healthy females. *British Journal of Nutrition* **78**, 215-22.

Poulter, N., Chang, C.L., Cuff. A., *et al.* (1993) Lipid profiles after the daily consumption of an oat-based cereal: a controlled crossover trial. *American Journal of Clinical Nutrition* **58**, 66-69.

Ranhotra, G.S., Gerolth, J.A. and Leinen, S.D. (1997) Hypolipidemic effect of resistant starch in hamsters is not dose dependent. *Nutrition Research*, **17** (2), 317-23.

Redard, C.L, Davis, P.A. and Schneeman, B.O. (1990) Dietary fiber and gender: effect on postprandial lipemia. *American Journal of Clinical Nutrition* **52**, 837-45.

Reiser, S., Powell, A.S., Scholfield, D.J., *et al.* (1989) Blood lipids, lipoproteins apoproteins and uric acid in man fed diets containing fructose or high-amylose cornstarch. *American Journal of Clinical Nutrition*, **49**, 832-39.

Ripsin, C.M., Keenan, J.M., Jacobs, D.R., *et al.* (1992) Oat products and lipid lowering- A metaanalysis. *Journal American Medical Association* **267**, 3317-25.

Roberfroid, M.B. and Delzenne, N.M. (1998) Dietary fructans, *Annual Review of Nutrition* **18**, (in press)

Spiller, G.A., Farquhar, J.W., Gates, J.E. *et al.* (1991) Effect of guar gum and an oat fiber source on plasma lipoproteins and cholesterol in hypercholesterolaemic adults. *Arteriosclerosis and Thrombosis* **11**, 1204-08.

Stark, A.H. and Madar, Z. (1993) In vitro production of short-chain fatty acids by bacterial fermentation of dietary fiber compared with effects of those fibers on hepatic sterol synthesis in rats. *Journal of Nutrition* **123** (12), 2166-73.

Swain, J.F., Rouse, I.L., Curley, C.B, *et al.* (1990) Comparison of the effects of oat bran and low fiber wheat on serum lipoprotein levels and blood pressure. *New England Journal of Medicine* **322**, 147-52.

Takase, S., Goda, T. and Watanabe, M. (1994) Monostearylglycerol-starch complex: its digestibility and effects on glycemic and lipogenic responses. *Journal of Nutrition and Vitaminology* **40**, 23-36.

Törrönen, R., Kansanen, L., Uusitupa, M., *et al.* (1992) Effects of an oat bran concentrate on serum lipids in free-living men with mild to moderate hypercholesterolaemia. *European Journal of Clinical Nutrition* **46**, 621-27.

Truswell, A.S. (1995) Dietary fibre and plasma lipids. *European Journal of Clinical Nutrition* **49**, S105-S109.

Vahouny, G.V., Satchithanadam, S., Chen, I., *et al.* (1988) Dietary fiber and intestinal adaptation: effects on lipid absorption and lymphatic transport in the rat. *American Journal of Nutrition* **47**, 201-06.

Van Horn, L., Moag-Stahlberg, A., Liu, K., *et al.* (1991) Effects on serum lipids of adding instant oats to usual american diets. *American Journal of Public Health* **81**, 183-88.

Vanhoof, K. and DeSchrijver, R. (1997) Consumption of enzyme resistant starch and cholesterol metabolism in normo and hypercholesterolemic rats. *Nutrition Research* **17** (8), 1331-40.

Whyte, J., McArthur, R., Topping, D. *et al.* (1992) Oat bran lowers plasma cholesterol in mildy hypercholesterolemic men. *Journal of American Dietetic Association* **92**, 446-49.

Williams, C.M. (1997) Postprandial lipid metabolism: effects of dietary fatty acids. *Proceedings Nutrition Society* **56**, 579-92.

Williams, C.M. (1998) Effects of inulin on blood lipids in humans. *Journal of Nutrition* (In press).

Wolever, T.M.S., Hegele, R.A., Connelly, P.W., *et al.* (1997) Long term effect of soluble fiber foods on postprandial fat metabolism in dyslipidemic subjects with apo E3 and apo E4 genotypes. *American Journal of Clinical Nutrition* **66**, 584-90.

Yamashita, K., Kawai, K. and Itakura, M. (1984) Effects of fructo-oligosaccharides on blood glucose and serum lipids in diabetic subjects. *Nutrition Research* **4**, 961-66.

Younes, H., Levrat, M.A., Demigne, C., *et al.* (1995) Resistant starch is more effective than cholestyramine as a lipid-lowering agent in the rats. *Lipids* **30** (9), 847-53.

Zhang, J.X., Hallmans, G., Andersson, H., *et al.* (1992) Dietary fibres: effects on lipid metabolism and mechanisms of action. *European Journal of Clinical Nutrition* **50**, 125-33.

Zilversmit, D.B. (1979) Atherosclerosis a postprandial phenomenon. *Circulation* **60**, 473-85.

CHAPTER 14

Bioavailability of Minerals

SUSAN J. FAIRWEATHER-TAIT AND IAN T. JOHNSON

Institute of Food Research, Norwich Research Park, Colney, Norwich, UK

1   Introduction

Fermentation is used throughout the animal kingdom as a strategy for the breakdown and assimilation of dietary macromolecules which resist hydrolysis by endogenous digestive enzymes. In ruminants this process occurs early in digestion and the main fermentation compartment lies between the oesophagus and the true stomach. In monogastric species the intestinal microflora reside primarily in the large bowel, the main functions of which are reabsorption of digestive secretions and the recovery of energy from undigested carbohydrates. Epithelial cells of the colon are specialised for this role and most of their transport pathways are adapted to facilitate the absorption of fermentation products, principally short chain fatty acids (SCFA), electrolytes and water. Nevertheless, a variety of other food-borne substances reach the large intestine. These include inorganic micronutrients which have escaped absorption in the small bowel.

Minerals associated with cell wall polysaccharides or phytate may be released into solution by bacterial degradation in the large bowel, and hence become available for absorption. There is considerable variation in the anatomy, function and nutritional role of the mammalian hindgut but mineral absorption seems to be an important function in many species. In humans, the major site of mineral absorption is the proximal small bowel but there is increasing evidence that the colon can absorb nutritionally significant quantities of some inorganic micronutrients, and this process may be susceptible to manipulation by fermentable substrates. In this chapter the physiological and nutritional significance of colonic transport processes will be outlined, and evidence that the colonic microflora play a role in their regulation will be reviewed.

G.R. Gibson and M.B. Roberfroid (eds.), Colonic Microbiota, Nutrition and Health, 233-244.
© 1999 *Kluwer Academic Publishers. Printed in the Netherlands.*

2   Water and electrolytes

One of the main functions of the colon is the absorption of water. When the intact human colon is perfused with physiological saline there is net absorption of water, sodium and chloride, and secretion of bicarbonate and potassium (Devroede and Phillips, 1969). Water transport is essentially a secondary phenomenon coupled to active absorption of electrolytes, and particularly sodium, which is absorbed against a substantial concentration gradient. If the healthy human colon is perfused with saline media containing varying concentrations of sodium, net absorption is obtained at luminal concentrations exceeding 25 mEq per liter (Billich and Levitan, 1969). In the small intestine, the minimum sodium concentrations which sustain net sodium absorption are 133 mEq per litre in the jejunum, and 35 mEq per litre in the ileum respectively (Fordtran et al., 1968). Evidently, the ability of the alimentary tract to absorb sodium from dilute solutions increases distally. This is consistent with the need to progressively reduce the water content of faecal material such that faeces can be stored and voided at will. Failure of this basic physiological function results in diarrhoea. Under normal circumstances the human colon probably absorbs about 2 litres of water per day, but the maximum absorptive capacity is thought to be between 5 and 6 litres per day (Debongnbie and Phillips, 1978).

Our current understanding of colonic electrolyte absorption is based upon knowledge gained through a variety of experimental techniques including perfusion studies in healthy human volunteers and in vitro studies using sheets of colonic mucosa obtained primarily from rats and rabbits. There are marked differences between species but some generalisations are possible. The distal colon of the rabbit transfers sodium from the lumin to the serosal compartments by an energy dependent electrogenic process which is independent of chloride and can be blocked by the drug amiloride. Chloride is also absorbed in the distal rabbit colon, but the process is independent of sodium, electroneutral, and coupled directly to secretion of bicarbonate. In the proximal rabbit colon, and in both proximal and distal segments of rat colon, sodium absorption is apparently electroneutral and dependent upon chloride. For obvious reasons there is far less information available on the characteristics of sodium chloride transport in different segments of human colon but the pattern seems to be similar to that of the rabbit, with electrogenic sodium uptake in the distal segments and electroneutral absorption in the proximal bowel.

3   Electrolytes and short chain fatty acids

The major products of carbohydrate fermentation by anaerobic bacteria of the colon are the SCFA acetate, propionate and butyrate. Much smaller quantities of lactate and succinate are present, together with branched chain fatty acids such as isobutyrate and isovalerate which are derived from amino acid fermentation. The importance of SCFA in the present context lies in the fact that they play an important role in the maintenance of sodium and water absorption in the large bowel and may also influence the transport or other cations including calcium.

Ruppin *et al.* (1980) intubated the intact caecum of healthy human volunteers, perfused with SCFA at 0, 30, 60 or 90 mmol/L, and measured water, sodium and potassium absorption and bicarbonate secretion. The absorption of water and electrolytes was enhanced at all three concentrations compared to the control perfusion, and bicarbonate secretion was increased. More recently, Bowling *et al.* (1993) conducted similar studies using a multilumen perfusion tube which allowed them to compare absorption of water and electrolytes in both the ascending and distal colon. The perfusate in these studies contained acetate (50 mmol/L) propionate (20 mmol/L) and butyrate (20 mmol/L). The short chain fatty acid infusions stimulated uptake of sodium, potassium and chloride. Water uptake was similarly enhanced, to such an extent that net water secretion caused by gastric infusion of enteral feed was reversed. This suggested that adequate SCFA production was an important factor in the maintenance of optimal colonic function and prevention of diarrhoea.

Short chain fatty acids are absorbed throughout the human small and large bowel (Ruppin *et al.*, 1980; Schmitt *et al.*, 1976; Binder and Mehta, 1989) but most of the mechanistic studies that have been carried out have used rabbit or rodent colon as the experimental model. The mechanism of SCFA uptake remains a matter of some controversy, and it may indeed vary from one region of the gut to another. One school of thought maintains that SCFA undergo protonation close to the mucosal surface and are then passively absorbed into the colonic epithelial cells by non-ionic diffusion. According to the model of Binder & Mehta (1989), which is based on rat distal colon, the SCFA molecules then dissociate in the more alkaline environment of the mucosal cell cytoplasm, liberating protons which then are then secreted back into the lumen via a Na^+/H^+ exchanger. Some SCFA are also partially recycled in exchange for chloride, and the net result is an enhanced absorption of sodium chloride.

It has consistently been observed however that absorption of SCFA in the colon is associated with an increase in the luminal concentration of bicarbonate. To account for this it has been proposed that SCFA are absorbed, at least partly, by a specific SCFA/bicarbonate exchange transporter. This hypothesis appears to have been recently strengthened through studies carried out with vesicles derived from the apical membranes of human colonocytes (Harig *et al.*, 1996). In this system, there is good evidence for the existence of a structurally specific SCFA transporter with no coupling to the movement of sodium or chloride. The mechanism by which SCFA enhance uptake of sodium chloride, and hence water, therefore remains unresolved.

4   Calcium and magnesium

Calcium is an essential micronutrient because of its structural role as a constituent of bone and its participation as a messenger in a variety of neurological and other cellular signalling events. As with other inorganic nutrients, the small intestine is the major site of calcium absorption in most mammalian species, including humans, in whom most dietary calcium is absorbed by a saturable, transcellular pathway that is subject to physiological and nutritional regulation via vitamin D. However, some calcium is also absorbed by a nonsaturable, concentration-dependent process along the whole intestine, independent of nutritional and physiological regulation. There is ample evidence that the hindgut, and particularly the caecum, plays an important role in calcium absorption in the rat (Karbac and Feldmeier, 1993; Ohta *et al.*, 1997; Younes *et al.*, 1996.) Studies with ileostomists suggest that the human colon also has an innate capacity to absorb calcium which is of metabolic significance, at least after major resection of the small intestine (Hylander *et al.*, 1980), and even in healthy human subjects, colonic absorption of calcium may account for a substantial part of the total calcium uptake. For example, when a solution of ^{47}Ca was introduced into the colon of adult men and women via a colonoscope, mean absorption was 14.1% at the caecum and 5.3% at the hepatic flexure (Sandstrom *et al.*, 1986). More recently Trinidad *et al.* (1996) have shown that significant uptake of calcium occurs in the distal colon and rectum of humans.

The metabolic activity of the colonic flora may influence the bioavailability of calcium in the colon by modifying its speciation, and hence solubility, or by exerting a direct effect on the mucosal transport pathway. Recent evidence suggests that both types of mechanism may be important in humans. To deal first with the issue of intraluminal speciation, it is well established that phytate (myoinositol hexaphosphate), the storage form of phosphorus in plants, forms insoluble complexes with divalent metals, the order of stability being: $Zn(2+) > Cu(2+) > Co(2+) > Mn(2+) > Ca(2+)$ (Maddaiah *et al.*, 1964). This renders the mineral unavailable for absorption in the intestine (McCance and Widdowson, 1942; Cummings *et al.*, 1979). For example, when introduced into the human colon via a colonoscope, phytate labelled with ^{47}Ca reduced calcium absorption from 14.1% to 4.0% (Sandstrom *et al.*, 1990). It is probable therefore that hydrolysis of phytic acid complexes in the small or large intestine would enable the released calcium to become available for absorption (Wise, 1983). Although phytase activity has been reported in the small intestinal mucosa of rats and humans (Bitar and Reinhold, 1972), its role in hydrolysing phytate is considered negligible compared to that of the gastrointestinal microflora (Moore and Veum, 1983). The fact that there appears to be no adaptation to prolonged consumption of high phytate diets in terms of altered mineral absorption (Brune *et al.*, 1989), or phytate excretion (Schwartz *et al.*, 1986), supports this contention.

Microbial hydrolysis of phytate is usually considered to be insignificant in the small intestine; thus the inhibitory effect of phytate is more pronounced for minerals with negligible absorption in the large intestine, such as iron and zinc (see below), than with calcium where there appears to be a mechanism for colonic uptake. Nevertheless, in a study with ileostomists Sandberg *et al.* (1982) found that phytate was partially digested in the stomach and small intestine, possibly indicating an adaptive response. Bacterial digestion of other plant constituents may also be important in terms of increased efficiency of absorption of calcium from high fibre diets. For example, the fermentation of uronic acids in pectin (found in wholegrain cereals, fruits and vegetables) and alginates (found in seaweed), which also bind calcium, may be a key step in releasing it in a form suitable for absorption (James *et al.*, 1978).

Using experimental animals it is possible to study directly the effects of fermentable carbohydrates on the concentration and availability for absorption of minerals in the lumen of the large bowel. Demigne *et al.* (1989), demonstrated that fermentable oligosaccharides and starch simultaneously increased the concentration of both short and the minerals calcium, magnesium and phosphorus in the caecal contents of rats. The larger intraluminal calcium pool was associated with increased uptake compared to controls. Indeed, the absorption of calcium was almost linearly related to the concentration of inulin in the diet. Other groups have recently shown that dietary supplementation with fructo-oligosaccharides similarly enhances the uptake of calcium (Delzenne *et al.,* 1995; Ohta *et al.,* 1995, 1998a), and alleviates the reduction in bone calcium content and bone mineral density which follows gastrectomy in rats (Ohta *et al.,* 1998b).

Human beings lack the large self-contained caecum which is a characteristic of the rat, but the limited information which is available on the effects of fermentable polysaccharides on mineral absorption in human subjects suggests that fermentable carbohydrates may also stimulate calcium absorption in man. Coudray *et al.,* (1997) conducted a balance study in which they fed control diets containing 18g of dietary fibre, or the same diets supplemented with sugar beet fibre (40g/d) or inulin (40g/d), for 28 d to 9 healthy human subjects in a crossover design with 3 repetitions. Ingestion of sugar beet fibre, which is relatively rich in calcium, increased both the intake and retention of calcium without influencing apparent absorption. However inulin, which is virtually mineral-free, also led to a significant increase in both absorption and retention of calcium. No significant effects on other aspects of mineral metabolism were observed, and the authors concluded that at these high levels of intake both types of fermentable carbohydrate could improve calcium balance in humans, with no adverse effects on other inorganic nutrients.

Inulin is well fermented in the human proximal colon. Forty grams of inulin per day, which must be close to the upper limits of human tolerance, will cause a substantial increase in the production of SCFA and a marked acidification of the luminal contents. It is proposed, though not yet proven, that this acidification of the luminal contents leads directly to increased calcium absorption. One possible explanation is that the reduction in pH simply leads to greater solubilisation of calcium so that the biologically available concentration in the luminal phase is increased (Remesy *et al.,* 1993). A second possibility is that SCFA contribute directly to the enhancement of calcium absorption via a cation exchange mechanism analogous to that discussed earlier in relation to sodium transport (Lutz and Scharrer, 1991). Trinidad *et al.,* (1993, 1996) conducted rectal infusion studies with solutions of SCFA and calcium in human volunteers and showed that both disappearance of calcium from the colonic lumen and its appearance in the circulation were enhanced by the presence of acetate and propionate (total concentration 75 mM) in the infusate. They concluded that their findings were consistent with a hypothetical ion-exchange mechanism in which SCFA in a protonated form were absorbed and underwent dissociation within the intracellular environment, yielding a proton which was then re-secreted into the lumen in exchange for calcium. Propionate was observed to cause a greater enhancement of calcium absorption than acetate, and they showed that this might be due to the fact that the rate of uptake of propionate was significantly faster than that of acetate.

Quite apart from its role as an essential micronutrient, calcium may play an important role as a regulator of conditions within the colonic lumen. Many studies in the field of colon carcinogenesis have shown that supplemental dietary calcium decreases the cytotoxicity of intestinal contents by precipitating cytoxic surfactants such as bile acids, thereby strengthening the barrier function of the intestinal mucosa and possibly surrounding the endogenous microflora with a less aggressive environment (Lapre et al., 1993). Intraluminal calcium also appears to down-regulate crypt cell proliferation by a direct effect on epithelial cell function (Appleton et al., 1991). This supposedly beneficial effect on the mucosa will be dependent upon the biological availability of intraluminal calcium to the cell, so that solubilisation of calcium in the acidic environment caused by fermentation of complex carbohydrate is likely to be important here too.

It has recently been shown that rats fed a low-calcium milk diet had a significantly impaired colonization resistance to Salmonella enteritidis compared with rats on high-calcium diets containing milk or yoghurt (Bovee-Oudenhoven et al., 1996). Although in vitro studies show that divalent cations, such as calcium and magnesium, play a regulatory role in the expression of virulence genes of salmonellae (Vescove et al., 1996), growth of the pathogen in vivo is not dependent upon the environmental calcium concentration. Further studies have demonstrated that dietary calcium increased the colonization resistance to Salmonella enteritidis and reduced translocation of this invasive pathogen to the systemic ciculation (Bovee-Oudenhoven et al., 1997). Several mechanism(s) of action have been proposed: (a) calcium increases the excretion of organic phosphate, which precipitates bile acids and fatty acids (b) calcium phosphate increases the faecal excretion of bacteria (stimulation of bacterial growth is recognised to be an important host defence mechanism) and (c) calcium phosphate co-precipitates soluble iron in the intestinal lumen, thereby minimising epithelial cell damage and maximising resistance against pathogens. The proposed trophic effect of dietary calcium phosphate on the endogenous microflora and its relevance for the resistance to salmonellae is under further investigation. However, it is widely accepted that when the endogenous flora is optimal, colonisation of pathogens is repressed through the production of antimicrobial substances and by competition for nutrients and adhesion sites on the intestinal epithelium. The current view is that an adequate supply of dietary calcium is effective in reducing the severity of gastrointestinal infections, which is particularly important for immuno-compromised individuals.

Magnesium is absorbed primarily in the small intestine, both by a facilitated process and by passive diffusion, but factors controlling absorption are poorly understood. The negative effect of high fibre diets on magnesium absorption has been investigated by a number of research groups (Kelsay et al., 1970). When a group of adult men were given a diet containing wheat bran (high in phytate) initially magnesium absorption was low but showed a consistent positive trend throughout the 48 day study, indicating intestinal adaptation. The same result was found for manganese, but not for calcium, copper or zinc (Drews et al., 1979). The active principals in high fibre foods that are believed to reduce magnesium absorption are hemicellulose (Brune et al., 1989), cellulose (Slavin and Marlett, 1980) and oxalate (Kelsay and Prather, 1980). Fermentable carbohydrates such as resistant starch and inulin, which are essentially free of these cell wall constituents, have been reported to enhance magnesium absorption in the rat (Younes et al., 1996; Demigne et al., 1989) but not, to judge from the recent balance study of Coudray et al., (1997), in humans.

5 Iron

Iron is present in the human diet in two forms, haem iron derived from meat, and non-haem iron present in a variety of plant foods. Both forms are absorbed, via different receptor mediated pathways, principally in the duodenum and proximal jejunum. Only a small fraction of dietary iron is absorbed, partly because the intraluminal pool contains ligands which bind non-haem iron and inhibit its uptake by mucosal receptors, and partly because absorption is down-regulated in individuals whose body iron stores are depleted (Hallberg, 1981; Baynes and Bothwell, 1990). The main sites of iron absorption probably differ significantly between species. There is evidence that under experimental conditions significant quantities of iron can be absorbed from the colon of the guinea pig (Conrad *et al.*, 1966) and the dog (Chernelch *et al.*, 1970) but despite early reports of colonic iron absorption in humans (Ohkawara *et al.*, 1963), this route is not thought to be of any nutritional significance in healthy humans.

Patients with ulcerative colitis or Crohn's disease who had been given an ileostomy were reported to be mildly iron deficient, probably due to pre-existing iron deficiency resulting from preceding illness and operation (Kennedy *et al.*, 1982). However, the abnormality was less pronounced in patients in whom the ileostomy had been established for more than 3 years, which may be related to the time required to replenish body iron or possibly evidence of intestinal adaptation with respect to iron absorption, as occurs with sodium and water, whereby the terminal ileum takes over some of the functions of the colon. This hypothesis supports the earlier suggestions that the colon may, under certain conditions, transport iron. If this were the case, then colonic fermentation of plant cell wall material that contains iron or the breakdown of iron-binding compounds by bacteria would provide an additional route of entry of iron into the body, as would hydrolysis of phytate-iron and polyphenol-iron complexes. At present, however, there is no evidence to suggest that colonic absorption of iron has any biological significance.

Although it is unlikely that the colonic microflora play any role in systemic iron metabolism they may nevertheless exert important effects on the intraluminal iron pool. Unabsorbed dietary iron enters the colon and is likely to become available for participation in a combination of Haber-Weiss and Fenton type reactions which can generate hydrogen peroxide and hydroxyl radicals (Blakeborough *et al.*, 1989; Babbs, 1990; Graf and Eaton, 1985). Such reactions require a supply of molecular oxygen and the presence of bacteria. Faecal microorganisms have been shown to generate oxygen free radicals under aerobic conditions in vitro (Babbs, 1990), and the appropriate conditions may exist in a microenvironment close to the mucosal surface. In principle, active oxygen species generated in this manner may interact with colonocytes, causing oxidative damage to lipids and DNA, thus increasing the risk of a mutation, either as an initiating event, or later in the adenoma-carcinoma sequence (Graf and Eaton, 1985). Alternatively iron mediated reactions may be involved in the conversion of pro-carcinogen to carcinogen within the faecal stream.

Interrelationships between intraluminal iron and certain types of bacteria have been implicated in various aspects of human health. For example, there has been particular interest in the iron-binding bifidobacteria found in the faeces of breast-fed infants.

The faecal flora of breast-fed infants is substantially different from that of a formula-fed infant. Among breast-fed infants the predominant organisms are bifidobacteria, lacto-bacilli and staphylococci, whereas in formula-fed infants the predominant organisms are enterococci, coliforms and bacteroides (Wharton *et al.*, 1994). It has been proposed that bifidobacteria can protect the gastrointestinal tract of the breast-fed infant by producing a large quantity of acetate buffer to maintain the pH at about 5, thus inhibiting the growth of pathogenic organisms, and by sequestering iron and thus making it unavailable to pathogenic organisms (Balmer and Wharton, 1991). Attempts have been made to modify infant formulas so as to encourage the development of a faecal flora similar to that found in breast-fed infants. There is some indication that the addition of iron to both whey and casein based infant formulas discourages colonisation and growth of staphylococci and bacteriodes but encourages clostridia and enterococci. However, the type of protein had more effect upon the growth of *Bifidobacterium*; whey formulas encouraged colonisa-tion by bifidobacteria but iron fortification had no modulating effect on these bacteria (Bezkorovainy, 1989). Modification of milk proteins by desialylation does not favour the development of a bifidus flora in one-month old infants (Chierici *et al.*, 1997).

The role of the iron-binding protein lactoferrin, found in high concentrations in breast but not cow's milk, is still not clear. The proposal that it optimises iron absorption from breast milk has not been confirmed (Fairweather-Tait *et al.*, 1987; Davidsson *et al.*, 1994). However, as it remains to a large extent undegraded in the gastrointestinal tract of the infant, it may play a role in control of the intestinal microflora by limiting the supply of iron to bacteria in the intestinal lumen (Brock, 1980).

6   Zinc

Intubation studies using marker perfusion of the upper gastrointestinal tract have established that the site of zinc absorption appears to be more distal than that of iron (Matseshe *et al.*, 1980). Furthermore, the levels of zinc in the chyme are well in excess of dietary levels, indicating that zinc of endogenous origin is secreted into the gut. However, unlike calcium, very little zinc is absorbed in the human colon. Sandstrom *et al.*, (1990) found that only 4.1% of a 30mmol (2mg) dose of ^{67}Zn introduced into the colon of adult men and women through a colonoscope was absorbed. As the efficiency of zinc absorption is dose-related, care was taken in using an appropriate quantity of zinc, namely a realistic level of nonabsorbed zinc reaching the colon after a single meal. When a similar dose of zinc was administered orally, fractional absorption was 65-74%, illustrating the insignificant role of the colon with respect to zinc absorption. As with other minerals absorbed primarily in the proximal small bowel, fermentable carbohydrates apparently have no effect on absorption (Coudray *et al.*, 1997).

Phytate has repeatedly been shown to be one of the most potent inhibitors of zinc absorption but some inconsistencies in results have been observed (Anon, 1983). The explanation lies in the modulating effect of calcium as this forms mixed complexes with phytate and zinc that are less soluble than simple phytate-zinc complexes.

As discussed above, these complexes can be hydrolysed by colonic bacteria thereby releasing bound calcium and zinc but with the very limited capacity for zinc absorption in the colon, the release of zinc by fermentation has no biological significance.

References

Anonymous (1983) Phytate and zinc metabolism. *Nutrition Reviews* **41**, 64-66.

Appleton G.V.N., Owen R.W., Wheeler E.E., Challacombe D.N. and Williamson, R.C.N. (1991) Effect of dietary calcium on the colonic luminal environment. *Gut* **32**, 1374-77.

Babbs, C.F. (1990) Free radicals and the etiology of colon cancer. *Free Radical Biology and Medicine* **8**, 191-200.

Balmer, S.E. and Wharton, B.A. (1991) Diet and faecal flora in the newborn: iron. *Archives Diseases in Children* **66**, 1390-94.

Baynes, R.D. and Bothwell, T.H. (1990) Iron deficiency. *Annual Review of Nutrition* **10**, 133-48.

Bezkorovainy, A. (1989) Iron transport and utilization by bifidobacteria. In: *Biochemistry and Physiology of Bifidobacteria* (Bexkorovainy A., Miller-Catchpole R., eds.) Florida: CRC Press pp. 147-76.

Billich, C.O. and Levitan, R. (1969) Effects of sodium concentration and osmolarity on water and electrolyte absorption from the intact human colon. *Journal of Clinical Investigation* **48**, 1336-47.

Binder, H.J. and Mehta, P. (1989) Short-chain fatty acids stimulate active sodium and chloride absorption in vitro in the rat distal colon. *Gastroenterology* **96**, 989-96.

Bitar, K. and Reinhold, J.G. (1972) Phytase and alkaline phosphatase activities in intestinal mucosae of rat, chicken, calf, and man. *Biochemistry Biophysica Acta* **268**, 442-52.

Blakeborough, M.H., Owen, R.W. and Bilton, R.F. (1989) Free radical generating mechanisms in the colon: Their role in the induction and promotion of colorectal cancer. *Free Radical Research Communication* **6**, 359-67.

Bovee-Oudenhoven, I., Termont, D., Dekker, R. and Van der Meer, R. (1996) Calcium in milk and fermentation by yoghurt bacteria increase the resistance of rats to Salmonella infection. *Gut* **38**, 59-65.

Bovee-Oudenhoven, I.M.J., Termont, D.S.M.L., Weerkamp, A.H., Faassen-Peters, M.A.W. and Van der Meer, R. (1997) Dietary calcium inhibits the intestinal colonzation and translocation of *Salmonella* in rats. *Gastroenterology* **113**, 550-57.

Bowling, T.E., Raimundo, A.H., Grimble, G.K. and Silk, D.B.A. (1993) Reversal by short-chain fatty acids of colonic fluid secretion induced by enteral feeding. *Lancet* **342**, 1266-68.

Brock, J.H. (1980) Lactoferrin in human milk: its role in iron absorption and protection against enteric infection in the newborn infant. *Archives Diseases in Children* **55**, 417-21.

Brune, M., Rossander, L. and Hallberg, L. (1989) Iron absorption: no intestinal adaptation to a high-phytate diet. *American Journal of Clinical Nutrition* **49**, 542-45.

Chernelch, M., Fawwaz, R., Sargent, T. and Winchell, H.S. (1970) Effect of phlebotomy and pH on iron absorption from the colon. *Journal of Nuclear Medicine* **11**, 25-27.

Chierici, R., Sawatzki, G., Thurl, S., Tovar, K. and Vigi, V. (1997) Experimental milk formulae with reduced protein content and desialylated milk proteins: influence on the faecal flora and the growth of term new-born infants. *Acta Paediatrica* 557-63.

Conrad, M.E., Weintraub, L.R., Sears, D.A. and Crosby, W.H. (1966) Absorption of hemoglobin iron. *American Journal of Physiology* **211**, 1123-30.

Coudray, C., Bellanger, J., Castiglia-Delavaud, C., Rémésy, C., Vermorel, M. and Rayssignuier, Y. (1997) Effect of soluble or partly soluble dietary fibres supplementation on absorption and balance of calcium, magnesium, iron and zinc in healthy young men. *European Journal of Clinical Nutrition* **51**, 375-80.

Cummings, J.H., Hill, M.J., Houston, H., Branch, W.J. and Jenkins, D.J.A. (1979) The effect of meat protein and dietary fiber on colonic function and metabolism. 1. Changes in bowel habit, bile acid excretion and calcium absorption. *American Journal of Clinical Nutrition* **32**, 2086-93.

Davidsson, L., Lastenmeyer, P., Yuen, M., Lonnerdal, B. and Hurrell, R.F. (1994) Influence of lactoferrin on iron absorption from human milk in infants. *Pediatric Research* **35**, 117-24.

Debongnie, J.C. and Phillips, S.F. (1978) Capacity of the human colon to absorb fluid. *Gastroenterology* **74**, 698-703.

Delzenne, N., Aertssens, J., Verplaetse, H., Roccaro, M. and Roberfroid, M. (1995) Effects of fermentable oligosaccharides on mineral, nitrogen, and energy digestive balance in the rat. *Life Sciences* **57**, 1579-87.

Demigné, C., Levrat, A.M. and Rémésy, C. (1989) Effect of feeding fermentable carbohydrates on cecal concentrations of minerals and their fluxes between the cecum and blood plasma in the rat. *Journal of Nutrition* **119**, 1625-30.

Devroede, G.J. and Phillips, S.F. (1969) Conservation of sodium, chloride and water by the human colon. *Gastroenterology* **56**, 421-26.

Drews, L.M., Kies, C. and Fox, H.M. (1979) Effect of dietary fiber on copper, zinc, and magnesium utilization by adolescent boys. *American Journal of Clinical Nutrition* **32**, 1893-97.

Fairweather-Tait, S.J., Balmer, S.E., Scott, P.H. and Minski, M.J. (1987) Lactoferrin and iron absorption in newborn infants. *Pediatric Research* **22**, 651-54.

Fordtran, J.S., Rector, F.C. and Carter, N.W. (1968) The mechanisms of sodium absorption in the human small intestine. *Journal of Clinical Investigation* **47**, 884-900.

Graf, E. and Eaton, J.W. (1985) Dietary suppression of colonic cancer-fiber or phytate? *Cancer* **56**, 717-18.

Hallberg, L. (1981) Bioavailability of dietary iron in man. *Annual Review of Nutrition* **1**, 123-47.

Harig, J.M., Ng, H.E., Dudeja, P.K., Brasitus, T.A. and Ramaswamy, K. (1996) Transport of n-butyrate into human colonic luminal membrane vesicles. *American Journal of Physiology* **271**, G415-22.

Hylander, E., Ladefoged, K. and Jarnum, S. (1980) The importance of the colon in calcium absorption following small-intestinal resection. *Scandinavian Journal of Gastroenterology* **15**, 55-60.

James, W.P.T., Branch, W.J. and Southgate, D.A.T. (1978) Calcium binding by dietary fiber. *Lancet* **1**, 638-39.

Karbach, U. and Feldmeier, H. (1993) The cecum is the site with the highest calcium absorption in rat intestine. *Digestive Diseases and Sciences* **38**, 1815-24.

Kelsay, J.L., Behall, K.M. and Prather, E.S. (1970) Effect of fiber from fruits and vegetables on metabolic responses of human subjects. II. Calcium, magnesium, iron, and silicon balances. *American Journal of Clinical Nutrition* **32**, 1876-80.

Kelsay, J.R. and Prather, E.S. (1983) Mineral balances of human subjects consuming spinach in a low-fiber diet and in a diet containing fruits and vegetables. *American Journal of Clinical Nutrition* **38**, 12-19.

Kennedy, H.J., Callender, S.T., Truelove, S.C. and Warner, G.T. (1982) Haematological aspects for life with an ileostomy. *British Journal of Haematology* **52**, 445-54.

Lapre', J.A., De Vries, H.T., Termont, D.S.M.L., Kleibeuker, J.H., De Vries, E.G.E. and Van derMeer, R. (1993) Mechanism of the protective effect of supplemental dietary calcium on cytolytic activity of fecal water. *Cancer Research* **53**, 248-53.

Levrat, A.M., Rémésy, C. and Demigné, C. (1991) High propionic acid fermentations and mineral accumulation in the cecum of rats adapted to different levels of inulin. *Journal of Nutrition* **121**, 1730-37.

Lutz, T. and Scharrer, F. (1991) The effect of SCFA on Ca absorption by the rat colon. *Experimental Physiology* **76**, 615-18.

Maddaiah, V.T., Kurnick, A.A. and Reid, B.L. (1964) Phytic acid studies. *Proceeding Society Experiment Biology Medicine* **115**, 391-93.

Matseshe, J.W., Phillips, S.F., Malagelada, J.R. and McCall, J.T. (1980) Recovery of dietary iron and zinc from the proximal intestine of h19ealthy man: studies of different meals and supplements. *American Journal of Clinical Nutrition* **33**, 1946-53.

McCance, R.A. and Widdowson, E.M. (1942) Mineral metabolism of healthy adults on white and brwon bread dietaries. *Journal Physiology* **101**, 44-85.

Moore, R.J. and Veum, T.L. (1983) Adaptive increase in phytate digestibility by phosphorus-deprived rats and the relationship of intestinal phytase (*EC* 3.1.3.8) and alkaline phosphatase (*EC* 3.1.3.1) to phytate utilization. *British Journal of Nutrition* **49**, 145-52.

Ohkawara, Y., Bamba, M., Nakai, J., Kinka, S. and Masuda, M. (1963) The absorption of iron from the human large intestine. *Gastroenterology* **44**, 611-15.

Ohta, A., Ohtsuki, M., Baba, S., Adachi, T., Sakata, T. and Sakaguchi, E. (1995) Calcium and magnesium absorption from the colon and rectum sre increased in rats fed fructo-oligosaccharides. *Journal of Nutrition* **125**, 2417-24.

Ohta, A., Ohtsuki, M., Baba, S., Hirayama, M. and Adachi, T. (1998) Comparison of the nutritional effects of fructo-oligosaccharides of different sugar chain lengths in rats. *Nutrition Research* **18**, 109-20.

Ohta, A., Ohtsuki, M., Hosono, A., Adachi, T., Hara, H. and Sakata, T. (1998) Dietary fructooligosaccharides prevent osteopenia after gastrectomy in rats. *Journal of Nutrition* **128**, 106-10.

Ohta, A., Baba, S., Ohtsuki, M., Takizawa, T., Adachi, T. and Hara, H. (1997) In vivo absorption of calcium carbonate and mafnesium oxide from the large intestine in rats. *Journal of Nutritional Science and Vitaminology* **43**, 35-46.

Rémésy, C., Levrat, M.-A., Gamet, L. and Demigné, C. (1993) Cecal fermentations in rats fed oligosaccharides (inulin) are modulated by dietary calcium level. *American Journal Physiology* **264**, G855-62.

Ruppin, H., Bar-Meir, S., Soergel, K.H., Wood, C.M. and Schmitt, M.G. (1980) Absorption of short chain fatty acids by the colon. *Gastroenterology* **78**, 1500-07.

Sandstrom, B., Cederblad A., Kivisto, B., Stenquist, B. and Andersson, H. (1986) Retention of zinc and calcium from the human colon. *American Journal of Clinical Nutrition* **44**, 501-04.

Sandstrom, B., Cederblad, A., Stenquist, B. and Andersson, H. (1990) Effect of inositol hexaphosphate on retention of zinc and calcium from the human colon. *European Journal of Clinical Nutrition* **44**, 705-08.

Sandberg, A.S., Hasselblad, C., Hasselblad, K. and Hulten, L. (1982) The effect of wheat bran on the absorption of minerals in the small intestine. *British Journal of Nutrition* **48**, 185-91.

Schmitt, M.G., Soergel, K.H. and Wood, C.M. (1976) Absorption of short chain fatty acids from the human jejunum. *Gastroenterology* **70**, 211-15.

Schmitt, M.G., Soergel, K.H., Wood, C.M. and Steff, J.J. (1977) Absorption of short chain fatty acids from the human ileum. *American Journal of Digestive Diseases* **22**, 340-47.

Schwartz, R., Apgar, B.J. and Wien, E.M. (1986) Apparent absorption and retention of Ca, Cu, Mg, Mn, and Zn from a diet containing bran. *American Journal of Clinical Nutrition* **43**, 444-55.

Slavin, J.L. and Marlett, J.A. (1980) Influence of refined cellulose on human bowel function and calcium and magnesium balance. *American Journal of Clinical Nutrition* **33**, 1932-39.

Trinidad, T.P., Wolever, T.M.S. and Thompson, L.U. (1993) Interactive effects of Ca and SCFA on absorption in the distal colon of man. *Nutrition Research* **13**, 417-25.

Trinidad, T.P., Wolever, T.M.S. and Thompson, L.U. (1996) Effect of acetate and propionate on calcium absorption from the rectum and distal colon of humans. *American Journal of Clinical Nutrition* **63**, 574-78.

Vescovi, E.G., Soncini, F.C. and Groisman, E.A. (1996) Mg2+ as an extracellular signal: environmental regulation of Salmonella virulence. *Cell* **84**, 165-74.

Wharton, B.A., Balmer, S.E. and Scott, P.H. (1994) Sorrento studies of diet and fecal flora in the newborn. *Acta Paediatrica* **36**, 579-84.

Wise, A. (1983) Dietary factors determining the biological activities of phytate. *Nutrition Abstracts and Reviews* **53**, 791-806.

Younes, H., Demigne, C. and Rémésy, C. (1996) Acidic fermentation in the caecum increases absorption of calcium and magnesium in the large intestine of the rat. *British Journal of Nutrition* **75**, 301-14.

CHAPTER 15

Diet and Biotransformation of Carcinogenic Compounds in the Gut by Enzymes of Microflora and of Intestinal Cells

BEATRICE L. POOL-ZOBEL

Institute for Nutrition and Environment, Friedrich-Schiller-University Jena, Dornburger Str. 25, 07743 Jena, GERMANY

1   Abstract

An unfavourable diet is expected to increase colon cancer risk by elevating exposure to both endogenous and exogenous toxic, genotoxic or carcinogenic compounds. Exogenous risk factors include food contaminants such as heterocyclic amines, polycyclic aromatic hydrocarbons, metals, or mycotoxins. Examples of endogenous risk compounds are bile acids, reactive oxygen species (peroxides, oxygen radicals), lipid peroxidation products (4-hydroxy-alkenals, α, ß-unsaturated aldehydes) and nitrosamines. All compounds may be subject to local metabolic conversion (by the microflora or by intestinal cells) and systematically (*e.g.* by the liver). Metabolic conversion pathways include the formation, activation and inactivation of toxic, genotoxic and carcinogenic substances. The effect of these compounds is dependent on the concentration of reactive intermediates in the specific target tissue, which in turn is governed by pharmacokinetic factors, exposure levels, and the balance between activating and deactivating enzymes. The colon may be more susceptible to carcinogens than the liver, since lower levels of deactivating enzymes are present in intestinal cells. Moreover, exposure times to toxic compounds may be longer. Enzymes from intestinal cells as well as from the microflora may be modulated by diet. Several studies have indicated that the ingestion of complex carbohydrates may favourably alter the balance of biotransformation towards inactivation of reactive compounds. Thus, diet has an impact on health not only by allowing a reduction of exposure to exogenous and endogenous factors, but also by modulating the microflora and the enzymes of epithelial cells in such a manner that a reduced load of reactive compounds is present in the colon and ultimately delivered to cells.

G.R. Gibson and M.B. Roberfroid (eds.), Colonic Microbiota, Nutrition and Health, 245-255.
© 1999 *Kluwer Academic Publishers. Printed in the Netherlands.*

The consequence of reduced exposure to carcinogenic substances is a lowering of cancer risk.

2   Introduction

Diet is one of the most important risk factors in human cancer development. An approximately 30% mortality rate of cancer can be attributed to nutritional factors (Doll and Peto, 1981). Diets high in energy or animal fats, red meat, or spicy foods have been linked to an elevated tumour risk in some epithelial tissues (colon, stomach, oesophagus), whereas high vegetable intake is associated with lower risk (Potter and Steinmetz, 1996). From a toxicological viewpoint, ingestion of a diet that has an elevated cancer risk may be associated with the higher intake of toxic, genotoxic or carcinogenic compounds. Additionally, higher levels of endogenously formed toxic metabolites may prevail. These could include compounds such as bile acids, reactive oxygen radicals (ROR), lipid peroxidation products as well as endogenously generated nitroso-compounds. Both endogenous and exogenous products could participate in the complex multistep process of carcinogenesis, one of the important mechanisms of which is induction of mutations in critical target genes, responsible for the process of carcinogenesis (Fearon and Vogelstein, 1990). Mutations occurring in tumour suppressor genes and proto-oncogenes have been identified in various stages of the neoplastic process (Boland *et al.*, 1995). In part, some of these are hereditary and can contribute to approximately 1% of all cancers, but the rest is most likely due to the acquisition of mutations in somatic cells of tumour target tissue (Fearon, 1997).

One of the basic concepts in toxicology is that the impact of these environmental factors is dependent on the concentration of reactive intermediates in specific target tissues. This is governed in turn by pharmacokinetic factors and exposure levels, as well as by the balance between activating and deactivating enzymes. The colon is an important tumour target tissue which may be affected by diet. Both enhancing and inhibiting effects of dietary factors have been identified (Hill, 1995). Colorectal tumours are the second to third most frequent human tumours in areas with "Western Style Diets", including several countries in Europe, North America and Australia (Levi *et al.*, 1995). The complex interactions of diet, exposure to risk factors, microflora and epithelial cell metabolism in the colon tissue will be reviewed here.

3   Susceptibility of the colon to suspected risk factors

3.1   PHARMACOKINETIC FACTORS

The colon is very susceptible to dietary factors, since toxic compounds ingested with food may reach the large intestine via several pathways. One is by direct passage through the gastrointestinal tract with the food bulk, others include systemic pathways either reaching the intestinal epithelia on the apical end with bile, or on the basolateral side with blood. Compounds reaching the colon in this manner are usually conjugates of directly resorbed compounds which have already been partially biotransformed in the liver. Additionally, it can also be speculated that reactive compounds may reach the colon within exfoliated cells originating from upper regions of the small intestine.

3.2 EXPOSURE LEVELS IN THE COLON

Colonic tissue may be considered particularly susceptible to endogenous and exogenous toxic compounds on account of two major properties. One is that particularly long exposure periods can be expected, since faecal retention times in the gut may be several hours or more. Secondly, compounds reaching the colon by systemic pathways, blood or bile, are usually partially activated intermediates easily reactivated by the cleavage of conjugating groups. On the other hand, apical exposure of colonic cells to certain types of reactive compounds may be reduced by the mucous barrier, which can scavenge electrophiles and thus protect intestinal cells.

3.3 BALANCE OF METABOLISING ENZYMES

Toxic compounds are subject to metabolic conversion by a variety of enzymes which, depending on the chemical nature of the compound, may ultimately yield a metabolite that is of higher, lower or similar reactivity as the parent compound. The yield of reactive intermediates in the gut will depend on the type of compounds and enzymes therein.

3.3.1 *Metabolising enzymes of the intestinal cells*
In mammalian cells, phase I or phase II metabolising enzymes have been identified. In this context, phase I enzymes usually refer to oxidising systems, like cytochrome P450 monooxygenases or those containing flavin which convert substrates to more hydrophilic intermediates by introducing oxygen. The resulting hydroxyl groups can be conjugated with glutathione, glucuronic acid, sulphate or other groups by phase II enzymes (glutathione S-transferases, UDPGA-transferases, sulfotransferases, acetyl transferases). Frequently, phase I oxidation reactions are accompanied by the formation of reactive electrophilic intermediates, which may react or covalently bind with macromolecules. The conjugation reaction however is commonly, but not always, associated with deactivation since most of the resulting compounds are less reactive than the previously formed intermediates.

The major site of such metabolism is the liver. However, the intestine also contains important metabolising enzymes such as cytochrome p4501A1 and glutathione S-transferase p (GSTP1). Previous studies have shown that the normal colonic epithelium of the human small and large intestine contains specific isoenzymes of these super families, which are different from those detected in the liver (Peters *et al.*, 1992). The presence of specific isoezymes of many enzymes involved in biotransformation are governed by genetic polymorphisms (Sivaraman *et al.*, 1994). Such polymorphisms are due to the deletion of genes coding for a whole protein or may affect a single base substitution, leading to an altered enzyme activity. Some genetic polymorphisms such as

GSTM1*0 can be frequent and affect up to 50% of a population. It is expected that individual susceptibility in cancer burden related to environmental exposure, is governed by the pattern of these genetic enzyme polymorphisms (Bartsch and Hietanen, 1996). Altogether, the metabolic capacity in the colon is several fold lower than in the liver. This lower capacity pertains to several enzymes largely responsible for activation as well as deactivation. Since the colon, however, is exposed to compounds already partially activated in the liver, the low deactivating capacity (*e.g.* lower levels of glutathione S-transferases; see Fig. 1) is one additional reason why this tissue is susceptible to risk factors.

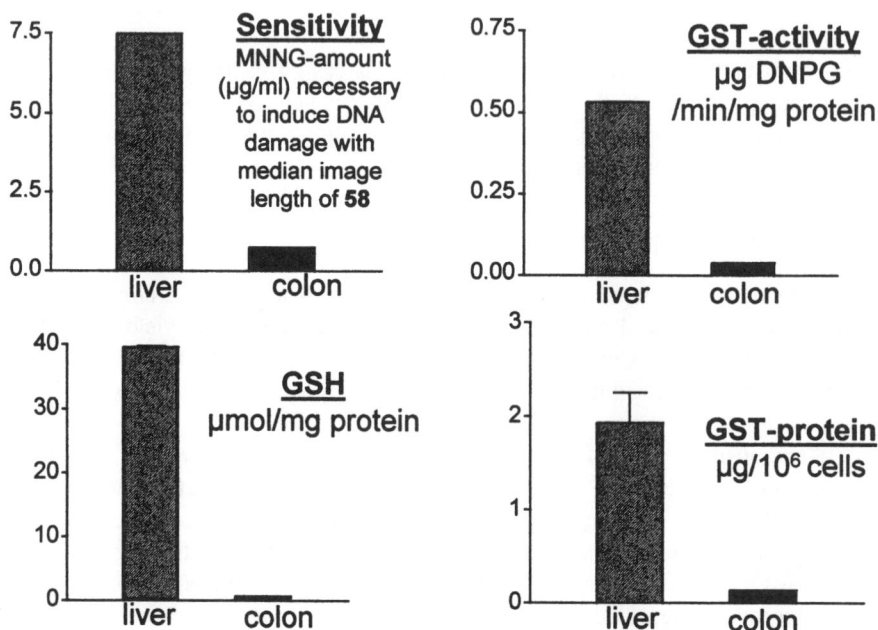

Figure 1. Metabolic deactivation capacity by glutathione S-transferase (GST) in liver and colon of rat. (Pool-Zobel *et al.*, (1996); and unpublished results).

3.3.2 *Metabolism by microbiota enzymes*

The intestinal microflora is composed of numerous bacterial species and strains which produce various enzymes, *e.g.* hydrolases (ß-glycosidases, ß-glucuronidases, sulfatases), reductases (of azo groups, aromatic nitro groups, ketone and aldehyde groups), demethylation, lyases (leading to deamination, decarboxylation,), and transferases (*e.g.* sulfotransferases). Depending on the type of substrate, these reactions can be considered to be toxic or beneficial to the host (Reddy, 1990). Thus, the reduction of carcinogenic 1-nitropyrene to 1-aminopyrene by nitroreductases of intestinal bacteria, or the cleavage of N-glucuronide conjugates of heterocyclic amines by ß-glucuronidases, can be considered to be metabolic activation steps (Howard *et al.*, 1983; Kaderlik *et al.*, 1994).

Bacterial enzymes can hydroxylate the heterocyclic amine 2-amino-3-methyl-3H-imidazo [4,5-f]quinoline (IQ) in the aromatic ring leading to a more reactive species (Rumney *et al.*, 1993). Also, deconjugation of cycasin to yield genotoxic intermediates by ß-glycosidases is an activation step, leading to more DNA-reactive intermediates. In contrast, the liberation of aglycones from secondary plant ingredients by ß-glycosidases may be considered to be associated with health protective beneficial properties. Similarly, methylmercury compounds can be synthesised from the biologically less reactive mercuric chloride by the intestinal flora (Mallett and Rowland, 1990). Thus, bacterial enzymes may participate both in activation and deactivation reactions. The consequence, *i.e.* extent of toxic effects is dependent on types of compounds and enzymes present in the gut at a given time.

4   Effect of suspected risk factors in human colon cells

Although there are several reports on the contents of different enzyme systems in colon cells, little is known about the consequences of these enzymes for suspected risk factors. The most important will be the extent of DNA damage and mutations induced by individual compounds at physiological exposure concentrations. Clearly, since much research is necessary to elucidate the potencies of damaging effects by suspected carcinogens in the colon, sophisticated techniques are needed to study these aspects. Accordingly, we have developed a novel method to investigate chemicals for genotoxicity in cells of human colon biopsies.

4.1   DETECTION OF DNA DAMAGE IN HUMAN COLON CELLS WITH THE MICROGEL ELECTROPHORESIS ASSAY ("COMET ASSAY")

The method is based on the determination of DNA damage in primary cells freshly isolated from human colon biopsies with the single cell microgel electrophoresis technique (Pool-Zobel *et al.*, 1996). Colonic cells are obtained from the biopsies. These are incubated with Proteinase-K and collagenase, suspended and centrifuged to form cell pellets. After incubation, cells are lysed, alkalinised and subjected to electrophoresis and staining with ethidium bromide and microscopical analysis. Images of each slide are scored and damaged DNA is visible as "comet-like" images. This method has been given the name of "Comet-Assay". Experimental details are given by Pool-Zobel *et al.* (1996). The potential of colon cancer risk compounds to induce genetic damage using the assay has been assessed as follows:

4.1.1 *Nitroso compounds*
Nitroso compounds include nitrosamines, which are mainly metabolised by cytochrome p450 monooxygenases, and nitrosamides which are directly active. Their present relevance for carcinogenesis in humans is seen mainly through the potential of indigenous formation, especially within the intestinal tract. A rough estimate indicates that 45 to 75 % of total exposure comes from the endogenous route (Tricker, 1997) In the colonic lumen,

nitroso compounds have been shown to be increased with higher intakes of red meat and nitrate (Rowland *et al.*, 1991; Bingham *et al.*, 1996). Although the chemical structures of these compounds have not yet been identified, their formation may be a relevant risk factor. Little is known about the metabolic conversion by human intestinal cells. However, early studies have shown that non-symmetrical nitrosamines may be transformed by jejunal and ileal tissues of rats (Richter *et al.*,1985). We have now investigated this in human colon cells with N-methyl-N-nitro-N-nitrosoguanidine (MNNG) and found this compound to be a potent inducer of DNA damage. N-methyl-N-nitro-N-nitrosoguanidine is a directly active compound that may be deactivated by GST. Liver cells of rats contain approximately 10 GST activities and GSH levels than do colon cells. As such, 10-fold more MNNG is needed to induce equal DNA damage in the liver compared to gut cells (Fig. 1). For directly acting compounds, the colon may be more susceptible than the liver, since lower levels of deactivating enzymes are present.

4.1.2 *Heterocyclic amines*
Heterocyclic amines are pyrolysis products of proteins formed during the extensive heating of meat and fish products. They are considered to be very potent endogenous risk candidates for human colon cancer (Eisenbrand and Tang, 1993). Two compounds, namely 2-amino-1-methyl-6-phenylimidazo[4,5-b]pyridine (PhIP) and IQ (discussed above), deserve particular attention. They induce tumours in the colons of rats and mice (Eisenbrand and Tang, 1993). In contrast to MNNG, PhIP and IQ have very complex modes of metabolic activation. N-hydroxylation in the liver is the first step followed by O-acetylation. The conjugate is transported to the intestine where cleavage yields the reactive species, or N-acetylation to give a form which may be transported and reactivated in colon cells (Kaderlik *et al.*, 1994). Apparently, the enzymes which carry out the first activation step are not present in human colon cells, (since neither PhIP nor IQ were genotoxic in the cells of several donors' Pool-Zobel and Leucht, 1997). Thus, for the proportion of ingested PhIP and IQ which are not resorbed but reach the colon lumen directly, these compounds may only be of toxicological significance if they are activated by the microflora, or resorbed at this later stage and passed into the liver. Non-metabolised substances may be excreted with the faeces.

4.1.3 *Hydrogen peroxide*
Hydrogen peroxide (H_2O_2) is an indigenous substance, representative of compounds arising during oxidative stress. Activated forms of oxygen may be increased in a pro-oxidant state due to nutritional imbalance mainly by different forms of high energy intake. There is convincing evidence that pro-oxidant states are involved in carcinogenesis by initiating primary DNA damage and promoting initiated cells to an increased neoplastic conversion (Ames *et al.*, 1993). Hydrogen peroxide decomposes in the presence of transition metals to yield the hydroxyl radical OH (Imlay *et al.*, 1988). This may induce lesions typical of oxidative DNA damage (Collins *et al.*, 1995). Cellular anti-oxidant defence systems such as glutathione peroxidases, catalases and iso-forms of glutathione S-transferases can deactivate peroxides (Bonorden and Pariza, 1994). Human cells are highly susceptible to H_2O_2 activities since the antioxidant defence systems are not always sufficient to protect against the dosages studied. Two other sources of H_2O_2,

in addition to the usual pathways of cellular metabolism, may be important in colon carcinogenesis. One is that certain growth factors such as insulin, epidermal growth factor (EGF) and platelet derived growth factor (PDGF) can mediate their effectiveness via the production of H_2O_2 (Sundaresan *et al.*, 1995). Also, the gut microflora can produce reactive oxygen species (Owen *et al.*, 1996). In human colonic cells, amounts of 50-500 mM are genotoxic (Pool-Zobel and Leucht, 1997). Endogenous oxidised DNA pyrimidine bases, probably also due to peroxide activity, can be efficiently detected in human colon cells (unpublished results), meaning that a steady state of exposure to oxidants and elimination of oxidative damage takes place in this tissue.

4.1.4 *Other risk factors*
In addition to the examples described above, other chemical carcinogens or toxins associated with food or with the metabolism of nutrients have been investigated in primary human colon cells. These compounds include contaminants such as benzo(a)pyrene, aflatoxin B1, which had only low genotoxic potency in human colon cells and others such as the pesticide lindane or mercuric chloride, which are toxic and genotoxic (Pool-Zobel and Leucht, 1997). Most indigenous risk compounds investigated so far, like dinitroso caffeidine and lithocholic acid have toxic potential. Also, hexenal which is either a representative of natural food ingredients, a food additive or an endogenous product of lipid peroxidation has been shown to be strongly genotoxic in human colon cells (Gölzer *et al.*, 1996). Thus, the human colon is susceptible to a variety of compounds with different chemical and toxicological properties. Since levels of phase II enzymes are relatively low, directly active compounds seem to be especially effective genotoxins in colon cells.

5 Modulation of intestinal enzymes by the diet

Dietary factors can either directly modulate the activities and formation of intestinal enzymes by action of individual micronutrients, or indirectly by altering the gut fermentation.

5.1 ENZYME INDUCTION IN INTESTINAL CELLS

The activities of micronutrients, or of fermentation products, may alter enzyme activity by direct interactions with catalytic systems. Conversely, the formation of enzymes may be genetically regulated, within colonic cells, by the action of micronutrients and single fermentation products. A few studies are available showing that diet can induce glutathione S-transferases (GST) in the gut. In humans, a high intake of glucosinolate-containing cruciferous vegetables (*e.g.* Brussel sprouts) has resulted in increased rectal GST-α and -π isoezyme levels (Nijhoff *et al.*, 1995). This enhancement, of detoxification enzymes, may partly explain the epidemiological association between high intake of glucosinolates and a decreased risk of colorectal cancer. The enhancement may be due to the glucosinolates or flavonoid glycosides, which show inducing capacity *in vitro*

(Ballongue *et al.*, 1997). However, Brussel sprouts also contain complex carbohydrates which can yield fermentation products with inducing potential. Our data indicate that GST amounts in rats with a conventional microflora are higher than in germ free equivalents and that rodents fed complex carbohydrates have more GSTP1 (unpublished results).

The fermentation of complex carbohydrates generates a higher yield of SCFA and the consequence may be a relatively higher proportion of butyrate (Cummings and Macfarlane, 1997). Butyrate has been shown to induce GSTP1 in human colon tumor cells *in vitro* (Stein *et al.*, 1996).

Numerous micronutrients have been shown to induce GST in model in vitro systems (Prestera *et al.*, 1993). It is possible that these, together with fermentation products, increase enzymes of the chemoprevention system in the colon. However, enzymes of the phase I system in the colon can also be modulated by known enzyme inducers (Rosenberg, 1991). In the rat, a high protein diet has been shown to increase levels of cytochrome p-450, in addition to increasing activities of GST, and UDPGA transferases (Tutelyan *et al.*, 1990). We have observed a two fold enhancement of cytochrome reductase in the rat gut, following ingestion of lactic acid bacteria (Pool-Zobel *et al.*, 1996). Also in the rat, it has been shown that dietary fibre decreases the cytochrome P4501A1 contents in the colonic mucosa, a mechanism associated with protective properties. (Kawata *et al.*, 1992). Thus, interactions of this type coupled to the induction of GST (as discussed above), strongly shifts the balance of metabolic conversion, for several types of chemicals, towards inactivation rather than activation.

5.2   MODULATION BY ENZYMES OF THE MICROFLORA

The composition of the colonic microflora is strongly dependent on the diet. Therefore, enzymes present in the gut flora are also a direct reflection of dietary factors. In particular, fibre, complex carbohydrates and other non-digestible food ingredients fermented by bacteria of the gut can impact on the composition and metabolic activities of intestinal contents (Cummings and Macfarlane, 1997). Certain observations have been made when switching from a high protein to a high fibre or carbohydrate diet. These are that species of bifidodobacteria or lactobacilli may increase. The formation of toxic amines, sulphides, and phenols (ammonia, phenol, indol, cresols, skatol) is generally lower whereas organic acid production (lactic acid, butyrate, acetic acid) is enhanced and the pH value decreases through stimulation of the activity of procarcinogenic enzymes such as azoreductase, 7-α-dehydroxylase, ß-glucuronidases, nitroreductases and urease have been shown to be decreased in rats and humans (Birkett *et al.*, 1996; Hara *et al.*, 1994). In contrast there are reports showing that ß-glycosidases can be increased (Rowland and Tanaka, 1993). As mentioned above, ß-glycosides are responsible for cleaving bioactive plant glycosides to yield aglycones. Some may lead to toxic intermediates (*e.g.* cycasine), others may have chemopreventive potential (*e.g.* glycosides of antioxidant flavonoids, monoterpenes, lignans). If adequate supplies of the latter are present in the gut, this may again have the consequence that balance of metabolic conversion is shifted more towards the production of protective rather than toxic intermediates.

6 Conclusion

Available evidence suggests that the metabolic capacity of the intestine, arising both from the gut lumen as well as from intestinal cells, has a profound input on the conversion of compounds to toxic, less toxic and protective intermediates. This conversion capacity may be modulated by diet. Most evidence suggests that dietary fibre and complex compounds, as well as some secondary plant ingredients, may favourably alter the conversion capacity towards protection. This type of protection has the consequence that fewer lesions such as DNA damage (Rowland *et al.*, 1996), preneoplastic lesions or even tumours (Reddy *et al.*, 1997) may result. These findings, which have been mainly obtained through animal experiments, are recently being supported more and more by model studies with human cells, body fluids or microflora preparations. More research in the future will be necessary to assess beneficial health effects by the multifold types of complex carbohydrates contained in foods, and to determine the quantitative impact of protective mechanisms in humans.

References

Ames, B.N., Shigenaga, M.K. and Hagen, T.M. (1993) Oxidants, antioxidants, and the degenerative diseases of aging. *Proceedings of the National Academy of Sciences* U.S.A. **90**, 7915-22.

Ballongue, J., Schumann, C. and Quignon, P. (1997) Effects of lactulose and lactitol on colonic microflora and enzymatic activity. *Scandanavian Journal of Gastroenterology* **32** Suppl. 222, 41-44.

Bartsch, H. and Hietanen, E. (1996) The role of individual susceptibility in cancer burden related to environmental exposure. *Environmental Health Perspectives* **104**, 569-77.

Bingham, S.A., Pignatelli, B., Pollock, J.R.A., Ellul, A., Malaveille, C., Gross, G., Runswick, S., Cummings, J.H. and O'Neill, I.K.O. (1996) Does increased endogenous formation of N-nitroso compounds in the human colon explain the association between red meat and colon cancer? *Carcinogenesis* **17**, 515-23.

Birkett, A., Muir, J., Phillips, J., Jones, G. and O'dea, K. (1996) Resistant starch lowers fecal concentrations of ammonia and phenols in humans. *American Journal of Clinical Nutrition* **63**, 766-72.

Boland, C.R., Sato, J., Appelman, H.D., Bresalier, R.S. and Feinberg, A.P. (1995) Microallelotyping defines the sequence and tempo of allelic losses at tumour suppressor gene loci during colorectal cancer progression. *Nature Medicine* **1**, 902-09.

Bonorden, W.R. and Pariza, M.W. (1994) Antioxidant nutrients and protection from free radicals. In *Nutritional Toxicology*, eds. Kotsonis, F.N., Mackey, M. and Hielle, J., pp. 19-48. New York: Raven Press Ltd.

Collins, A.R., Duthie, S.J. and Dobson, V.L. (1995) Direct enzymic detection of endogenous oxidative base damage in human lymphocyted DNA. *Carcinogenesis* **14**, 1733-35.

Cummings, J. and Macfarlane, F.T. (1997) Role of intestinal bacteria in nutrient metabolism. *Clinical Nutrition* **16**, 3-11.

Doll, R. and Peto, R. (1981) The causes of cancer: Quantitative estimates of avoidable risks of cancer in the United States today. *Journal of the National Cancer Institute* **66**, 1191-1308.

Eisenbrand, G. and Tang, W. (1993) Food-borne heterocyclic amines. Chemistry, formation, occurrence and biological activities. A literature review. *Toxicology* **84**, 1-82.

Fearon, E.R. (1997) Human cancer syndromes: Clues to the origin and nature of cancer. *Science* **278**, 1043-50.

Fearon, E.R. and Vogelstein, B. (1990) A genetic model for colorectal tumorigenesis. *Cell* **61**, 759-67.

Gölzer, P., Janzowsky, C., Pool-Zobel, B.L. and Eisenbrand, G. (1996) Hexenal-induced DNA damage and formation of cyclic 1,N^2-(1,3-Propano)-2^l-deoxygunanosine adducts in mammalian cells. *Chemical Research in Toxicology* **9**, 1207-13.

Hara, H., Li, S.T., Sasaki, M., Maruyama, T., Terada, A., Ogata, Y., Fuyimori, I. and Mitsuoka, T. (1994) Effective dose of lactosucrose on fecal flora and fecal metabolites of humans. *Bifidobacteria Microflora* **13**, 51-63.

Hill, M.J. (1995) Diet and cancer: A review of scientific evidence. *European Journal of Cancer Prevention* **4**, Supplement 2, 3-42.

Howard, P.C., Beland, F.A. and Cerniglia, C.E. (1983) Reduction of the carcinogen 1-nitropyrene to 1-aminopyrene by rat intestinal bacteria. *Carcinogenesis* **4**, 985-90.

Imlay, J., Chin, S.M. and Linn, S. ((1988)) Toxic DNA damage by hydrogen peroxide through the Fenton reaction in vivo and in vitro. *Science* 640-42.

Kaderlik, K.R., Minchin, R.F., Mulder, G.J., Ilett, K.F., Daugaard-Jenson, M., Teitel, C.H. and Kadlubar, F.F. (1994) Metabolic activation pathway for the formation of DNA adducts of the carcinogen 2-amino-1-methyl-6-phenylimidazo[4,5-ß]pyridine (PhIP) in rat extrahepatic tissues. *Carcinogenesis* **15**, 1703-09.

Kawata, S., Tamura, S., Matsuda, Y., Ito, N. and Matsuzawa A., Y. (1992) Effect of dietary fiber on cytochrome P450IAI induction in rat colonic mucosa. *Carcinogenesis* **13**, 2121-25.

Levi, F., Lavecchia, C., Lucchini, F. and Negri, E. (1995) Cancer mortality in Europe. *European Journal of Cancer Prevention* **4**, 389-417.

Mallett, A.K. and Rowland, I.R. (1990) Bacterial enzymes: Their role in the formation of mutagens and carcinogens in the intestine. *Drug Dispositon* **8**, 71-79.

Nijhoff, W.A., Mulder, T.P.J., Verhagen, H., Van Poppel, G. and Peters, W.H.M. (1995) Effects of consumption of Brussels sprouts on plasma and urinary glutathione S-transferase class -α and -π in humans. *Carcinogenesis* **16**, 955-57.

Owen, R.W., Wimonwatwatee, T., Spiegelhalder, B. and Bartsch, H. (1996) A high performance liquid chromatography system for quantification of hydroxyl radical formation by determination of dihydroxy benzoic acids. European *Journal of Cancer Prevention* **5**, 233-40.

Peters, W.H.M., Boon, C.E.W., Roelofs, H.M.J. and Wobbes, T. (1992) Expression of drug-metabolizing enzymes and P-170 glycoprotein in colorectal carcinoma and normal mucosa. *Gastroenterology* **103**, 448-55.

Pool-Zobel, B.L. and Leucht, U. (1997) Induction of DNA damage in human colon cells derived from biopsies by suggested risk factors of colon cancer. *Mutation Research* **375**, 105-16.

Pool-Zobel, B.L., Neudecker, C., Domizlaff, I., JI, S., Schillinger, U., Rumney, C.J., Moretti, M., Villarini, M., Scassellati-Sforzolini, G. and Rowland, I.R. (1996) *Lactobacillus-* and *Bifidobacterium*-mediated antigenotoxicity in colon cells of rats: Prevention of carcinogen-induced damage in vivo and elucidation of involved mechanisms. *Nutrition and Cancer* **26**, 365-80.

Potter, J.D. and Steinmetz, K.A. (1996) Vegetables, fruit and phytoestrogens as preventive agents. In *Principles of Chemoprevention*, eds. Stewart, B.W., Mcgregor, D.B. and Kleihues, P., pp. 61-90. Lyon: International Agency for research on Cancer.

Prestera, T., Holtzclaw, W.D., Zhang, Y. and Talalay, P. (1993) Chemical and molecular regulation of enzymes that detoxify carcinogens. *Proceedings of the National Academy of Sciences U.S.A.* **90**, 2965-69.

Reddy, B.S. (1990) Intestinal microflora and carcinogenesis. *Bifidobacteria Microflora* **9**, 65-76.

Reddy, B.S., Hamid, R. and Rao, C.V. (1997) Effect of dietary oligofructose and inulin on colonic preneoplastic aberrant crypt foci inhibition. *Carcinogenesis* **18**, 1371-74.

Richter, E., Feng, X. and Wiessler, M. (1985) Biotransformation of symmetric dialkylnitrosamines in the rat intestine. *Biochemical Pharmacology* **34**, 445-46.

Rosenberg, D.W. (1991) Tissue specific induction of the carcinogen inducible cytochrome P450 isoform, P450IA1 in colonic epithelium. *Archives Biochemistry Biophysics* **284**, 223-26.

Rowland, I.R., Bearne, C.A., Fischer, R. and Pool-Zobel, B.L. (1996) The effect of lactulose on DNA damage induced by 1,2-dimethylhydrazine in the colon of human-flora-associated rats. *Nutrition and Cancer* **26**, 38-47.

Rowland, I.R., Granli, T., Bockman, O.C., Key, P.E. and Massey, R.C. (1991) Endogenous N-nitrosation in man assessed by measurement of apparent total N-nitroso compounds in feces. *Carcinogenesis* **12**, 1395-1401.

Rowland, I.R. and Tanaka, R. (1993) The effects of transgalactosylated oligosaccharides on gut flora metabolism in rats associated with a human faecal microflora. *Journal of Applied Bacteriology* **74**, 667-74.

Rumney, C.J., Rowland, I.R. and O'Neill, I.K. (1993) Conversion of IQ to 7-OHIQ by gut microflora. *Nutrition and Cancer* **19**, 67-76.

Sivaraman, L., Leatham, M.P., Yee, J., Wilkens, L.R., Lau, A.F. and Marchand, L.L. (1994) CYP1A1 genetic polymorphisms and in situ colorectal cancer. *Cancer Research* **54**, 5692-95.

Stein, J., Schr der, O., Bonk, M., Oremek, G., Lorenz, M. and Caspary, W.F. (1996) Induction of glutathione-S-transferase-pi by short-chain fatty acids in the intestinal cell line CaCo-2. *European Journal of Clinical Investigation* **26**, 84-87.

Sundaresan, M., Yu, Z.X., Ferrans, V.J., Irani, K. and Finkel, T. (1995) Requirement for generation of H_2O_2 for platelet-derived growth factor signal transduction. *Science* **270**, 296-99.

Tricker, A.R. (1997) N-nitrosocompounds and man. Sources of exposure, endogenous formation and occurrence in body fluids. *European Journal of Cancer Prevention* **6**, 226-68.

Tutelyan, V.A., Kravchenko, L.V., Avrenyeva, L.I. and Kuzmina, E.E. (1990) The activity of xenobiotic-metabolizing enzymes in the liver and small intestine of rats fed low and high levels of protein. *Nutrition Research* **10**, 1119-29.

CHAPTER 16

Large Bowel Cancer and Colonic Foods

A.V. RAO
Department of Nutritional Sciences, University of Toronto, CANADA

1   Introduction

Large bowel cancer along with breast, prostate, lung, endometrium and ovarian tumours are the leading causes of cancer mortality in the Western countries (The American Institute of Cancer Research, 1997). In addition to these cancers, other Western diseases that also contribute significantly to morbidity and sometimes mortality, include other gastrointestinal diseases (constipation, hiatus hernia, appendicitis, diverticular disease and haemorrhoids), cardiovascular diseases (coronary heart disease, stroke, essential hypertension deep vein thrombosis, pulmonary embolism, varicose veins), and metabolic diseases (obesity, diabetes, gallstones, renal stones, osteoporosis, gout). The development of these disorders appears to be associated with life style and environmental factors of which diet is a major component. (Potter, 1995; 1996).

2   Diet and large bowel cancer

Large bowel cancer is initiated through the proliferation of healthy epithelial cells that line the colon and rectum. The abnormal cell proliferation results in the formation of dysplastic crypts that progress towards aberrant crypts, polyps and eventually tumours and cancers (Bingham and Cummings, 1997). Although genetic predisposition is associated with these cellular changes, dietary factors are being recognised as playing an important role in the overall outcome. Of the several dietary factors, high intakes of energy and fat, and low intake of dietary fibre, characteristic of Western diets, have shown the strongest association with increased risk of large bowel cancer (Bingham and Cummings, 1997). Other cross-sectional studies have shown that dietary fat and meat increase the risk, while cereal foods, fruits and vegetables have the opposite effect. A majority of the case-control studies point to the consumption of red meats as increas-

G.R. Gibson and M.B. Roberfroid (eds.), Colonic Microbiota, Nutrition and Health, 257-265.
© 1999 *Kluwer Academic Publishers. Printed in the Netherlands.*

ing risk. Dairy products and eggs are shown to have no effect on relative risk but consumption of fruits and vegetables are consistently shown to reduce cancer risk. Fibre content of cereals, vegetables and fruits are suggested as ingredients which provide protection (Bingham, 1990).

Among the mechanisms of diet-induced carcinogenesis associated with intake is the bacterial metabolism of bile acids in the colon (Hill *et al.,* 1971; 1975). Secondary bile acids, which stimulate cell proliferation in the colon, are known tumour promoters. Recently, it has been suggested that bile acids stimulate phospholipid turnover in the cell membrane and release diacylglycerol, which in turn activates protein kinase C and causes cell proliferation (Reddy *et al.,* 1994). Increased production of reactive oxygen species (ROS), as a result of lipid peroxidation, and oxidative damage to DNA is also recognised as another possible mechanism of action of dietary fats (Ames *et al.,* 1993; 1995, Wiseman, 1996). The association of increased cancer risk with protein and meat intakes are attributed to the formation of mutagenic and carcinogenic compounds, such as heterocyclic amines that are produced during cooking, formation of N-Nitroso compounds, amines and amides from the bacterial fermentation of protein, accumulation of ammonia in the colon during high-meat diets and the induction of bile-acid degrading enzymes (Bingham and Cummings, 1997).

Major component of the fruits and vegetables providing protection against colorectal cancer are the non-starch polysaccharides (NSP). However, other minor constituents include υ3 and υ6 fatty acids, antioxidant micronutrients, flavonoids and phenols, lycopenes, sulphur compounds, isoflavones, lignans and saponins. Suggested mechanisms of action of these protective factors include immune stimulation, hormonal action, antioxidants and metabolic detoxification of genotoxic compounds.

3   Colonic bacteria

With increased recognition of the role of environmental and dietary factors in the aetiology of several cancers, in particular cancer of the large bowel, the relationship between gastrointestinal microbial flora and nutritional factors in humans and laboratory animals are now being more intensely investigated (see Chapters 4, 8 and 15).

The intestinal microbial flora of humans represents a rich ecosystem composed of a wide range of metabolically active microorganisms. The large bowel is the most heavily colonised part of the gastrointestinal tract, yielding up to 10^{12} bacteria per gram of intestinal contents in healthy human subjects. Major genera include bacteroides, clostridia, fusobacteria, streptococci, bifidobacteria, enterobacteria, eubacteria, and coliforms. Traditionally, the liver is considered a major site of metabolic activity in the body. However, due to the presence of the microorganisms in the colon and a ready supply of substrate, a very high degree of metabolic activity takes place in the human large intestine. It has been shown that intestinal microbial flora play a crucial role in influencing the physiological state of the host (see Chapter 1). However, it can also contribute significantly to the genesis of various disease states, including cancer of the large bowel, but also provides a major defence against infection. There is much current interest in understanding the nature of the substrate that enters the large intestine, nature of the microorganisms colonising the gastrointestinal tract, microbial enzymes, metabolic end products, interactions between the metabolites and the gastrointestinal mucosal cells and their health consequences.

The composition of the microbial flora in the gastrointestinal tract is dependent on a number of factors including diet. Diet influences the composition of gastrointestinal microflora, and as a result, alters other physiological states of the colon such as lumenal pH, colonic immune function and bacterial enzymatic activity which in turn affects the host health.

Several epidemiological studies have shown differences in the nature of intestinal microbial flora between various population groups at risk of large bowel cancer. The faecal microflora composition of Americans, who are at high risk of colon cancer, have been observed to be significantly different from a lower risk Japanese population (Finegold *et al.*, 1974). Similarly, the faecal bacterial composition of colonic polyp patients is different from that of control populations (Finegold *et al.*, 1975). The nature of microbial activity depends, to a great extent, on the composition and amount of sustrate being delivered to the large intestine. A diet rich in complex carbohydrates will support microbial activity consistent with a 'healthier' fermentation and the formation of several organic acids (see Chapter 4). Faecal samples from healthy humans consuming a diet rich in dietary fibre typically show a low pH, and a predominance of anaerobic bacteria, such as bifidobacteria.

4 'Colonic foods' and microflora

In response to the consumer need for foods which are beneficial to health, the concept of functional foods has evolved. These foods target specific functions in the body in a positive way due to the presence of health enhancing ingredients (Roberfroid, 1995). Colonic foods are an example of such functional foods that target the large intestine. These are foods that contain an ingredient that does not undergo significant modification during transit through the small intestine, but reach the colon where they are utilised by the resident bacteria producing metabolites that influence the physiological and biochemical processes in a beneficial manner. Recently, the terms probiotics, prebiotics and synbiotics have been used to describe various colonic foods (Gibson and Roberfroid, 1995).

Dietary fibre was one of the earliest food components shown to influence the total fermentable load entering the colon (Jenkins *et al.*, 1986). It has been suggested to play a beneficial role in a number of metabolic disorders including hyperlipidemia, diabetes, chronic renal failure, diverticular diseases and cancer (Bingham, 1990; Chen *et al.*, 1979; Jenkins *et al.*, 1976; Rampton *et al.*, 1984; Trowell, 1976). Dietary fibre from various plant sources influences colonic metabolism and bowel habit differently (Jenkins *et al.*, 1987). Insoluble dietary fibres, such as cellulose and lignin, are mainly responsible for faecal bulking. Soluble fibres, including pectin and guar gum, are readily fermented by intestinal bacteria thereby contributing little to faecal-bulking but being responsible for other metabolic effects of fibre (Cummings, 1984). Some of the effects may result from colonic fermentation, the nature of which may in turn depend on the type of linkage between individual sugar monomer components (Salyers *et al.*, 1978). The main products of bacterial fermentation of dietary fibre are organic acids and gases including hydrogen, carbon dioxide, and in some individuals, methane. Short chain fatty acids (SCFA) constitute the predominant organic acids (Cummings, 1981). Lactic acid, produced by lactobacilli and bifidobacteria, is present in high concentrations in infants. In adults, lactate is rapidly converted to propionate and acetate by other intestinal bacteria such as the propionibacteria (Wolin and Miller, 1983). Besides acting as a source of energy for intestinal mucosal cells (Roediger, 1980, Kripke *et al.*, 1989), SCFA have been claimed to reduce serum lipids (Ultrich, 1987) and improve carbohydrate metabolism (Wolever *et al.*, 1991). A major interest has been the reduction in pH of colonic contents resulting from the generation of SCFA as such a low faecal pH has been associated with reduced risk of colonic cancer (Samelson *et al.*, 1985, Walker *et al.*, 1986).

Ingesting high insoluble and soluble fibre diets for a period of 15 weeks has been shown to significantly lower faecal pH and the number of total aerobes. Faecal anaerobes and bifidobacterial counts increased significantly in two weeks but returned to their base levels at week 15. Similarly, the logarithmic ratio of anaerobes, to aerobes used as an index of fermentation activity, was increased significantly at the end of 15 weeks (Rao *et al.*, 1994). These data suggested the possibility that the intestinal microflora can be altered through intake of dietary fibre.

In addition to dietary fibre, resistant starch also reaches the large intestine and is fermented by resident bacteria (Asp, 1997). In a four week feeding study (Hylla *et al.*, 1998), it was shown that resistant starch increased faecal wet and dry weights significantly compared to a low starch intake. Furthermore, during the high resistant starch regime, bacterial ß-glucosidase activity and concentrations of total and secondary bile acids were significantly reduced. These results suggested that resistant starch had potentially important effects on bacterial metabolism in the human colon that may be relevant for cancer prevention. There is also an indication that resistant starch may be a good source of butyrate making it potentially important for colonic health (Christl *et al.*, 1997). In an animal feeding study, resistant starch was shown to stimulate bifidobacteria, lactobacilli and the production of SCFA. However, in healthy men who supplemented their habitual diet for one week with 32 g resistant starch, there was no effect on putative risk factors for colon cancer when compared to glucose (Heijnen *et al.*, 1998).

Another group of carbohydrates not digested in the human small intestine, but fermented in the colon, are the oligosaccharides. Depending upon the type and sequence of the monosaccharide units they are resistant to hydrolysis by digestive enzymes in the small intestine and reach the colon intact. These non-digestible oligosccharides can include in their chain glucose, fructose, xylose and galactose (Roberfroid *et al.*, 1998). One of the most common are the inulin-type fructans. These fructooligosaccharides (FOS) are a mixture of oligosaccharides containing glucose linked to fructose units, which are present in many commonly consumed fruits and vegetables including bananas, tomatoes, garlic and onion (Spiegel *et al.*, 1994). It is estimated that the consumption level of FOS ranges from 806 mg per day (Spiegel *et al.*, 1994) to 10 g per day (van Loo *et al.*, 1995). Commercially, FOS is either synthesised from sucrose, through the action of fungal ß-fructofuranosidase, or extracted from chicory root in the form of inulin and hydrolysed under controlled conditions. Several human feeding studies have been conducted using FOS and are reviewed by Roberfroid *et al.*, (1998) and elsewhere in this book. Chicory FOS was shown to significantly modify the composition of the faecal microflora with bifidobacteria numbers showing the highest increase in both in vivo (Gibson *et al.*, 1995) and in vitro studies (Wang and Gibson, 1993, Gibson and Wang, 1994). Predominance of bifidobacteria in the large intestine of humans and animals is now recognised as being beneficial for the maintenance of good health and prevention of chronic diseases including cancer of the large bowel (Mitsuoka, 1990). In this context, FOS can constitute important colonic food ingredients that may markedly affect host health.

5   Colonic foods and experimental carcinogenesis

Epidemiological and experimental evidence suggests a strong relationship between diet, intestinal microflora and the incidence of colon cancer. The selective growth stimulation of bifidobacteria by FOS has generated interest in studying their role in colon carcinogenesis. Studies have demonstrated that bifidobacterial cultures, when administered to animals, enhanced their immune response (Sekine *et al.*, 1995), and inhibited colon, liver and mammary carcinogenesis in rats (Reddy and Rivenson, 1993, Singh *et al.*, 1997). In another study, oral administration of indigenous strains of bifidobacteria to CF1 mice induced intestinal colonisation and selective proliferation of these bacteria (Koo and Rao,1991). In the same study, when mice were treated with 1,2-dimethylhydrazine and maintained on a diet supplemented with neosugar at the rate of 5% (w/v) for 38 weeks they showed a significantly lower incidence of aberrant crypts and foci, the precursor lesions of colonic carcinogenesis, compared to mice treated with the carcinogen alone.

A recent study also demonstrated that feeding diets containing 10% (w/v) oligofructose and inulin to azoxymethane treated rats significantly inhibited the formation of aberrant crypt foci and their multiplication (Reddy *et al.*, 1997). In this work, inulin was shown to cause a greater degree of inhibition when compared to oligofructose. Another study has demonstrated an inhibitory effect on abberant crypt foci development by *Bifidobacterium longum* and inulin: there was also a synergistic effect of the synbiotic combination (Rowland *et al.*, 1998). The effect of ingesting FOS by healthy human subjects on faecal bifidobacteria and selected indexes of colonic carcinogenesis has been studied by Bouhnik *et al.*, (1996). They concluded that ingesting 12.5 g/day for 12 days significantly increased faecal bifidobacteria counts but did not alter faecal total anaerobes, pH, the activities of nitroreductase, azoreductase, ß-glucuronidase, and the concentrations of bile acids and neutral sterols. In another study, it was shown that the administration of inulin and oligofructose significantly suppressed the growth of transplanted tumours, TLT and EMT6 in mice (Taper *et al.*, 1997). These studies suggest:
1) stimulation of the immune system
2) inhibition of bile acid metabolism

3) suppression of the metabolic activity of bacterial enzymes that are responsible for the formation of potential carcinogenic compounds in the colon, as potential mechanisms of action of the non-digestible carbohydrates in reducing the incidence of colon cancer. Stimulation of butyrate production in the large gut may also have positive effects with respect to tumourigenesis (Scheppach *et al.*, 1995).

Conclusion

Evidence in support of the hypothesis that diet is an important contributory factor responsible for the causation of 'Western' diseases is compelling. Several components of the diet can, and do, influence the nature of intestinal microorganisms either beneficially or adversely. Amongst the beneficial ingredients are the non-digestible carbohydrates. They provide substrates for a 'healthy' fermentation by colonic bacteria producing metabolites, such as SCFA, that can have a positive effect in the gut. Of the multitude of bacterial species which colonise the large intestine, a predominance of bifidobacteria may now be considered as being relevant for the maintenance of good health and prevention of chronic diseases, including cancer of the large bowel. A group of prebiotic oligosaccharides, that include inulin and FOS, have been shown to selectively stimulate the growth and activity of bifidobacteria in the colon. Several human trials have demonstrated that these nondigestible carbohydrates can be tolerated at fairly high levels of ingestion without any adverse physiological effects. In addition to stimulating bifidobacteria they may also lower serum triglycerides, phospholipids and cholesterol and VLDL of rats (Fiordaliso *et al.*, 1995). Due to the importance of colonic diseases there is interest in developing 'functional' foods that are capable of maintaining a healthier colon. In this context, the concept of probiotics (containing beneficial microorganisms) and prebiotics (containing food ingredients that selectively support the growth of indigenous beneficial microorganisms) may have significance for prevention of large bowel cancers and other diseases.

Although considerable advancements have been made during the last decade in understanding the role of the probiotics and prebiotics in colonic health, further studies are required to understand the long term implications of ingesting these foods on the course of cancer and other degenerative diseases and to elucidate their mechanisms of action.

References

Ames, B.N., Shigenaga, M.K. and Hagan, T.M. (1993) Oxidants, antioxidants and the degenerative diseases of aging. *Proceedings of the National Academy of Sciences* **90**, 7915-22.

Ames, B.N., Gold, L.S. and Willet, W.C. (1995) Causes and prevention of cancer. *Proceedings of the National Academy of Sciences* **92**, 5258-65.

Asp, N.G. (1997) Resistant starch - an update on its physiological effects. *Advances in Experimental Biology* **427**, 201-10.

Bingham, S.A. (1990) Mechanisms and experimental and epidemiological evidence relating dietary fibre (non-starch polysaccharides) and starch to protection against large bowel cancer. *Proceedings of the Nutrition Society* **49**, 153-71.

Bingham, S.A. and Cummings, J.H. (1997) Diet and large bowel cancer. In *Diet, Nutrition and Chronic Disease. Lessons from Contrasting Worlds*. ed. Shetty, P. S and McPherson, K. John Wiley and Sons. New York.

Bouhnik, Y., Flourie, B., Riottot, M., Bisetti, N., Gailing, M.F., Guibert, A., Bornet F. and Rambaud, J.C. (1996) Effects of fructo-oligosaccharides ingestion on fecal bifidobacteria and selected metabolic indexes of colon carcinogenesis in healthy humans. Nutrition. *Cancer* **26**, 21-29.

Chen, W.J.L. and Anderson, J.W. (1979) Effects of plant fibres in decreasing plasma total cholesterol and increasing high density lipoprotein cholesterol. *Proceedings Society Experimental Biology Medicine* **162**, 310-13.

Christl, S.U., Katzenmaier, U., Hylla, S., Kasper, H. and Scheppach, W. (1997) In vitro fermentation of high-amylose cornstarch by a mixed population of colonic bacteria. *Journal of Parenteral Eteral Nutrition* **21**, 290-95.

Cummings, J.H. (1981) Short chain fatty acids in the human colon. *Gut* **22**, 763-79.

Cummings, J.H. (1984) Microbial digestion of complex carbohydrates in man. *Proceedings Nutrition Society* **43**, 35-44.

Finegold, S.M, Attebery, H.R. and Sutter, V.L. (1974) Effect of diet on human fecal flora, comparison of Japanese and American diets. *American Journal Clinical Nutrition* **27**, 1456-69.

Finegold, S.M., Flora, D.J., Attebery, H.R and Sutter, V.L. (1975) Fecal bacteriology of colonic polyp patients and control patients. *Cancer Research* **35**, 3407-17.

Fiordaliso, M., Kok, N., Desager, J.P., Goethals, F., Deboyser, D., Roberfroid, M. and Delzenne, N. (1995) Dietary oligofructose lower triglycerides, phospholipids and cholesterol in serum and very low density lipoproteins of rats. *Lipids* **30**, 163-67.

Gibson, G.R. and Wang, X. (1994) Enrichment of bifidobacteria from human gut contents by oligofructose using continuous culture. *FEMS Microbiology Letters* **118**, 121-27.

Gibson, G.R., Beatty, E.R., Wang, X. and Cummings, J.H. (1995) Selective stimulation of bifidobacteria in the human colon by oligofructose and inulin. *Gastroenterology* **108**, 975-82.

Gibson, G.R. and Roberfroid, M.B. (1995) Dietary modulation of the human colonic microbiota: introducing the concept of prebiotics. *Journal of Nutrition* **125**, 1401-12.

Heijnen, M.L, van Amelsvoort, J.M., Deurenberg, P. and Beynen, A.C. (1998) Limited effect of consumption of uncooked (RS2) or retrograded (RS3) resistant starch on putative risk factors for colon cancer in healthy men. *American Journal of Clinical Nutrition* **67**, 322-31.

Hill M.J, Drasar, B.S, Hawksworth, G., Aries, V., Crowther, J.S. and Williams, R.E. (1971) Bacteria and aetiology of cancer of the large bowel. *Lancet* **I**, 95-100.

Hill, M.L., Drasar, B.S. and Williams, R.E. (1975) Faecal bile acids and clostridia in patients with cancer of the large bowel. *Lancet* **I**, 535-39.

Hylla, S., Gostner, A., Dusel, G., Anger, H., Bartram, H.P., Christl, S.U, Kasper, H. and Scheppach, W. (1998) Effects of resistant starch on the colon in healthy volunteers: possible implications for cancer prevention. *American Journal of Clinical Nutrition* **67**, 136-42.

Jenkins, D.J.A., Goff, D.V., Leeds, A.R., Alberti, K.G., Wolever, T.M., Gassulol, M.A. and Hockaday, T.D. (1976) Unabsorbed carbohydrates and diabetes: decreased post-prandial hyperglycaemia. *Lancet* **2**, 172-74.

Jenkins, D.J., Jenkins, A.L. and Wolever, T.M. (1986) Fiber and starchy foods: gut function and implications in disease. *American Journal Gastroentrology* **81**, 920-30.

Jenkins, D.J., Jenkins, A.L., Rao, A.V. and Thompson, L.U. (1987) Starchy foods, type of fiber, and cancer risk. *Preventative Medicine* **16**, 545-53.

Koo, M. and Rao, A.V. (1991) Longterm effect of bifidobacteria and neosugar on precursor lesions of colonic cancer in CF1 mice. Nutrition. *Cancer* **16**, 249-57.

Kripke, S.A., Fox, A.D., Berman, J.M., Settle, R.G. and Rombeau, J.L. (1989) Simulation of intestinal mucosal growth with intracolonic infusion of short-chain fatty acids. *Journal of Parenteral and Enteral Nutrition* **13**, 109-16.

Mitsuoka, T. (1990) Bifidobacteria and their role in human health. *Journal of Industrial Microbiology* **6**, 263-68.

Potter, J.D. (1995) Risk factors for colon neoplasia - Epidemiology and biology. *European Journal of Cancer* **31**, 1033-38.

Potter, J.D. (1996) Nutrition and colorectal cancer. *Cancer Causes and Control* **7**, 127-46.

Rampton, D.S., Cohen, S.L., Crammond, V.B., Gibbons, J., Lilburn, M.F, Rabet, J.Y., Vince, A.J., Wager, J.D. and Wrong, O.M. (1984) Treatment of chronic renal failure with dietary fiber. *Clinical Nephrology* **21**, 159-63.

Rao, A.V., Shiwnarain, N., Koo, M. and Jenkins, D.J.A. (1994) Effect of fiber-rich foods on the composition of intestinal microflora. *Nutritional Research* **14**, 523-35.

Reddy, B.S. and Rivenson, A. (1993) Inhibitory effect of *Bifidobacterium longum* on colon, mammary and liver carcinogenesis induced by 2-amino-3-methylimidazol[4,5-f]quinoline, a food mutagen. *Cancer Research* **53**, 3914-18.

Reddy, B.S., Simi, B. and Engle, A. (1994) Biochemical epidemiology of colon cancer: effect of types of dietary fiber on colonic diacylglycerols in women. *Gastroenterology* **106**, 883-89.

Reddy, B.S., Hamid, R. and Rao, C.V. (1997) Effect of dietary oligofructose and inulin on colonic preneoplastic aberrant crypt foci inhibition. *Carcinogenesis* **18**, 1371-74.

Roediger, W.E.W. (1980) The colonic epithelium in ulcerative colitis: An energy-deficiency disease? *Lancet* **2**, 712-15.

Roberfroid, M.B. (1995) A. non-digestible oligosaccharides and prebiotics: two new concepts in nutrition *Proceedings First ORAFTI Research Conference* pp. 9-23.

Roberfroid, M.B., van Loo, J.A.E. and Gibson, G. R. (1998) The bifidogenic nature of chicory inulin and its hydrolysis products. *Journal of Nutrition* **128**, 11-19.

Rowland, I.R., Rumney, C.J., Coutts, J.T. and Lievense, L.C. (1998) Effect of *Bifidobacterium longum* and inulin in gut bacterial metabolism and carcinogen-induced aberrant crypt foci in rats. *Carcinogenesis* **19**, 281-85.

Salyers, A.A., Palmer, J.K. and Wilkins, T.D. (1978) Degradation of polysaccharides by intestinal bacterial enzymes. *American Journal of Clinical Nutrition* **31**, 128-30.

Samelson, S.L., Nelson, R.L. and Nyhus, L.M. (1985) Protective role of faecal pH in experimental colon carcinogenesis. *Journal of Royal Society of Medicine* **78**, 230-33.

Scheppach, W., Bartram, J.K. and Richter, F. (1995) Role of short chain fatty acids in the prevention of colorectal cancer. *European Journal of Cancer* **31**, 1077-80.

Sekine, K., Ohta, J., Onishi, M., Tatsuki, T., Shimokawa, Y., Toida, T., Kawashima, T. and Hashimoto, Y. (1995) Analysis of antitumor properties of effector cells stimulated with a cell wall preparation (WPG) of *Bifidobacterium infantis*. *Biological Pharmaceutical Bulletin* **18**, 148-53.

Singh, J., Riverson, A., Tomita, M., Shimamura, S., Ishibashi, N. and Reddy, B.S. (1997) *Bifidobacterium longum*, a lactic acid-producing intestinal microflora inhibits colon cancer and modulates the intermediate biomarkers of colon carcinogenesis. *Carcinogenesis* **18**, 833-41.

Spiegel, J.E., Rose, R., Karabell, P. and Frankos, V.H. Schmitt DF. (1994) Safety and benefits of fructooligosaccharides as food ingredients. *Food Technology* **48**, 85-89.

Taper, H.S., Delzenne, N.M. and Roberfroid, M.B. (1997) Growth inhibition of transplantable mouse tumors by non-digestible carbohydrates. *International Journal of Cancer* **71**, 1109-12.

The American Institute of Cancer Research. Washington, (1997) Food, nutrition and the prevention of cancer: A global perspective. 35-52.

Trowell, H. (1976) Definition of dietary fiber and hypotheses that it is a protective factor in certain diseases. *American Journal of Clinical Nutrition* **29**, 417-27.

Ultrich, I.H. (1987) Evaluation of high-fiber diet in hyperlipidemia: A review. *Journal of American Nutrition* **6**, 19-25.

van Loo, J., Coussement, P., de Leenheer, L., Hoebregs, H. and Smits, G. (1995) On the presence of inulin and oligofructose as natural ingredients in the Western diet. *Critical Reviews in Food Science and Nutrition* **35**, 525-52.

Walker, A.R., Walker, B.F. and Walker, A.J (1986) Faecal pH, dietary fibre intake, and proneness to colon cancer in four South African populations. *Br. J. Cancer* **53**, 489-95.

Wang, X. and Gibson, G.R. (1993) Effects of the in vitro fermentation of oligofructose and inulin by bacteria growing in the human large intestine. *Journal of Applied Bacteriol* **75**, 373-80.

Wiseman, H. (1996) Dietary influences on membrane function: importance in protection against oxidative damage and disease. *Journal of Nutritional Biochemistry* **7**, 2-15.

Wolever, T.M.S., Spadafora, P. and Eshuis, H. (1991) Interaction between colonic acetate and propionate in humans. *American Journal of Clinical Nutrition* **53**, 681-87.

Wolin, M.J. and Miller, T.L. (1983) Carbohydrate fermentation. In Hentges D.J. (ed.) *Human Intestinal Microflora in Health and Disease*. New York, Academic Press, 147-65.

CHAPTER 17

Gastrointestinal Infections

ERIKA ISOLAURI[1], HEIKKI ARVILOMMI[2]
AND SEPPO SALMINEN[3]
[1]*Department of Paediatrics,University of Turku, 20520 Turku, FINLAND*
[2]*National Public Health Institute, Turku, FINLAND*
[3]*Department of Biochemistry and Food Chemistry, University of Turku, FINLAND*

1   Introduction

Gastrointestinal infection has been estimated to cause significant mortality in African, Asian and Latin American infants and young children. In populations with good nutrition and sanitary conditions, death rates are lower, but infectious diarrhoea remains a major world-wide cause of illness in early childhood. Acute infections of the gastrointestinal tract are usually self-limiting diseases. Diarrhoea begins suddenly, accompanied usually by vomiting and fever. The principal pathogens are viruses, such as rotavirus and adenovirus; and enteroinvasive bacteria such as shigellae, salmonellae, *Yersinia enterocolitica, Campylobacter jejuni, Escherichia coli*; *Vibrio cholerae* and *Clostridium perfringens* and a number of protozoa such as *Giardia lamblia* and *Entamoeba histolytica*.

Rotavirus is recognised as the leading cause of severe gastro-enteritis in children (Steinhoff, 1980). Due to improved diagnostic tools, data are also accumulating on other viruses such as intestinal adenoviruses, astroviruses, caliciviruses, Norwalk and Norwalk-like viruses and atypical rotaviruses as causes of acute gastro-enteritis. Bacterial aetiology of acute gastro-enteritis during childhood can be established in 10-15% of intestinal infections, and their incidence rises in warm climates and under poor sanitary conditions. The most commonly detected bacterial pathogens are *Yersinia enterocolitica* and *Campylobacter jejuni, Escherichia coli*, whether toxigenic, invasive or of pathogenic serotype, with *Salmonella* spp. and *Shigella* spp. appearing infrequently. Enterotoxigenic *Escherichia coli* (ETET) are the most commonly identified pathogens in travellers diarrhoea, while enteropathogenic *Escherichia coli* (EPEC) strains have been responsible for infantile summer diarrhoea and nursery epidemics. Cholera is

G.R. Gibson and M.B. Roberfroid (eds.), Colonic Microbiota, Nutrition and Health, 267-279.
© 1999 *Kluwer Academic Publishers. Printed in the Netherlands.*

a problem in Asia. Chronic infections in the gastrointestinal tract are less frequent. Antibiotic-associated diarrhoea is usually due to the activities of toxin producing bacteria such as *Clostridium difficile*.

Most microbes gain access to the body via the mucosae. Continuous exposure to potentially pathogenic antigens imposes high demands on the mucosal surfaces of the body, including the respiratory, gastrointestinal and urogenital tracts. Mucosae of the gastrointestinal tract comprise an important organ of defence, providing a well functioning barrier against antigen challenge, from diet and microorganisms (Sanderson and Walker, 1993).

2   Gut mucosal defence against gastrointestinal infections

The gastrointestinal barrier controls antigen transport in the gut. In order to establish an infection, the pathogen must circumvent an impressive array of intestinal mucosal defences (Table 1). This defence depends on a number of factors in both intestinal lumen and mucosa which restrict colonisation by pathogenic bacteria in the gut, and interfere with the adherence of microorganisms to the mucosal surface (Sanderson and Walker, 1993; Schreiber and Walker, 1988).

During postnatal development, major maturational events occur in the gut defence barrier, such as the appearance of mucosal proteins, digestive enzymes and development of the intestinal flora. Gastric acidity is an important defence against microorganisms. Secretion of hydrochloric acid by the gastric mucosa develops slowly during the first few weeks of life, reaching the lower limit of normal for adults by three months of age (Grand *et al.*, 1976). Goblet cell mucus covering the epithelial surface of the gastrointestinal tract is an important physical barrier that interferes with intestinal attachment of luminal antigens and microorganisms (Schreiber and Walker, 1988). Low proteolytic activity and lower mucous coat binding of antigens may partly explain enhanced absorption of intact macromolecules and antibodies during the neonatal period (Grand *et al.*, 1976). Maturational changes are also known to affect epithelial cell membranes. Studies in experimental animals have demonstrated greater microvillous membrane binding of cows milk protein and cholera toxin postnatal than in adult animals (Schreiber and Walker, 1988).

Lysosomal degradation also provides a significant barrier to the passage of intact macromolecules across cells. Lysosomes contain acid hydrolases, such as lipases, proteases, nucleases and carbohydrases that break down all types of macromolecules. Breakdown in the immature intestine is less effective than in later life (Sanderson and Walker, 1993; Schreiber and Walker, 1988).

Normal bacterial and viral populations can prevent overgrowth of potential pathogens in the gastrointestinal tract. Germ-free mice have been shown to exhibit larger weight losses and more marked and prolonged enhancement of intestinal permeability after rotavirus infection than rotavirus-infected conventional mice (Heyman *et al.*, 1987). Microbial colonisation begins immediately after birth. The maternal intestinal flora is a source of bacteria which colonise the newborn intestine. Colonisation is also determined by contact with the surroundings. However, the development of intestinal microflora is a gradual process (reviewed in Alderberth, 1998). Consequently, some of these host defence functions are not fully established during infancy, which may make infants vulnerable to infection. Breast feeding encourages the growth of bifidobacteria, which are thought to be favourable for preventing gastrointestinal infection.

The demonstration that during absence of the intestinal microflora, antigen transport is increased indicates that gut microorganisms are important constituents of the intestinal defence barrier. Moreover, bacterial colonisation provides maturational signals for the gut-associated lymphoid tissue (Shroff *et al.*, 1995).

The intestine is the largest lymphoid organ in the body. Hence, the gut-associated lymphoid tissue comprises an important element of the full immunological capacity of the host (Brandtzaeg, 1995). Regulatory events of the intestinal immune response take place in organised lymphoepithelial tissue and secretory sites. Firstly, there are organised lymphoid tissues which comprise Peyer's patches. Secondly, lymphocytes and plasma cells are distributed throughout the lamina propria. Thirdly, intraepithelial lymphocytes are located above the basal lamina, in the intestinal epithelium. Peyer's patches play an essential role in intestinal immune function. These aggregations of lymphoid follicles are covered by a unique epithelium comprising cuboidal epithelial cells, very few goblet cells and specialised antigen sampling cells, the M cells. Antigen transport across this epithelium is characterised by rapid uptake and reduced degradation (Ducroc *et al.*, 1983).

While blood-borne and tissue immunity has a predominance of IgG antibodies compared to IgA and IgM, IgA antibody production is abundant at mucosal surfaces. IgG-, IgM- and IgE-secreting cells also function albeit at a significantly lower capacity. Furthermore, in contrast to IgA in serum, secretory IgA is present in dimeric or polymeric form. Secretory IgA antibodies in the gut are part of the common mucosal immune system which includes respiratory tract and lacrimal, salivary and mammary glands. Consequently, an immune response initiated in the gut-associated lymphoid tissue can affect immune response at other mucosal surfaces (Brandtzaeg, 1995). Lymphocytes activated within the Peyer's patches disseminate via mesenteric lymph nodes, thoracic ducts and blood back to the lamina propria, but also traffic between other secretory tissues.

3   Immune response to gastrointestinal infection

The immune response to intestinal infection is diverse, and includes cellular components such as B- and T- lymphocytes, dendritic cells and epithelial cells, together with secreted products like antibodies and cytokines. The exact role of the components in combating infection and providing protection against future attacks is not entirely clear. However, secretory immunoglobulins, particularly IgA, appear crucial in maintaining the defence. Secreted by plasma cells in lamina propria, dimeric IgA and pentameric IgM are actively transported with the aid of polymeric Ig receptor through epithelial cells into the gut lumen, where it functions as an immunological barrier (Table 1). This occurs by preventing the adherence of bacteria and viruses to epithelial cells. In addition to a lumenal barrier function, in vitro studies suggest that dimeric IgA could also clean microbes from the lamina propria as excreted complexes through epithelial cells. Furthermore, viruses inside the epithelial cells may be neutralised by IgA in transport (Mazenac *et al.*, 1993). Whatever the exact mechanisms by which secretory IgA works, it alone, without other arms of immunity, can protect against challenge with enteric pathogens including species of salmonellae (Michetti *et al.*, 1992).

Table 1. Factors controlling antigen transport in the gut (Schreiber and Walker, 1998)

Non-immunological	Immunological
Saliva	Secretory antibodies
Gastric acid	Peyer's patches
Intestinal microflora	Cells in lamina propria
Intestinal proteolysis	Intraepithelial lymphocytes
Peristalsis	
Mucus	
Epithelial membrane	

Observing and dissecting the local gut immune response in patients with enteric infection has been difficult for obvious reasons. The majority of available data are derived from studies on B cell response to enteric infections, measured as circulating antibody-secreting cells (Kantele and Takamen, 1988; Kaila *et al.*, 1992; Orr *et al.*, 1992). Similar relevant data have been obtained from research on oral vaccines. The lymphocyte maturation cycle involves antigen transport across M cells and presentation of antigens to helper T cells, which proliferate and induce B cells to respond by switching from IgM surface bearing to IgA. The specific antibody-secreting lymphocytes appear in peripheral blood 2-4 days after antigen exposure, reach a maximum after 6-8 days and persist for 2-3 weeks. The potential for these cells to home in the gut is clear. Migration of lymphocytes into tissues is targeted by homing receptors on that interact with their ligands on endothelial cells. In diarrhoea patients (Kantele *et al.*, 1996) or orally vaccinated volunteers (Quiding-Jarbrink *et al.*, 1995; Kantele *et al.*, 1997) it has been shown that all cells secreting specific antibody expressed $\alpha 4\beta 7$, the gut homing receptor. By contrast, parenteral vaccination induced cells, only half of which exhibited these receptors. This is in line with the notion that the majority of circulating antibody secreting cells induced in the gut are on their way back to gut mucosae.

Cells that are induced by enteric infection or oral vaccination predominantly secrete IgA as would be expected. There are substantial numbers of cells secreting IgM and, more strikingly, IgG. No active transport system across the epithelium is known for IgG.

However, IgG-secreting cells have similar homing properties to IgA-cells (Kantele *et al.*, 1996). Another, perhaps unexpected, feature of the response is that, in addition to cells secreting pathogen-specific antibodies, there are large amounts of non-specific cells. This may be explained by polyclonal stimulation of B cells, but the biological purpose is unclear.

Unlike studies on B cells as described above, investigations on T cell responses to intestinal infections are rare. Levine's group has demonstrated cytotoxic T cells following oral *Salmonella* vaccine and suggests that this response may be important in protection during typhoid fever (Sztein *et al.*, 1995). The role of intraepithelial T lymphocytes in human intestinal infections is unknown.

In understanding how probiotics might amplify the immune response in the gut, it may be relevant to consider the role of epithelial cells and cytokine network. Epithelial cells seem to express a variety of cytokine receptors (Reinecker and Podolsky, 1995) and also produce cytokines (Brandtzaeg, 1995). These properties together with the capacity to communicate with intraepithelial lymphocytes via MHC molecules - both classical and non-classical - make epithelial cells probably important in sensing the gut microbiota and infiltration.

4   Pathomechanisms of gastrointestinal infections

Most pathogens disturb the intestinal barrier function and cause diarrhoea either by invading the mucosa or by producing an enterotoxin in the lumen. The capacity of the pathogen to bind to the intestinal surface is an important determinant of its virulence. Attachment to the mucosa enables the microorganism to resist gut defence mechanisms, such as peristaltic clearing (Table 1). Adherence is mediated by bacterial surface binding to cell receptors. Bacterial adhesins are associated with projections on their surface, pili or fimbriae.

Rotaviruses invade the highly differentiated absorptive columnar cells of the small intestinal epithelium, where they replicate causing defective sodium and chloride transport. The invasion results in partial disruption of the intestinal mucosa with loss of microvilli and decrease in the villus/crypt ratio. Diarrhoea is mainly due to failure of the epithelium to differentiate during rapid migration to repair the disruption.

Bacteria causing infection are classified according to whether they secrete an entero-toxin or invade the bowel wall. Enteroinvasive bacteria, such as EIEC and shigellae, are capable of epithelial penetration and replication. Enterotoxigenic *Escherichia coli* do not invade enterocytes nor replicate in the lamina propria, but colonise the mucosal surface and elaborate enterotoxin. This results in net fluid and electrolyte secretion into the lumen of the small intestine: coupled sodium and chloride transport across the microvillus membrane is decreased and chloride secretion by crypt cells increased. The pathogenic mechanism is consequently similar to secretory diarrhoea induced by *Vibrio cholerae*.

Clostridium difficile is the aetiologic agent of pseudomembranous colitis, one of the common causes of antibiotic associated diarrhoea and the most common cause of nosocomial diarrhoea. Infection is commonly associated with prior or concomitant antimicrobial therapy, which disrupts the gut barrier function, enabling *Clostridium difficile* to establish more effectively and produce toxin.

The main clinical manifestations of acute gastrointestinal infection are due to water and electrolyte losses. Characteristics of rotavirus infection include watery diarrhoea, fever and vomiting, with resulting dehydration often severe enough to warrant hospitalisation. Rotavirus causes approximately half of the hospital cases of childhood diarrhoea during the year, and more that 80 % of cases in winter months, *i.e.* the epidemic season of rotavirus infection (Steinhoff, 1980). Some infectious agents, particularly invasive bacteria, are capable of causing dysenteric forms of diarrhoea, with passage of scanty stools containing mucus and blood. Dysentery may be accompanied by malaise and toxaemia. Diarrhoeal deaths are mainly due to severe dehydration from persistent diarrhoea. The current accepted guidelines for treatment are based on correcting the dehydration by oral rehydration solutions. In addition, immediately after completion of oral rehydration, full feeding of a previously tolerated diet can be reintroduced (Sandhu *et al.*, 1997). There is no need for elimination or restriction of milk or milk-based products.

During infection the microecological balance is disturbed and microorganisms adhering to the epithelium override the gut defence barrier (Table 1). Gastrointestinal infection results in a partial disruption of the intestinal mucosa with increased intestinal permeability, which leads to aberrant absorption of intraluminal antigens (Jalonen *et al.*, 1991). Moreover, disturbed microbial balance has been demonstrated during gastrointestinal infection (Isolauri *et al.*, 1994). These changes may be identified as novel targets for the management of gastrointestinal infection (Fig. 1).

5   Specific dietary foods for the treatment and prevention of gastrointestinal infections

The demonstration that the gut microbiota is an important constituent in the intestinal mucosal barrier has led to the development of a range of novel therapeutic interventions. One strategy is based on the consumption of mono- and mixed cultures of beneficial live microorganisms as probiotics. A probiotic has been defined as a live microbial feed supplement which beneficially affects the host animal by improving its intestinal microbial balance (Fuller, 1991). Some important criteria for a probiotic include the strain being of human origin, safe for human use, stable to acid and bile, able to adhere to the intestinal mucosa and produce antimicrobial components.

The beneficial effects of lactobacilli have been attributed to their ability to suppress the growth of pathogens, probably by secretion of antibacterial substances like lactic acid, peroxide, and bacteriocins. Amongst the possible mechanisms is promotion of the immunological and nonimmunological defence barrier in the gut (Fig. 1): normalisation of increased intestinal permeability (Isolauri *et al.*, 1993) and gut microflora interactions (Isolauri *et al.*, 1994). Another explanation for the gut-stabilising effect could be improvement of the intestine's immunological barrier, particularly intestinal IgA responses (Kaila *et al.*, 1992).

Several clinical studies have investigated the use of probiotics as dietary supplements for the prevention and treatment of certain gastrointestinal infections (Salminen *et al.*, 1998). Early results have not been consistent however, possibly due to the variety of preparations used, methodological problems and varying bacterial strains used in fermented products. Therefore, it is difficult to draw definitive conclusions on the precise contribution of probiotic bacteria to clinical results obtained, since test and placebo preparations often differ vastly in physical and chemical properties. Moreover, the strains have not always been well defined. However, more recently conducted studies can be used for a more accurate assessment of probiotic efficacy.

Recent work has indicated that probiotic bacteria can have a beneficial effect on the clinical course of rotavirus diarrhoea. Among the potentially successful therapeutic organisms have been *Lactobacillus acidophilus* NCFB 1748, *Lactobacillus acidophilus* LA1 (recently reclassified as *Lactobacillus johnsonii*), *Lactobacillus* GG (ATCC 53103) and *Lactobacillus casei* Shirota.

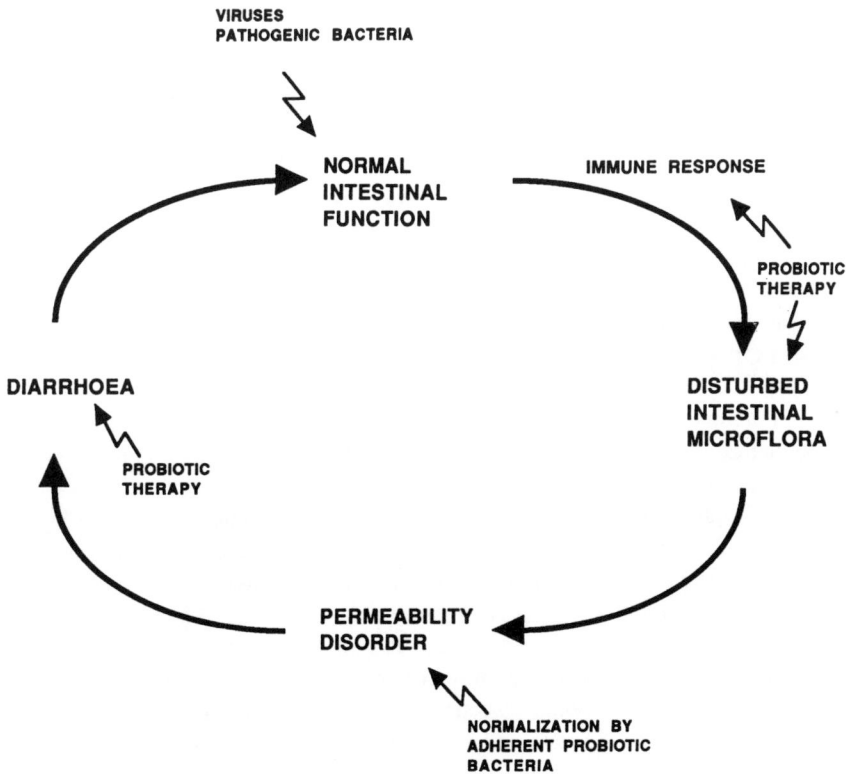

Figure 1. Targets for probiotics in gastrointestinal infection.

In patients hospitalised for acute rotavirus diarrhoea, *Lactobacillus* strain GG as a fermented milk or freeze-dried powder significantly reduced the duration of diarrhoea compared to a placebo group given pasteurised yoghurt (Isolauri *et al.*, 1991). This result has been confirmed in studies carried out in a similar population (Kaila *et al.*, 1992; Majamaa *et al.*, 1995) as well as in different populations (Pant *et al.*, 1996; Shornikova *et al.*, 1997; Canani *et al.*, 1997). Moreover, *Lactobacillus* GG was shown to reduce the duration of rotaviral excretion (Canani *et al.*, 1997). A reduction in the number of diarrhoeal episodes and the duration of diarrhoea was observed in infants given *Lactobacillus helveticus* and *Streptococcus thermophilus*-fermented formula (Brunser *et al.*, 1989) or *Lactobacillus acidophilus* and *Lactobacillus casei*-fermented milk (Gonzalez *et al.*, 1980) compared to a group given non-fermented formula or milk. *Enterococcus faecium* is one of the enterococcal species found in the faeces of healthy adults. *Enterococcus faecium*, strain 68 (SF68), has been investigated for the treatment of acute enteritis and other gut disorders (Mitra and Rabbani, 1990).

More recently, Saavedra *et al.* (1994) conducted a double-blind, placebo-controlled trial in hospitalised infants who were randomised to receive a standard infant formula or the same formula supplemented with *Bifidobacterium bifidum* and *Streptococcus thermophilus*. Over a 17-month follow-up, 31% of the patients given the standard infant formula, but only 7% of those receiving the probiotic-supplemented formula developed diarrhoea. The prevalence of rotavirus shedding was significantly lower in those receiving the probiotic.

In a study by Sepp *et al.* (1995) the clinical course and quantitative composition of faecal microflora was followed in 31 children treated for shigellosis using antimicrobial preparations with or without a probiotic *Lactobacillus* strain GG. In 6 children, *Lactobacillus* GG was administered alone, in 13 children with trimetoprim sulfamethoxazole and 12 children received trimetoprim sulfamethoxazole alone. By day 10, significantly more children had no diarrhoea in the probiotic treated groups when compared to the trimetoprim sulfamethoxazole-treated children

Several preparations have been tested for prevention of antibiotic-associated diarrhoea (Table 2). Significant positive effects have been reported for *Saccharomyces boulardii*, *Lactobacillus* GG and some bifidobacteria. Evidence from the available studies indicates that probiotic fermented milks may be of some value in the prevention of antibiotic associated diarrhoea.

The value of probiotic preparations in prophylaxis for traveller's diarrhoea has been studied using *Lactobacillus acidophilus*, *Bifidobacterium bifidum*, *Lactobacillus bulgaricus* and *Streptococcus thermophilus*, but the results have been conflicting (see Table 3 and references therein). This is not surprising in view of differences in probiotic species and vehicle used, the dosage schedule, as well as in travel destinations for which the studies were conducted. It may be however that there is evidence that some strains of lactic acid bacteria may provide protection against traveller's diarrhoea. Future studies should provide dose-response data and information on the aetiological agents causing diarrhoea in the destinations studied.

Table 2. Prevention of antibiotic-associated diarrhoea by probiotic bacteria: double-blind, placebo-controlled studies

Preparation Antibiotic	Number of subjects (reference)	test / control	Clinical effect
Lactobacillus acidophilus + *Lactobacillus bulgaricus*	Neomycin (Clements *et al.*, 1983)	20 / 19 16 / 16	No effect No effect
Bifidobacterium longum	Erythromycin (Colomel *et al.*, 1987)	10 / 10	Reduction in diarrhoea
Enterobacterium faecium SF68	several (Wunderlich *et al.*, 1989)	23 / 22	Reduction in diarrhoeal episodes
Lactobacillus GG yoghurt	Erythromycin (Siitonen *et al.*, 1990)	8 / 8	Reduction in diarrhoeal episodes
Lactobacillus acidophilus + *Lactobacillus bulgaricus*	Amoxicillin (Tankanow *et al.*, 1990)	30 / 30	No effect

6   Conclusions

There is now evidence that probiotic microorganisms have potential in the prevention and treatment of gastrointestinal infections. Best documented studies are in the treatment and prevention of infant diarrhoea with *Lactobacillus* GG and *Bifidobacterium animalis* preparations. Selected microbes, including *Saccharomyces boulardii*, some lactobacilli and bifidobacteria, may prevent antibiotic-associated diarrhoea and colitis related to *Clostridium difficile*. Several probiotic bacteria can influence the intestinal microflora by increasing numbers of bifidobacteria and lactic acid bacteria in faeces. Thus, a stabilising effect in the normal microflora therefore intestinal barrier mechanisms may result (Fig. 1).

There is, however, still a need for randomised placebo-controlled, double-blinded studies in human volunteers and in patients with specific intestinal infections. When the efficacy and safety, as well as mechanisms of action of candidate probiotic strains have been identified, biotherapeutic microbes may become important tools in maintaining balanced intestinal microflora. This could lead to effective gut barrier functions providing the basis of functional food, clinical food and pharmaceutical development through probiotics.

Table 3. The efficacy of probiotic preparations in preventing travellers diarrhoea: double-blind, placebo-controlled studies (adapted from Salminen et al., 1997)

Preparation	Number of subjects (reference)	Origin / destination	Clinical effect
Enterococcus faecium	81 (Black et al., 1989)	Denmark / Egypt	Protection rate 39.4 %
Lactobacillus	319 (Kollaritsch et al., 1990)	Austria / several	No effect
Enterococcus faecium	2144 (Kollaritsch et al., 1990)	Austria / several	No effect
Lactobacillus GG	331 (Oksanen et al., 1990)	Finland / Marmaris Turkey	No effect
Lactobacillus GG	146 (Oksanen et al., 1990)	Finland / Alanya Turkey	Protection rate 39.5 %
Saccharomyces boulardii	1016 (Kollaritsch et al., 1993)	Austria / several	Incidence reduced from 39.1 % to 28.7 %
Lactobacillus fermentum Lactobacillus acidophilus	282 (Katalaris et al., 1995)	UK / Belize	No effect
Lactobacillus GG	249 (Hilton et al., 1997)	USA / several	Protection rate 47 %

References

Adlerberth, I. (1998) The establishment of a normal intestinal microflora in the newborn infant. In: *Nestlé Nutrition Workshop: Probiotics, other Nutritional Factors and Intestinal Microflora.* Lippincott-Raven Publ: New York, in press.

Black, F.T., Andersen, P.L., Ørskov, F., Gaarslev, K. and Laulund, S. (1989) Prophylactic efficacy of lacto-bacilli on travellers diarrhea. *Travellers Medicine* **8**, 333-35.

Brandtzaeg, P. (1995) Molecular and cellular aspects of the secretory immunoglobulin system. *APMIS* **103**, 1-19.

Brunser, O., Araya, M, Espinoza, J., Guesry, P.R., Secretin, M.C. and Pacheco, I. (1989) Effect of an acidified milk on diarrhoea and the carrier state in infants of low socio-economic stratum. *Acta Paediatric Scandanavia* **78**, 259-64.

Canani, R.B., Albano, F., Spagnuolo, M.I., Di Benedetto, L., Stabile, A. and Guarino, A. (1997) Effect of oral administration of *Lactobacillus* GG on the duration of diarrhea and on rotavirus excretion in ambulatory children. *Journal of Pediatric Gastroenterology Nutrition* **24**, 469.

Clements, M.L., Levine, M.M., Ristaino, P.A., Daya, V.E. and Hughes, T.P. (1983) Exogenous lactobacilli fed to man - their fate and ability to prevent diarrheal disease. *Progress in Food and Nutrition Science* **7**, 29-37.

Colomel, J.F., Cortot, A. and Neut, C. (1987) Yoghurt with *Bifidobacterium longum* reduces erythromycin induced gastrointestinal effects. *Lancet* **2**, 43.

Ducroc, R., Heyman, M., Beaufrere, B., Morgat, J.L. and Desjeux, J.F. (1983) Horseradish peroxidase transport across rabbit jejunum and Peyer's patches in vitro. *American Journal Physiology*, **245**, G54-58.

Fuller, R. (1991) Probiotics in human medicine. *Gut* **32**, 439-42.

Gonzalez, S., Albarracin, G., Locascio de Ruiz Pesce, M., Male, M., Apella, M.C., Pesce de Ruiz Holgado, A. and Oliver, G. (1990) Prevention of infantile diarrhoea by fermented milk. *Microbiologie-Aliments-Nutrition* **8**, 349-54.

Grand, R.J., Watkins, J.B. and Torti, F.M. (1976) Development of the human gastrointestinal tract. *Gastroenterology* **70**, 790-810.

Heyman, M., Gorthier, G., Petit, A., Meslin, J-C, Moreau, C. and Desjeux, J-F. (1987) Intestinal absorption of macromolecules during viral enteritis: an experimental study on rotavirus-infected conventional and germ-free mice. *Pediatric Research* **22**, 72-78.

Hilton, E., Kolakowski, P., Smith, M. and Singer, C. (1997) Efficacy of *Lactobacillus* GG as a diarrhea preventative. *Journal of Travellers Medicine* **4**, 41-43.

Isolauri, E., Juntunen, M., Rautanen, T., Sillanaukee, P. and Koivula, T. (1991) A human *Lactobacillus* strain (*Lactobacillus* GG) promotes recovery from acute diarrhea in children. *Pediatrics* **88**, 90-97.

Isolauri, E., Kaila, M., Arvola, T., Majamaa, H., Rantala, I., Virtanen, E. and Arvilommi, H. (1993) Diet during rotavirus enteritis affects jejunal permeability to macromolecules in suckling rats. *Pediatric Research* **33**, 548-53.

Isolauri, E., Kaila, M., Mykkänen, H., Ling, W.H. and Salminen, S. (1994) Oral bacteriotherapy for viral gastro-enteritis. *Digestive Disease Science* **39**, 2595-600.

Jalonen, T., Isolauri, E., Heyman, M., Crain-Denoyelle, A.-M., Sillanaukee, P. and Koivula, T. (1991) Increased ß-lactoglobulin absorption during rotavirus enteritis in infants: relationship to sugar permeability. *Pediatric Research* **30**, 290-93.

Kaila, M., Isolauri, E., Soppi, E., Virtanen, E., Laine, S. and Arvilommi, H. (1992) Enhancement of the circulating antibody secreting cell response in human diarrhea by a human *Lactobacillus* strain. *Pediatric Research* **32**, 141-44.

Kaila, M., Isolauri, E., Virtanen, E. and Arvilommi, H. (1992) Preponderance of IgM from blood lymphocytes in response to infantile rotavirus gastro-enteritis. *Gut* **33**, 639-42.

Kantele, J.M., Arvilommi, H., Kontiainen, S., Salmi, M., Jalkanen, S., Savilahti, E., Westerholm, M. and Kantele, A. (1996) Mucosally activated circulating human B-cells in diarrhea express homing receptors directing them back to the gut. *Gastroenterology* **110**, 1061-67.

Kantele, A., Kantele, J.M., Savilahti, E., Westerholm, M., Arvilommi, H., Lazarovits, A., Butcher, E.C. and Mäkelä, P.H. (1997) Homing potentials of circulating lymphocytes in humans depend on the site of activation: oral, but not parenteral, typhoid vaccination induces circulating antibody-secreting cells that all bear homing receptors directing them to the gut. *Journal of Immunology* **158**, 574-79.

Kantele, A., Takanen, R. and Arvilommi, H. (1988) Immune response to acute diarrhea seen as circulating antibody secreting cells. *Journal of Infections Diseases* **158**, 1011-16.

Katalaris, P.H., Salam, I. and Farthing, M.J.G. (1995) Lactobacilli to prevent traveller's diarrhea? *New England Journal of Medicine* **333**, 1360-61.

Kollaritsch, H., Holst, H., Grobara, P. and Wiedermann, G. (1993) Prophylaxe der Reisediarrhö mit *Saccharomyces boulardii. Fortsch Medicine* **111**, 153-56.

Kollaritsch, H. and Wiedermann, G. (1990) Traveller's diarrhea among Austrian tourists: epidemiology, clinical features and attempts at nonantibiotic prophylaxis. In Pasini W. (ed.): *Proceedings of the Second International Conference on Tourist Health* WHO: Rimini., 74-82.

Majamaa, H., Isolauri, E., Saxelin, M. and Vesikari, T, (1995) Lactic acid bacteria in the treatment of acute rotavirus gastroenteritis. *Journal of Pediatric Gastroenterology Nutrition* **20**, 333-39.

Mazanec, M.B., Nedrud, J.G., Kaetzel, C.S. and Lamm, M.E. (1993) A three-tiered view of the role of IgA in mucosal defence. *Immunology Today* **14**, 430-35.

Michetti, P., Mahan, M.J., Slauch, J.M., Mekalanos, J.J. and Neutra, M.R. (1992) Monoclonal secretory immunoglobulin A protects mice against oral challenge with the invasive pathogen *Salmonella typhimurium. Infection and Immunology* **60**, 1786-92.

Mitra, A.G. and Rabbani, G.H. (1990) A double blind, controlled trial of Bioflorin (*Streptococcus faecium* S.F. 68) in adults with acute diarrhoea due to *Vibrio cholerae* and enterotoxigenic *Escherichia coli. Gastroenterology* **99**, 1149-52.

Oksanen, P., Salminen, S., Saxelin, M., *et al.* (1990) Prevention of traveller's diarrhea by *Lactobacillus* GG. *Annual Review of Medicine* **22**, 53-56.

Orr, N., Robin, G., Lowell, G. and Cohen, D. (1992) Presence of specific immunoglobulin A-secreting cells in peripheral blood after natural infection with *Shigella sonnei. Journal of Clinical Microbiology* **30**, 2165-68.

Pant, A.R., Graham, S.M., Allen, S.J., Harikul, S., Sabcharoen, A., Cuevas, L. and Hart, C.A. (1996) *Lactobacillus* GG and acute diarrhoea in young children in the tropics. *Journal of Tropical Pediatrics* **42**, 162-65.

Quiding-Järbrink, M., Lakew, M., Nordstrom, I., Banchereau, J., Butcher, E., Holmgren, J. and Czerkinsky, C. (1995) Human circulating specific antibody-forming cells after systemic and mucosal immunizations: differential homing commitments and cell surface differentiation markers. *European Journal of Immunology* **25**, 322-27.

Reinecker, H.C. and Podolsky, D.K. (1995) Human intestinal epithelial cells express functional cytokine receptors sharing the common gamma c chain of interleukin 2 receptor. *Proceedings National Academy of Science USA* **92**, 8353-57.

Saavedra, J.M., Bauman, N.A., Oung, I., Perman, J.A. and Yolken, R.H. (1994) Feeding of *Bifidobacterium bifidum* and *Streptococcus thermophilus* to infants in hospital for prevention of diarrhoea and shedding of rotavirus. *Lancet* **344**, 1046-49.

Salminen, S., Deighton, M., Benno, Y. and Gorbach, S.L. (1998) Lactic acid bacteria in the gut in normal and disordered states. In Salminen S. and von Wright A. (eds.): *Lactic Acid Bacteria: Functions and Technology.* Marcel Dekker Inc, New York, in press.

Sanderson, I.R. and Walker, W.A. (1993) Uptake and transport of macromolecules by the intestine: possible role in clinical disorders (an update) *Gastroenterology* **104**, 622-39.

Sandhu, B.K., Isolauri, E., Walker-Smith, J.A. *et al.* (1997) Early feeding in childhood gastroenteritis. A multicentre study on behalf of the European Society of Paediatric Gastroenterology and Nutrition (ESPGAN) working group on acute diarrhoea. *Journal of Pediatric Gastroenterology Nutrition* **24**, 522-27.

Schreiber, R.A. and Walker, W.A. (1988) The gastrointestinal barrier: antigen uptake and perinatal immunity. *Annual Allergy* **61**, 3-12.

Sepp, E., Tamm, E., Torm, S., Lutsar, I., Mikelsaar, M. and Salminen, S. (1995) Impact of a *Lactobacillus* probiotic on the fecal microflora in children with shigellosis. *Microecology Therapy* **23**, 74-80.

Shornikova, A.V., Isolauri, E., Burkanova, L., Lukovnikova, S. and Vesikari, T. (1997) A trial in the Karelian Republic of oral rehydration and *Lactobacillus* GG for treatment of acute diarrhoea. *Acta Paediatric* **86**, 460-65.

Shroff, K.E., Meslin, K. and Cebra, J.J. (1995) Commensal enteric bacteria engender a self-limiting humoral mucosal immune response while permanently colonizing the gut. *Infection and Immunology* **63**, 3904-13.

Siitonen, S., Vapaatalo, H., Salminen, S., Gordin, A., Saxelin, M., Wikberg, R., Kirkkola, A.L. (1990) Effect of *Lactobacillus* GG yoghurt in prevention of antibiotic associated diarrhea. *Annual Review of Medicine* **22**, 57-59.

Steinhoff, M.C. (1980) Rotavirus: the first five years. *Journal of Pediatrics* **96**, 611-22.

Sztein, M.B., Tanner, M.K., Polotsky, Y., Orenstein, J.M. and Levine, M.M. (1995) Cytotoxic T lymphocytes after oralimmunization with attenuated vaccine strains of *Salmonella typhi* in humans. *Journal of Immunology* **155**, 33987-93.

Tankanow, R.M., Ross, M.B., Ertel, I.J., Dickinson, D.G., McCormick, L.S. and Garfinkel, J.F. (1990) A double blind, placebo-controlled study of the efficacy of Lactinex in the prophylaxis of amoxycillin-induced diarrhea. *DICP* **24**, 382-84.

Wunderlich, P.F., Braun, L., Fumagalli, I., D'Apuzzo, V., Heim, F., Karly, M., Lodi, R., Politta, G., Vonbank F. and Zeltner, L. (1989) Double-blind report on the efficacy of lactic acid-producing *Enterococcus* SF68 in the prevention of antibiotic-associated diarrhoea and in the treatment of acute diarrhoea. *Journal of International Medical Research* **17**, 333-38.

CHAPTER 18

Probiotics in Consumer Products

R. KORPELA AND M. SAXELIN
Valio Ltd, P.O. Box 30, FIN-00039 Valio, FINLAND

1   Introduction

The **probiotic** approach towards an improved microflora composition involves adding live microorganisms to the gastrointestinal tract whilst **prebiotics** enhance certain components of the existing flora. This chapter will focus on the former in the context of consumer products. The effects of orally-administrated lactic acid bacteria (*e.g.* fermented milk products) on the gastrointestinal ecosystem were extensively researched at the beginning of this century by Professor Metchnikoff at the Pasteur Institute, Paris (Bibel, 1988). The rapid development of pharmaceutical chemistry and antibiotics draws from these ideas. During subsequent decades, development of new types of fermented milk products in the dairy industry has given rise to new interest in the role of probiotics in foodstuffs. This is especially related to lactic acid bacteria and bifidobacteria, in foods. Product development has been intensive during the past two decades (Saxelin, 1996).

2   Selection criteria for lactic acid bacterial strains as probiotics

During the selection of microbial strains for probiotic use, it is necessary that they are representative of microorganisms Generally Recognized As Safe (GRAS) or have good documentation on their safety. A probiotic strain should meet at least most of the *in vitro* criteria given in Table1. However, the final and most important factor is the *in vivo* effect of a strain or a product. The most important *in vitro* criteria are acid and bile tolerance. Studies *in vitro* do not take into account the kinetics of stomach emptying, gastric secretion, bile concentration or intestinal transit. There are a few studies on the pharmacokinetics of lactic acid bacteria. These indicate poor resistance on the part of yoghurt starter bacteria to gastric and bile acids. Species of *Lactobacillus acidophilus* and

G.R. Gibson and M.B. Roberfroid (eds.), Colonic Microbiota, Nutrition and Health, 281-289.
© 1999 *Kluwer Academic Publishers. Printed in the Netherlands.*

Bifidobacterium are somewhat resistant, but many differences exist between individual strains (Berrara *et al.*, 1991; Huis in't Veld *et al.*, 1994). Berrara *et al.* (1991), using both *in vitro* and *in vivo* methods, compared the acid tolerance of two commercial bifidobacterial strains. One strain was tolerant *in vitro*, whereas the other was sensitive to low pH. Survival in the human stomach *in vivo* gave comparable results during a 90 min follow-up period after consumption of the product. The gastric emptying rate was similar for both commercial products.

Table 1.   Properties of good probiotic strains

1) Technological aspects: -
   Sensory properties
   Growth properties
   Effects on food texture
   Viability in foods
   Stability in industrial processes

2) Physiological properties: -
   Resistance to;
   > pH
   > gastric juices
   > bile
   > pancreatic juices

3) Functional properties: -
   Adherence to the intestinal epithelium
   Colonization of the human intestinal tract (temporarily)
   Antagonism to pathogens
   Antimicrobial activity
   Enhancement/balancing effect on immune response

Adhesion is supposed to be important for the effects of probiotics but this is not an unambiguous question. Although most of the well-known strains are adhesive, some less or non-adhesive strains have shown certain probiotic effects. Adhesion may be important in some physiological situations, *e.g.* in enhancement of the immune system, but not necessarily in all effects.

The adhesive properties of lactic acid bacterial strains are considered to be host specific, but there also seems to be adhesion properties in non-host species. The adhesion of lactic acid bacteria and bifidobacteria appears to vary from strain to strain, and the common microorganisms used in dairy technology are not always among the most adhesive (Saxelin, 1996).

Colonisation resistance is mediated not only by the intestinal microflora but also by anatomical and physiological factors including an intact mucosal barrier, the secretion and composition of saliva, swallowing, the secretion of immunoglobulin A, lysozyme, hormones, the production of gastric and bile acid, desquamation of the mucous membranes cells, and normal gastrointestinal motility. These factors limit the adhesion of food micro-organisms to mucous membranes.

3   Types of probiotic product on the market

3.1   STARTERS

Bacterial cultures used in milk fermentations are called starter cultures. Today, they are manufactured in specialised production plants under strictly controlled conditions. Cultures are in concentrated forms, such as freeze-dried powders or frozen concentrates, and are inoculated into bulk starters (500-2000 litres) or directly to the final product milk, *e.g.* to 10,000 litres of milk.

The main function of starter bacteria in fermented milk products is the fermentation of lactose to lactic acid, giving rise to coagulation of casein and whey proteins. This gives consistency to the milk, and enables separation of the casein and whey fractions. Starter bacteria also degrade proteins in milk, making them nutritionally more available. All starter bacteria have their own typical sensory effects on the product and some genera are used mainly to produce aroma compounds in milk, *e.g.* diacetyl from citrate in butter. The formation of gas gives rise to eye formations in cheese, and polysaccharide formation contributes towards a thicker consistency of fermented milks. The most common bacterial species used in fermented milk products are shown in Table 2.

3.2   FERMENTED MILK PRODUCTS

The term "lactic acid bacteria" in the context of fermented foods mainly applies to the genera *Lactobacillus, Lactococcus, Leuconostoc* and *Streptococcus*, sometimes also to *Pediococcus* and *Enterococcus*. The genera *Vagococcus, Carnobacterium, Terageno-coccus* and *Aerococcus* also belong to lactic acid bacteria (Axelsson, 1993), but they are not used in fermented foods. Lactic acid bacteria do not form a homogeneous group of bacteria. They belong to separate phylogenetic branches, but the common features are that all have a fermentative metabolism and gain energy from the breakdown of carbo-hydrates, mainly into lactic acid. Homofermentative lactic acid bacteria form lactic acid to more than 85% of their metabolites, whereas heterofermentative lactic acid bacteria produce less than 50% lactic acid and also form CO_2 and ethanol (Hammes and Tichaczek, 1994).

Bifidobacterium do not belong to the lactic acid bacteria, since they have a distinct metabolic ('bifidus') pathway, the main fermentation end products being acetic and lactic acids. They are, however, widely used in fermented milk products and probiotic bacterial preparations, together with lactic acid bacteria (Ballongue, 1993).

In western countries, lactic acid bacteria are mainly consumed in milk products, but also in fermented vegetables and sausages such as salami. In Asia and Africa, milk is not a daily food ingredient and lactic acid bacterial intake is mainly through fermented vegetables and meat or fish products, from spontaneous or controlled fermentations. Traditionally, many foods of plant origin, such as fermented olives and vegetables, sauerkraut, pickled cucumber, sour dough, malolactic fermented wine and soy sauce, are prepared with lactic acid bacteria (Hammes and Tichaczek, 1994).

Table 2. Commonly used lactic acid bacteria and bifidobacteria and their functions in various fermented milk products (Saxelin, 1996)

Bacterial species	Fermented product	Main functions
Lactococcus lactis ssp. *lactis* and *cremoris*	Buttermilk (fermented milk) Butter, quarg, cheese *e.g.* gouda and edam	Lactic acid formation Milk coagulation
Lactococcus lactis ssp. *lactis* var. *diacetylactis*	-"-	Lactic acid formation Aroma formation (diacetyl from citrate) CO_2 formation
Leuconostoc mesenteroides ssp. *cremoris*	-"-	Aroma formation (diacetyl from citrate) CO_2 formation
Streptococcus thermophilus	Yoghurt Cheese (*e.g.* emmenthal)	Lactic acid formation Milk coagulation
Lactobacillus delbrückii ssp. *bulgaricus*	Yoghurt	Lactic acid formation Milk coagulation Yoghurt aroma formation (acetaldehyde)
Lactobacillus helveticus *Lactobacillus lactis*	Cheese (*e.g.* emmenthal)	Lactic acid formation Aroma formation (peptides etc.)
Lactobacillus acidophilus *L. casei* *L. rhamnosus*	Fermented milks, yoghurt, cheese	Acid formation Probiotics strains*
Bifidobacterium bifidum *B. breve* *B. infantis* *B. adolescentis* *B. longum*	Fermented milks, yoghurt	Lactic and acetic acid formation Probiotic strains*

* Health benefits documented for a few strains.

The viability of traditional yoghurt starter strains does not appear to be a major problem. Levels of both streptococci and *L. bulgaricus* appear to be stable or even to increase slightly during two weeks of storage. On the other hand, levels of *L. acidophilus* and bifidobacteria may decrease slightly over two weeks (Kneifel *et al.*, 1993). Kneifel *et al.* (1993) did not observe any dramatic decrease in the bifidobacterial content, in contrast to the results of Iwana *et al.* (1993), who failed to find any viable bacteria in three out of the eight bifidobacterial yoghurts manufactured in Europe. On the other hand, Samona and Robinson (1994) demonstrated the viability of three commercial *Bifidobacterium* strains, *B. bifidum, B. longum,* and *B. adolescentis,* to be excellent in fermented products during a three-week storage period. Bifidobacterial counts were influenced by co-inoculation with different yoghurt cultures, but no significant decline in numbers was observed. Obviously, strain selection is very important in bifidobacterial products (Nighswonger *et al.*, 1996). This is also the case with probiotic bifidobacterial powders, due to the low pH of gastric juices (Clark *et al.*, 1993). The viability of lactobacilli in fermented milk products is usually good, but stability of bifidobacteria is highly strain dependent (Nighswonger *et al.*, 1996).

Stability, in this context, usually refers to weeks at refrigerator temperatures. Ice cream may also form a possible vehicle for administering lactic acid bacteria or bifidobacteria in the human diet. In this case, viability can be as high as several months in the freezer (Modler *et al.*, 1990).

One way of administering probiotic bacteria is to freeze-dry them, and to use the lyophilised powder, or else to produce capsules, tablets or granules. Production of such pharmaceutical preparations makes it easier to consume the probiotics when travelling, or to ingest large quantities of bacteria without consuming milk (for strictly lactose-intolerant subjects). Unfortunately, in many countries such preparations are not subject to full quality control by the authorities, which makes it possible for products to be sold that do not fulfil the necessary declarations. It is possible to buy products on the market with too few, or no, viable cells, or containing other species than those which are claimed (Brennan *et al.*, 1983; Khedekar *et al.*, 1989; Majamaa *et al.*, 1995). The quality of products is improving, and preparations with poor quality will disappear as soon as quality standards are accepted and a stringent approval system introduced (Laulund, 1994).

4 Effect of dose and administration medium on intestinal viability

The use of bacteria in food fermentations is based on the extended shelf-life of food, the improvement of sensory characteristics, and better conditions for hydrolytic reactions, precipitation, coagulation etc. The end products of bacterial fermentation may possess antimicrobial potential, *e.g.* reduced pH, low oxidation-reduction potential and antimicrobial substances. Many metabolites prevent the growth of spoilage bacteria, and the exclusion of O_2 reduces growth of acid-tolerant yeasts and fungi.

Milk substances seem to protect bacteria in the stomach effectively. *Lactobacillus GG* could be recovered in stool samples during consumption of 100 million cfu/day administered as a freeze-dried powder (Saxelin *et al.*, 1991, 1993), while 100 million bacteria in sweet milk was an effective dose for a successful faecal recovery of the strain. When dried bacteria are taken with a meal, it is possible that survival could be improved. With a daily dose of 10-100 billion bacteria, all volunteers carried *Lactobacillus GG*, independent of the administration medium (Saxelin, 1996, 1998). Administration of probiotic bacteria in a milk base favours survival, because of buffering the stomach pH. Martini *et al.* (1987) detected the gastric pH to be > 2.7 for 3 h after yoghurt or milk ingestion. Many strains, die rapidly at pH 1 but not at pH > 2 (*e.g.* Pochart *et al.*, 1992). In principle, a meal ingested together with probiotic bacteria should have a similar effect.

Intestinal colonisation with probiotic bacteria does not seem to be permanent in most humans. In adult subjects, *Lactobacillus GG* was recovered in about 30% of faecal samples studied 2 weeks after discontinuation of administration in the adults (Ling *et al.*, 1992, 1994), and in about 50% of samples in newborns 3 to 4 weeks after intake (Sepp *et al.*, 1993). Faecal *L. reuteri* also decreased 1 week after discontinuation, reaching the original levels after 2 months (Wolf *et al.*, 1995). These results indicate that probiotic bacteria should be consumed several times weekly to maintain their effect on the intestinal microecology.

Traditionally, fermented milk products contain living bacteria at the level of about 10^{10}-10^{11} cfu/portion. However, some new, probiotic strains do not grow well in milk. Generally they are used with traditional starter strains, in which case the probiotic forms only part of the microbial flora of the milk product. Since the survival of lactic acid bacterial strains in the human intestinal tract is variable, it is important to know the exact amount of bacteria needed to ensure colonisation or passage of living bacteria through the intestine. In the human intestinal tract, lactic acid bacteria are common right from the upper part of the small intestine through to the colon and faeces. Therefore, it is laborious to recover an orally-administered strain from faecal samples, especially when it is not dominant in the lactic acid bacterial flora.

At present, evidence to indicate the minimum amount of probiotic required to give a probiotic response is scarce. Only in certain strains, has the dose response been studied. In *Lactobacillus GG*, lowest doses in powder form were 5×10^9 cfu twice daily for the successful treatment of rotavirus diarrhoea in children (Majamaa *et al.*, 1995). This is the minimal level to reach faecal recovery of *Lactobacillus GG* in powder form, but in milk-based products, colonisation is achieved with even lower doses (Saxelin, 1996).

Consumption of fermented dairy products has a long tradition especially in Nordic countries, where intake is higher than other parts of the world. Probiotic bacteria with documented clinical effects in large-scale consumption include *Lactobacillus acidophilus* (NCFB 1748), *Lactobacillus casei* strain Shirota, *Lactobacillus GG* (ATCC 53103), and *Lactobacillus acidophilus* LA1. More recently, a few newer strains have entered large-scale consumption and may be classified as 'successful'. New and novel strains and also species have emerged and are likely to be included in our diet. These include *Lactobacillus reuteri* strains, *Bifidobacterium animalis*, *Lactobacillus casei*, *Bifidobacterium bifidum* and strains of *Enterococcus faecium*. Food products which at present contain probiotics with documented health effects in the market are shown in Table 3, whilst Table 4 gives some properties associated with certain commercial strains.

5   Conclusion

Market research seems to show that consumers are more and more looking for healthy products, that may help prevent illness. In marketing, there is a gradual change towards more positive messages, but the current food laws do not allow any health claims, which would be needed to satisfy consumer requirements. Probiotic bacteria do not have special characteristics or flavour that can be tasted or detected by consumers. The only way to get the health message to the consumers is first to convince medical and scientific professionals with firm evidence from good clinical and/or nutritional studies. The development of new strains of bacteria that could either improve current health use or be used for clinical problems in which probiotics have not previously been used must be based on good and well controlled human volunteer trials. Each strain should be documented and tested independently, on their own merit. In this context, the food industry has a great responsibility.

Table 3. Food products with documented health effects

1) Milk based
    yoghurts and drinking yoghurts
    cultured milk
    quark
    milk-based drinks
    sweet milk

2) Others
    juices and juice soups
    whey-based products

3) Potential products
    fresh cheese
    ice cream
    etc.

Table 4. Comparative properties of certain commercially available probiotics

Property	*Lactobacillus casei* Shirota	*Lactobacillus GG (ATCC 53103)*	*Lactobacillus johnsonii (LA1)*	*Lactobacillus acidophilus (NCFB 1748)*	*Lactobacillus acidophilus (NCFM)*
Origin	human	human	human	?	human
Safety	verified	verified	verified	verified	verified
Acid stability	good	good	good	good	good
Bile stability	resistant	resistant	resistant	resistant	resistant
Colonisation	-	+	+	-	-
Bacteriocins	no	yes	yes	no	yes
Adherence (Caco-2)	no	yes	yes	no	yes
Adherence (mucosa)	?	yes	yes	yes	?

References

Axelsson, L.T. (1993) Lactic acid bacteria: classification and physiology. In *Lactic Acid Bacteria*. Salminen, S. and von Wright, A., eds. Marcel Dekker, Inc. New York. pp.1-63.

Ballongue, J. (1993) Bifidobacteria and probiotic action. In *Lactic Acid Bacteria*. Salminen S. and von Wright A., eds. Marcel Dekker, Inc. New York. pp. 357-28.

Berrara, N., Lemeland, J.F., Laroche, G., Thouvenot, P. and Piaia, M. (1991) *Bifidobacterium* from fermented milk: survival during gastric transit. *Journal of Dairy Science* **74**, 409-13.

Bibel, D.J. (1988) Elie Metchnikoff's bacillus of long life. *ASM News* **54**(12), 661-65.

Brennan, M., Wanismail, B. and Ray, B. (1983) Prevalence of viable *Lactobacillus acidophilus* in dried commercial products. *Journal Food Prot* **46**(10), 887-92.

Clark, P.A., Cotton, L.N. and Martin, J.H. (1993) Selection of bifidobacteria for use as dietary adjuncts in cultured dairy foods: II - Tolerance to simulated pH of human stomachs. *Cultured Dairy Products Journal* November, 11-14.

Hammes, W.P. and Tichaczek, P. (1994) The potential of lactic acid bacteria for the production of safe and wholesome food. Z. *Lebens Unters Forsch* **198**, 193-201.

Huis In't Veld, J.H.J., Havenaar, R. and Marteau, P. (1994) Establishing a scientific basis for probiotic R. & D. *Tibtech* **12**, 6-8.

Iwana, H., Masuda, H., Fujisawa, T., Suzuki, H. and Mitsuoka, T. (1993) Isolation and identification of *Bifidobacterium ssp.* in commercial yogurts in Europe. *Bifidobacteria Microflora* **12**, 39-45.

Khedekar, C.D., Sannabhadti, S.S. and Dave, J.M. (1989) Prevalence of lactobacilli in some of the commercial therapeutic preparations available in India. *Indian Journal of Dairy Science* **42**(4), 741-43.

Kneifel, W., Jaros, D. and Erhard, F. (1993) Microflora and acidification properties of yogurt and yogurt-related products fermented with commercially available starter cultures. *International Journal of Food Microbiology* **18**, 179-89.

Laulund, S. (1994) Commercial aspects of formulation, production and marketing of probiotic products. In *Human Health: The Contribution of Microorganisms*. S.A.W. Gibson, ed., Springer-Verlag London Ltd., London, pp. 159-73.

Ling, W.H., Hänninen, O., Mykkänen, H., Heikura, M., Salminen, S. and von Wright, A. (1992) Colonization and fecal enzyme activities after oral *Lactobacillus GG* administration in elderly nursing home residents. *Annual Nutrition Metabolism* **36**, 162-66.

Ling, W.H., Korpela, R., Mykkänen, H., Salminen, S. and Hänninen, O. (1994) *Lactobacillus strain* GG supplementation decreases colonic hydrolytic and reductive enzyme activities in healthy female adults. *Journal of Nutrition* **124**, 18-23.

Majamaa, H, Isolauri, E, Saxelin, M. and Vesikari, T. (1995) Lactic acid bacteria in the treatment of acute rotavirus gastroenteritis. *Journal of Pediatric Gastroenterology Nutrition* **20**, 333-38.

Martini, M.C., Bollweg, G.L., Levitt, M.D. and Savaiano, D.A. (1987) Lactose digestion by yoghurt ß-galactosidase: influence of pH and microbial cell integrity. *American Journal of Clinical Nutrition* **45**, 432-36.

Modler, H.W., McKellar, R.C., Goff, H.D. and Mackie, D.A. (1990) Using ice cream as a mechanism to incorporate bifidobacteria and fructooligosaccharides into the human diet. *Cultured Dairy Products Journal* August, 4-9.

Nighswonger, B.D., Brashears, M. and Gilliland, S.E. (1996) Viability of *Lactobacillus acidophilus* and *Lactobacillus casei* in fermented milk products during refrigerated storage. *Journal of Dairy Science* **79**, 212-19.

Pochart, P., Marteau, P., Bouhnik, Y., Goderle, I., Bourlioux, P. and Rambaud, J.P. (1992) Survival of bifidobacteria ingested via fermented milk during their passage through the human small intestine: an *in vivo* study using intestinal perfusion. *American Journal of Clinical Nutrition* **55**, 78-80.

Samona, A. and Robinson, R.K. (1994) Effect of yogurt cultures on the survival of bifidobacteria in fermented milks. *Journal of Society Dairy Technology* **47**, 58-60.

Saxelin, M. (1998) Development of dietary probiotics: estimation of optimal *Lactobacillus GG* concentrations. Valio Ltd, Research and Development, Helsinki and Department of Biochemistry of food Chemistry, University of Turku, Finland. Turku.

Saxelin, M. (1996) Colonization of the human gastrointestinal tract by probiotic bacteria. *Nutrition Today* **31**, 55-85.

Saxelin, M., Ahokas, M. and Salminen, S. (1993) Dose response on the fecal colonization of *Lactobacillus strain GG* administered in two different formulations. *Microbial Ecolology in Health and Disease* **6**, 119-22.

Saxelin, M., Elo, S., Salminen, S. and Vapaatalo, H. (1991) Dose-response colonization of feces after oral administration of *Lactobacillus casei* strain GG. *Microbial Ecolology in Health and Disease* **4**, 209-14.

Sepp, E., Mikelsaar, M. and Salminen, S. (1993) Effect of administration of *Lactobacillus casei* strain GG on the gastrointestinal microbiota of newborns. *Microbial Ecolology in Health and Disease* **6**, 309-14.

Wolf, B.W., Garleb, K.A., Ataya, D.G. and Casas, I.A. (1995) Safety and tolerance of *Lactobacillus reuteri* in healthy adult male subjects. *Microbial Ecolology in Health and Disease* **8**, 41-50.

CHAPTER 19

Prebiotics in Consumer Products

ANNE FRANCK
ORAFTI, Tienen, BELGIUM

1   Introduction

Amongst those components most likely to be used in functional foods, prebiotics show very interesting properties with some already being recognised and used as improved food ingredients (Coussement, 1996). A prebiotic has been defined as "a non-digestible food ingredient that beneficially affects the host by selectively stimulating the growth and/or activity of one or a limited number of bacteria in the colon, and thus improves host health" (Gibson and Roberfroid, 1995). To be classified as a prebiotic, a food ingredient should be neither hydrolysed, nor absorbed in the upper part of the gastrointestinal tract, be selectively fermented by a limited number of potentially beneficial bacteria in the colon, and alter the composition of the colonic microbiota towards a healthier community. A prebiotic may also induce systemic effects which can be beneficial to the host.

2   Natural occurrence and production

Most prebiotics now identified are non-digestible oligosaccharides (NDO), (Delzenne and Roberfroid, 1994). They are obtained either by extraction from plants (*e.g.* chicory inulin, soybean-oligosaccharides) eventually followed by an enzymatic hydrolysis (*e.g.* oligofructose from inulin, maltooligosaccharides from starch, xylooligosaccharides from xylan), by synthesis (transglycosylation reactions) from mono- or disaccharides such as maltose (isomaltooligosaccharides), glucose syrup (gentiooligosaccharides), sucrose (fructooligosaccharides) and lactose (transgalactosylated oligosaccharides or galactooligosaccharides). Their production processes are illustrated in Figs 1-3. They are available at different purity grades, either as white powders or viscous syrups (Crittenden and Playne, 1996).

G.R. Gibson and M.B. Roberfroid (eds.), Colonic Microbiota, Nutrition and Health, 291-300.
© 1999 *Kluwer Academic Publishers. Printed in the Netherlands.*

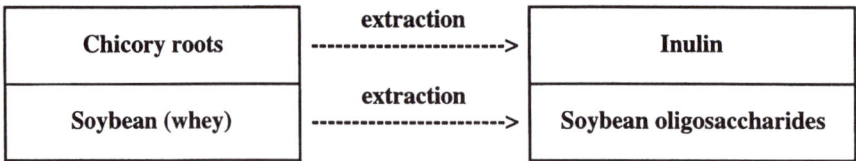

Figure 1. Production of non-digestible oligosaccharides by extraction of naturally occurring compounds.

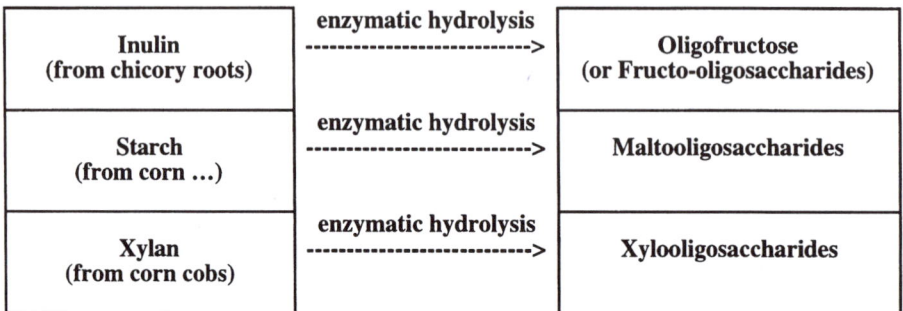

Figure 2. Production of non-digestible oligosaccharides by specific enzymatic hydrolysis.

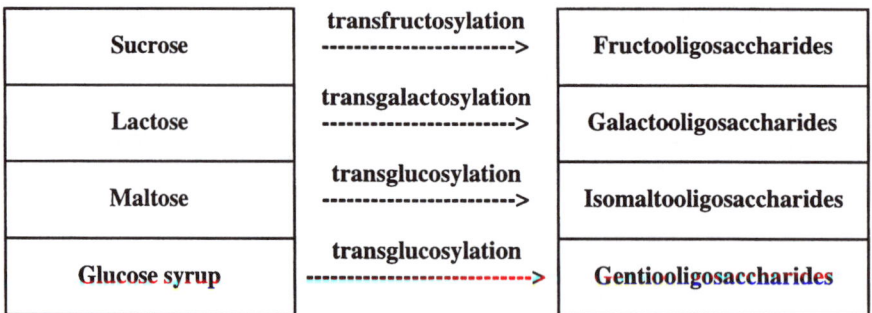

Figure 3. Production of non-digestible oligosaccharides by enzymatic transglycosylation.

The NDO which contain fructose (inulin, oligofructose) or galactose (soyoligosaccharides, galactooligosaccharides) are selectively fermented by the microflora in the human colon leading to a bacterial composition that is dominated by bifidobacteria, a perceived health promoting genus (Salminen *et al.*, 1998). Other oligosaccharides, available in Japan but not yet allowed for food use in Europe or the United States, include malto-, isomalto-, palatinose-, xylo- and gentio-oligosaccharides (Crittenden and Playne, 1996). Not all have been shown to have prebiotic properties. As such, this chapter will emphasise the fructose- and galactose-based oligosaccharides, as these are the best recognised prebiotics. An overview of both is shown in Table 1.

The ß(1-2) fructans inulin and oligofructose (or fructo-oligosaccharides, FOS) are, hitherto, the best studied prebiotics. Fructans are extremely abundant natural oligosaccharides. Inulin and oligofructose are common constituents of many foods such as leek, onion, garlic, artichoke, salsify, asparagus, banana and wheat. There has been widespread and common knowledge of their natural occurence and consumption in human and animal foods for many years (see Van Loo *et al.*, 1995). The average consumption in the normal human diet has been evaluated at several grams per day. Inulin and oligofructose have been recognised as dietary fibres in most countries. A new analytical method to measure fructans in foods was recently adopted by AOAC International (method number 997.08) (Hoebregs, 1997).

Inulin is industrially obtained from chicory roots (*Cichorium intybus*) by hot water extraction, followed by refining and spray-drying. Inulin is a polydisperse mixture of linear molecules, all with the same basic chemical structure which is symbolised as G-Fn (with G = glucosyl moiety, F = fructosyl moiety and n = number of fructose units linked through ß(1-2) bonds). The degree of polymerisation (DP) of native chicory inulin ranges between 3 and 60, with an average value of about 10 (De Leenheer, 1996). Inulin, from which the lower DP-fraction has been physically removed and having an average DP of about 25, is available as a high performance fat replacement (Coussement and Franck, 1997). High purity (> 90%) inulin is commercially available as a white powders.

Oligofructose is obtained by a partial enzymatic hydrolysis of inulin using a specific endo-inulinase. It is composed of linear G-Fn and Fn chains with a DP ranging from 2 to 8 (with an average value about 4). Fructooligosaccharides are also produced by synthesis from sucrose through a transfructosylation reaction giving G-Fn molecules with a DP from 3 to 5 (Bornet, 1994). Oligofructose products are available with different purity grades (up to 95% oligofructose) as viscous syrups or white powders.

Soybean oligosaccharides are directly extracted and purified from soybean whey, a by-product from the production of soy protein concentrate (Crittenden and Playne, 1996). Their chemical structure consists of 1 or 2 galactose units linked to one sucrose molecule (raffinose and stachyose). The galactosyl moieties are linked together through α (1-6) bonds. Soybean oligosaccharides are commercially available as a syrup containing raffinose (6%) and stachyose (18%), as well as sucrose, glucose and fructose (52%).

Some components of galactooligosaccharides have been reported to occur naturally in human breast milk at levels of about 3 mg/L. Industrially, transgalactosylated oligosaccharides are produced by synthesis from lactose using a ß-galactosidase. They have the chemical structure G-Galn (with G = glucosyl moiety, Gal = galactosyl moiety and n = number of galactose units linked together). The DP ranges from 2 to 8, with a mean of about 3. Some of the chains are branched and the galactosyl moieties are mainly linked together through ß(1-6) and/or ß(1-4) bonds (Ekhart and Timmermans, 1996).

Inulin, oligofructose, soybean oligosaccharides and transgalactosylated oligosaccharides have been evaluated by the Health Authorities of different countries and confirmed as 'safe' in many cases. Inulin and oligofructose (or FOS) can be used in food

Table 1. Overview of prebiotics currently available in Europe and/or the USA

	Inulin	Oligofructose (or fructo-oligosaccharides)	Galacto-oligosaccharides	Soybean oligosaccharides
Industrial production	Water extraction from chicory roots	Enzymatic hydrolysis of chicory inulin or Enzymatic synthesis from sucrose	Enzymatic synthesis from lactose	Extraction from soybeans
Chemical structure(*)	$G - F_n$	$G - F_n$ and F_n	$G - Gal_n$	$Gal_n - G - F$
Osidic bond between the fructose or galactose units	$\beta (1 - 2)$	$\beta (1 - 2)$	$\beta (1 - 6)$ $\beta (1 - 4)$	$\alpha (1 - 6)$
Degree of polymerisation	3 - 60 (average 10)	2 - 8 (average 4)	2 - 8 (average 3)	3 - 4 (average 4)

(*) G = glucosyl, F = fructosyl, Gal = galactosyl, n = number of fructose or galactose units linked together

products in Europe, the USA and Japan, soybean oligosaccharides in the USA and Japan, and transgalactosylated oligosaccharides in Japan and some European countries. Studies conducted to evaluate potential toxic effects in animals and humans revealed no adverse effects (Clevenger *et al.*, 1988). The only noted side effects were occasional occurence of flatulence and soft stools after the ingestion of large quantities (Briet *et al.*, 1995; Hata *et al.*, 1991). These effects are comparable with those observed for most soluble dietary fibres. In practice, the usual levels of prebiotic use (typically 2-4 g/ serving) are below the amounts at which intestinal discomfort occurs. The caloric value of NDO has been estimated between 1 and 2 kcal/g (Livesey, 1992; Roberfroid *et al.*, 1993). Their physiological properties have been reviewed by Tomomatsu (1994) and are described in Chapter 7. They are also summarised in Table 2.

3   Technological properties

Prebiotics offer a unique combination of nutritional benefits (Chapter 7) and technological advantages. Some properties of currently used prebiotics are summarised in Table 3.

Inulin has a neutral taste, without any off-flavour or aftertaste, and combines easily with other ingredients without modifying flavours (Coussement and Franck, 1997). It is moderately soluble in water (about 10% at room temperature), which allows its incorporation into systems where some other fibres would precipitate. This is relevant for table spreads, milk products and drinks. Inulin behaves as a bulk ingredient, contributing to body and mouthfeel. Its viscosity in water is however rather low (less than 2 mPa.s for a 5% w/w solution in water). On the other hand, inulin has good fat replacing capacity. When mixed with water or other aqueous liquids, it forms a gel offering a creamy structure with a suitable texture and mouthfeel.

An inulin gel exhibits a visco-elastic rheological behaviour and can easily be incorporated into foods to replace fat by up to 100% (Franck, 1993). Such a gel is composed of a multi-dimensional network of insoluble sub-micron crystalline inulin particles in water. Large amounts of water can be immobilised in the network, which assures physical stability of the gel. Furthermore, inulin stabilises foams and emulsions, *e.g.* in aerated dairy desserts, ice creams, table spreads and salad-dressings. Inulin can also work in synergy with most gelling agents like gelatine, gums, alginate, carrageenan and maltodextrin.

Oligosaccharides such as oligofructose, soybean oligosaccharides and transgalactosylated oligosaccharides are soluble (about 80% in water at room temperature). They can combine well with delicate tastes and may enhance certain fruit flavours. In food formulations, sweetness can be increased by adding high potency sweeteners as required. In combination with aspartame and acesulfame K, NDO provide mixtures with a goodmouthfeel, stability and sustained (fruit) flavour with reduced aftertaste, thus exhibiting good synergy (Wiedmann and Jager, 1997).

Non-digestible oligosaccharides show stability during food processing (*e.g.* heat treatment). They can also contribute towards texture and mouthfeel, have humectant

Table 2. Nutritional properties of currently used prebiotics

Prebiotic effect : modulation of the gut flora, promoting beneficial bacteria
   (e.g. bifidobacteria) and repressing harmful groups (e.g. clostridia)

Non-digestibility and low caloric value (1-2 kcal/g)

Suitable for diabetics

Soluble dietary fibres

Stool bulking effect : increase in stool weight and frequency

Improvement of calcium bioavailability

Reduction in serum levels of cholesterol and triglycerides (in
   experimental animals)

Prevention of colon carcinogenesis (in rats and mice)

---

properties, reduce water activity ensuring high microbiological stability, affect boiling and freezing points and can have a moderate reducing power, depending on the chemical composition. As such, NDO possess technological properties closely related to those of sugar and glucose syrups (Crittenden and Playne, 1996). This makes NDO suitable ingredients as sugar replacers while, at the same time, decreasing caloric content of the end product and adding prebiotic properties.

4   Applications in food products

Prebiotics can be used either for their nutritional advantages, technological properties or both. Potential food applications are illustrated in Table 4.

The use of NDO as fibre-like ingredients can lead to an improved taste and texture (Coussement, 1995). Non-digestible oligosaccharides can give increased crispiness and expansion to extruded snacks and cereals. They may also help maintain breads and cakes moist and fresh. Their solubility allows incorporation into fluid systems such as drinks, dairy products and table spreads.They are also increasingly used in functional foods as prebiotic ingredients to stimulate the growth of beneficial intestinal bacteria (see Chapter 7).

Because of specific gelling characteristics, inulin allows the development of low-fat foods (Coussement and Franck, 1997). This is especially useful for products like table spreads, butter-like products, cream cheeses and processed cheeses. Inulin allows the replacement of significant amounts of fat and stabilisation of the emulsion, while providing a short spreadable texture. It gives good results in water-in-oil spreads with a fat content from 20% to 60%, as well as in water continuous formulations containing 15% fat or less. In low-fat dairy products such as milk drinks, fresh cheeses, yoghurts,

Table 3. Technological properties of currently used prebiotics

Property	FOS / GOS / SOS (*)	Inulin
Aspect	Colourless viscous syrup (75°Bx) or white powder	White powder
Taste	Slightly sweet, synergy with high potency sweeteners	Neutral, without off-flavour
Sweetness versus sucrose	25 - 35%	< 10%
Solubility in water (room temperature)	About 80% W/W	About 10% W/W
Viscosity at 30% W/W in water (10°C)	About 5 mPa.s	About 100 mPa.s
Freezing point depression at 10% W/W	- 0.6°C	- 0.3°C
Others	Sugar replacement Moisture retention / humectant Water activity close to sugar	Fat replacement Gelling capacity (at high concentration) Foam and emulsion stabilisation

(*) FOS = fructooligosaccharides, GOS = galactooligosaccharides, SOS = soybean oligosaccharides

Table 4. Food applications of prebiotics

Application	Functionality
Dairy products (yoghurts, cheeses, desserts, drinks)	Fat or sugar replacement, body & mouthfeel, foam stabilisation, fibre & prebiotic
Frozen desserts	Fat or sugar replacement, texture & mouthfeel, melting behaviour
Fruit preparations	Sugar replacement, synergy with intense sweeteners, body & mouthfeel, fibre & prebiotic
Breakfast cereals & extruded snacks	Sugar replacement, crispiness & expansion, fibre & prebiotic
Baked goods & breads	Sugar replacement, moisture retention, fibre & prebiotic
Fillings	Fat or sugar replacement, texture & mouthfeel
Tablets & confectionery	Sugar replacement, fibre & prebiotic
Chocolate	Sugar replacement, heat resistance
Dietetic products & meal replacers	Fat or sugar replacement, synergy with intense sweeteners, body & mouthfeel, fibre & prebiotic
Table spreads & butter products	Fat replacement, texture & spreadability, stability, replacement of gelatine, fibre & prebiotic
Salad-dressings	Fat replacement, mouthfeel & body
Meat products	Fat replacement, texture & stability, fibre

creams and dairy desserts, the addition imparts improved flavour and a creamier mouthfeel. In frozen desserts, inulin provides easy processing, melting properties, as well as freeze-thaw stability. Fat replacement can also be applied in meat products, sauces and soups, low-fat meat products. In dairy mousses (*e.g.* chocolate, fruit, yoghurt or fresh cheese-based), incorporation of inulin improves processability and upgrades quality. The resulting products retain their typical structure for longer. Inulin has also found application in chocolates without added sugar, often in combination with sugar-alcohols. In several food products, inulin may also replace certain stabilisers such as gelatine.

Fructose-based oligosaccharides are already applied in several well-known products, for instance yoghurts, fermented milks, fresh cheeses, dairy drinks, desserts and meal replacers. The incorporation into baked goods allows replacement of sugar, fibre enrichment and good moisture retention properties. They also offer binding characteristics in cereal bars. Their use requires only minor adaptation of the production process. Non-digestible oligosaccharides allow bulk without calories and provide nutritional benefits without compromising taste and mouthfeel.

Nutrition Guidelines, as established by the World Health Organization (WHO), have put a strong emphasis on increasing the dietary fibre intake and decreasing fat. Moreover, reduction of energy intake to reduce obesity risk has been proposed. Addition of a few grams of NDO to foods can help to reach these goals. The prebiotic properties even offer a new dimension for the development of functional foods. One approach that may be also encouraged for future research is the combination of both probiotics and prebiotics (as synbiotics) to exert synergistic effects.

Reference

Bornet, F.R.J. (1994) Undigestible sugars in food products. *American Journal of Clinical Nutrition* **59** (suppl), 763S-69S.

Briet, F., Achour, L., Flourie, B., Beaugerie, L., Pellier, P., Franchisseur, C., Bornet, F. and Rambaud, J.C. (1995) Symptomatic response to varying levels of fructo-oligosaccharides consumed occasionally or regularly. *European Journal of Clinical Nutrition* **49**, 501-07.

Clevenger, M.A., Turnbull, D., Inoue, H., Enomoto, M., Allen, J.A., Henderson, L.M. and Jones, E. (1988) Toxicological evaluation of neosugar; genotoxicity, carcinogenicity and chronic toxicity. *Journal American Toxicity* **7** (5), 643-62.

Crittenden, R.G. and Playne, M. (1996) Production, properties and applications of food-grade oligosaccharides. *Trends in Food Science & Technology* **7**, 353-61.

Coussement, P. (1996) Pre- and synbiotics with inulin and oligofructose. *Food Technology in Europe* January, 102-04.

Coussement, P. (1995) A new generation of dietary fibres. *European Dairy Magazine* **3**, 22-24.

Coussement, P. and Franck, A. (1997) Multi-functional inulin. *Food Ingredients and Analysis International* October, 8-10.

De Leenheer, L. (1996) Production and use of inulin: industrial reality with a promising future, in *Carbohydrates as Organic Raw Materials III* (eds.. H. Van Bekkum, H. Roper and A.G.J. Voragen), VCH Publ. Inc., New York, pp. 67-92.

Delzenne, N.M. and Roberfroid, M.R. (1994) Physiological effects of non-digestible oligosaccharides. Lebensm *Wiss u Technology*, **27**, 1-6.

Ekhart, P.F. and Timmermans, E. (1996) Techniques for the production of transgalactosylated oligosaccharides (TOS) *Bulletin of the International Dairy Federation* **313**, 59-64.

Franck, A. (1993) Rafticreming: the new process allowing to turn fat into dietary fiber, in *FIE Conference proceedings* (1992), Expoconsult Publishers, Maarssen, pp. 193-97.

Hata, Y., Yamamoto, M. and Nakajima, K. (1991) Effects of soybean oligosaccharides on human digestive organs: estimation of fifty percent effective dose and maximum non-effective dose based on diarrhea. *Clinical Biochemistry and Nutrition* **10**, 135-44.

Gibson, G.R. and Roberfroid, M.B. (1995) Dietary modulation of the human colonic microbiota: introducing the concept of prebiotics. *Journal of Nutrition* **125**, 1401-12.

Hoebregs, H. (1997) Fructans in foods and food products, ion-exchange chromatograhic method: collaborative study. *Journal of AOAC International* **80** (5), 1029-37.

Livesey, G. (1992) The energy values of dietary fibre and sugar alcohols for man. *Nutrition Research Reviews* **5**, 61-84.

Roberfroid, M., Gibson, G.R. and Delzenne, N. (1993) The biochemistry of oligofructose, a nondigestible fiber: an approach to calculate its caloric value. *Nutrition Reviews* **51** (5), 137-46.

Salminen, S., Bouley, C., Boutron-Ruault, M.C., Cummings, J.H., Franck, A., Gibson, G.R., Isolauri, E., Moreau, M.C., Roberfroid, M.B. and Rowland, I. (1998) Gastrointestinal physiology and function. *British Journal of Nutrition* **80**, S147-71

Tomomatsu, H. (1994) Health effects of oligosaccharides. *Food Technology* October, 61-65.

van Haastrecht, J. (1995) Oligosaccharides: promising performers in new product development. *International Food Ingredients* **1**, 23-27.

Van Loo, J., Coussement, P., De Leenheer, L., Hoebregs, H. and Smits, G. (1995) On the presence of inulin and oligofructose as natural ingredients in the Western diet. *Critical Reviews in Food Science and Nutrition* **35** (6), 525-52.

WHO Study Group (1990) Diet, Nutrition, and the Prevention of Chronic Diseases. *World Health Organization Technical Report Series* **797**, Geneva.

Wiedmann, M. and Jager, M. (1997) Synergistic sweeteners. *Food Ingredients and Analysis International*, November-December, 51-56.

Index